Sustainable Potato Production and the Impact of Climate Change

Sunil Londhe
International Centre for Research in Agroforestry (ICRAF), India

A volume in the Practice, Progress, and Proficiency in Sustainability (PPPS) Book Series

IGI Global
DISSEMINATOR OF KNOWLEDGE
www.igi-global.com

Published in the United States of America by
IGI Global
Information Science Reference (an imprint of IGI Global)
701 E. Chocolate Avenue
Hershey PA 17033
Tel: 717-533-8845
Fax: 717-533-8661
E-mail: cust@igi-global.com
Web site: http://www.igi-global.com

Library of Congress Cataloging-in-Publication Data

Library of Congress Cataloging-in-Publication Data

Names: Londhe, Sunil, 1973- editor.
Title: Sustainable potato production and the impact of climate change / Sunil Londhe, editor.
Description: Hershey, PA : Information Science Reference, [2017] | "This book is an authoritative publication that provides the latest research on potato production in the future climate change scenario. Featuring exhaustive coverage on a variety of topics associated with food fundamentals such as availability, stability, utilization, and accessibility"--Provided by publisher.
Identifiers: LCCN 2016044777| ISBN 9781522517153 (hardcover) | ISBN 9781522517160 (ebook)
Subjects: LCSH: Potatoes. | Potatoes--Climatic factors. | Climatic changes.
Classification: LCC SB211.P8 S895 2017 | DDC 635/.21--dc23 LC record available at https://lccn.loc.gov/2016044777

This book is published in the IGI Global book series Practice, Progress, and Proficiency in Sustainability (PPPS) (ISSN: 2330-3271; eISSN: 2330-328X)

British Cataloguing in Publication Data
A Cataloguing in Publication record for this book is available from the British Library.

Practice, Progress, and Proficiency in Sustainability (PPPS) Book Series

ISSN:2330-3271
EISSN:2330-328X

Mission

In a world where traditional business practices are reconsidered and economic activity is performed in a global context, new areas of economic developments are recognized as the key enablers of wealth and income production. This knowledge of information technologies provides infrastructures, systems, and services towards sustainable development.

The **Practices, Progress, and Proficiency in Sustainability (PPPS) Book Series** focuses on the local and global challenges, business opportunities, and societal needs surrounding international collaboration and sustainable development of technology. This series brings together academics, researchers, entrepreneurs, policy makers and government officers aiming to contribute to the progress and proficiency in sustainability.

Coverage

- ICT and knowledge for development
- Knowledge clusters
- Technological learning
- Global Content and Knowledge Repositories
- Innovation networks
- Socio-Economic
- Global Business
- Outsourcing
- Strategic Management of IT
- Sustainable Development

IGI Global is currently accepting manuscripts for publication within this series. To submit a proposal for a volume in this series, please contact our Acquisition Editors at Acquisitions@igi-global.com or visit: http://www.igi-global.com/publish/.

The Practice, Progress, and Proficiency in Sustainability (PPPS) Book Series (ISSN 2330-3271) is published by IGI Global, 701 E. Chocolate Avenue, Hershey, PA 17033-1240, USA, www.igi-global.com. This series is composed of titles available for purchase individually; each title is edited to be contextually exclusive from any other title within the series. For pricing and ordering information please visit http://www.igi-global.com/book-series/practice-progress-proficiency-sustainability/73810. Postmaster: Send all address changes to above address.

Titles in this Series

For a list of additional titles in this series, please visit: www.igi-global.com

Handbook of Research on Green Economic Development Initiatives and Strategies
M. Mustafa Erdoğdu (Marmara University, Turkey) Thankom Arun (University of Essex, UK) and Imran Habib Ahmad (Global Green Growth Institute, South Korea)
Business Science Reference • copyright 2016 • 747pp • H/C (ISBN: 9781522504405) • US $335.00 (our price)

Impact of Water Pollution on Human Health and Environmental Sustainability
A. Elaine McKeown (Independent Researcher, USA) and George Bugyi (Pennsylvania State University, USA)
Information Science Reference • copyright 2016 • 422pp • H/C (ISBN: 9781466695597) • US $220.00 (our price)

Impact of Meat Consumption on Health and Environmental Sustainability
Talia Raphaely (Curtin University, Australia) and Dora Marinova (Curtin University, Australia)
Information Science Reference • copyright 2016 • 410pp • H/C (ISBN: 9781466695535) • US $210.00 (our price)

Handbook of Research on Pedagogical Innovations for Sustainable Development
Ken D. Thomas (Auburn University, USA) and Helen E. Muga (University of Mount Union, USA)
Information Science Reference • copyright 2014 • 800pp • H/C (ISBN: 9781466658561) • US $345.00 (our price)

Sustainability Science for Social, Economic, and Environmental Development
Nilanjan Ghosh (Multi Commodity Exchange of India Limited, India) and Anandajit Goswami (The Energy and Resources Institute, India)
Information Science Reference • copyright 2014 • 324pp • H/C (ISBN: 9781466649958) • US $195.00 (our price)

E-Innovation for Sustainable Development of Rural Resources During Global Economic Crisis
Zacharoula Andreopoulou (Aristotle University of Thessaloniki, Greece) Vagis Samathrakis (Alexander Technological Educational Institute of Thessaloniki, Greece) Soulla Louca (University of Nicosia, Cyprus) and Maro Vlachopoulou (University of Macedonia, Greece)
Business Science Reference • copyright 2014 • 317pp • H/C (ISBN: 9781466645509) • US $185.00 (our price)

Editorial Advisory Board

Table of Contents

Detailed Table of Contents

Potato is one of the most important vegetable crops in India accounting for 20-25% of area under cultivation of vegetables and grown in a wide range of climatic conditions. It is grown in almost all states under diversified agro-climatic conditions. Nearly 80% of the crop is grown in Indo- Gangetic plains comprising Uttar Pradesh, West Bengal, Punjab, Haryana, Bihar and other parts of India like Gujarat and Karnataka. Moreover, within the country, there is a lot of heterogeneity in potato productivity depending upon mostly on management and climatic conditions. The viability of commercial potato production is influenced by spatial and temporal variability in soils, agro climate, and the availability of water resources. The inter and intra-regional variations in productivity within the country are attributed to the variations in bio-physical factors vis-a-vis specific soil-climatic requirements of the crop. The present chapter discusses the impact of climate change on the land resources requirement for potato crop with reference to Indian context.

Potato is important crop for solving food and nutritional security problem of growing population of India. Application of N in two split dose i.e. half at planting time and rest at time of earthing up produce higher yields and higher N recovery. At the time of planting, calcium ammonium nitrate or ammonium sulphate should be preferred by furrow application. Selection of suitable variety may play major role beside time and method of application in improving nutrient use efficiency. Balanced use of major and micronutrients plays an important role in improving quality of produce besides good yield. Potato based cropping system mostly shows build up of P and negative balance of N and K which may be overcame by organic residues recycling. Intensive cropping system has resulted in wide spread deficiency of secondary and micro nutrients particularly Zn and these must be applied on soil test basis. Integrated nutrient management is a must for an exhaustive and responsive crop like potato.

Potato is one of the main staple foods in West Bengal, where it ranks second in production after Uttar Pradesh. There is lots of variation in productivity of the crop. It is due to climatic variability which causes widespread disease infection in potato crop. The shifting of onset and withdrawal of monsoons has also proved to be a barrier in the productivity of the crop. The farmers are habituated to plant the crop within 15th of November; however this is being disrupted because of the shifting of withdrawal of monsoons. Potato is a thermo sensitive crop. The crop growth rate of potato is significantly affected by cumulative maximum and minimum temperatures. Leaf area index significantly decreased with the increase in cumulative maximum and minimum temperatures. Rainfall and relative humidity are two crucial factors that determine the incidence of late blight in potato, the most devastating disease in Bengal. Rainfall increased productivity by lowering soil temperature and reducing hydrolysis of starch respiratory losses from tubers.

Potato is a temperate crop and higher day temperatures cause some areas to less suitable for potato production due to lower tuber yields and its quality. Tuber growth and yield can be severely reduced by temperature fluctuations outside 5-30 °C. The rate of warming in last 50 years is double than that for the last century. Increase in temperature and atmospheric CO2 are interlinked occurring simultaneously under future climate change and global warming scenarios. If CO2 is elevated to 550 ppm the temperature rise is likely to be 3 °C with decline in potato production by 13.72% in the year 2050. The changing climate will affect the potato production adversely due to drought, salinity, frost, flooding, erratic unseasonal rains etc. It may reduce seed tuber production, impact storage facility and potato processing industries. Therefore, the quantification of regional vulnerability and impact assessment is very important for the development of early warning on disease forecasting systems, breeding of short duration and heat, drought, salinity tolerant and disease resistant cultivars.

In the emerging global economic order in which agricultural crop production is witnessing a rapid transition to agricultural commodity production, potato is appearing as an important crop, poised to sustain and diversify food production in this new millennium. Temperature and unpredictable drought are two most important factor affecting world food securities and the catalyst of the great famines of the past. Decreased precipitation could cause reduction of irrigation water availability and increase in evapo-transpiration, leading to severe crop water-stress conditions. Increasing crop productivity in unfavourable environments will require advanced technologies to complement traditional methods which are often unable to prevent yield losses due to environmental stresses. Various crop management practices such as improved nutrient application rate, mulching, raised beds and other improved technology help to raise the productivity. Conservation farming practices play important role to restore soil and enhancing soil health and play important role to combat climate change issue.

Soil salinity is a major constrain to crop production and climate change accelerates it. It reduces plant water potential, causes ion imbalance, reduce plant growth and productivity, and eventually leads to death of the plant. This is the case in potato. However, potato has coping strategies such as accumulation of proline, an osmoregulator and osmoprotector. In addition, leaching of salts below the root zone is preferred, exogenous application of ascorbic acid and growth hormones are practiced to combat salinity. Breeding and genetic engineering also play key roles in salinity management of potato. Varieties such as: Amisk, BelRus, Bintje, Onaway, Sierra, and Tobique were tolerant in North America, variety Cara in Egypt, Sumi in Korea and varieties Vivaldi and Almera in Mediterranean region. Transgenic lines of Kennebec variety, lines S2 and M48 also proved tolerance due to transcription factor MYB4 encoded by rice Osmyb4 gene.

Potato (Solanum tuberosum L.) is one of the most important non-traditional tuber crops of Rajasthan. The potato tuber is a modified stem developed underground on a specialized structure called stolen. It contributes to food and nutritional security and provide cheap source of vegetable. It is used either alone or intermingled with other vegetables. It is also consumed as many fried salted food items. Potato is a highly nutritious, easily digestible, wholesome food. In Rajasthan, where varied climatic conditions promoting cultivation of almost every crops and vegetables, the economic conditions of growers, lack of storage facilities and lack of improved technologies for the state remain as bottleneck for its cultivation. In this chapter I tried to elaborate the constraints and possible suggestion for increasing cultivation of potato which is fairly to highly responsive to inputs supplied and gave cash returns in short periods.

In a field trial (2012), simulated aerosols: NH4Cl (reduced) and NaNO2 (oxidised) @ 10 & 20 kg ha-1y-1 (≈ 100 ppm & ≈ 200 ppm respectively), 1000 cm3m-2 of each along with a control were misted to population of Kufri Jyoti at different growth stages viz., vegetative (10-60 DAS), tuber initiation (60-90 DAS) and tuber bulking >90DAS). The higher dose of aerosols lowered nitrate reductase activity, nitrogen use efficiency, cell membrane stability, tuber yield, but increased photosynthesis, peroxidise activity significantly. The mechanisms of injury in terms of higher peroxidase activity and lower membrane stability of leaf cells have been elucidated. Foliar feeding of nitrogenous pollutant in the form of aerosols to plants at juvenile stage is important in addition to basal use of recommended fertilizers.

Climate change is a global problem affecting agricultural production, a good adaptation strategy for this phenomena should be sought for increase agricultural production. The study was conducted in Nigeria to assess the Impact of Climate Change on root and tuber crops production among farmers in Nigeria. Secondary data were used for the study, they were collected from NRCRI Umudike and other individual publications. The result shows that climate change had negative impact on root and tubers crops production including potato. Adaptation of Agriculture to climate change in the areas of crop and animal production, post harvest activities and capacity building, divers friction of livelihood sources through the use of different farming methods and improved agricultural practices will help to reduce the impact of climate change. Examples are establishment of forestry, generation of improved and disease resistance crop varieties addition of value into agricultural products and post harvest activities for climate change adaptation and sustainable development.

The quality and quantity of the potatoes produced is directly affected by the climatic factors that prevailed during the crop season. It is also well established that abiotic and biotic stresses cause tremendous losses to the crop. Host plants and their pathogens are prone to various climatic factors like temperature, relative humidity, rainfall and CO2 which are behaving in erratic manner. Phytophthora infestans has adapted itself at higher temperature so there are chances to spread at a larger area. The other potato diseases like early blight, bacterial wilt, soft rot and viral diseases may also behave differently at elevated temperature and high rainfall. Viral diseases of potato are serious threat to potato industry as most of the viruses are transmitted by vectors and vector populations are bound to increase with these changed climatic conditions. Therefore, potato researchers need to simulate these conditions and devise mitigation strategies for sustained potato production.

Potato production is seriously compromised due to prevalence of a number of diseases and they are the major constraints in potato production resulting in significant yield reduction. Integrated disease management of potato includes regular inspection for healthy seed or nursery, crop production, correct identification of the problem, cultural practices (crop rotation, sanitation etc.), biological control, soil fumigation (if necessary), seed or nursery stock treatment and disinfestations of cutting tools. Due to the ever increasing number of new fungicide resistant fungal pathogens, proper and timely diagnosis of potato diseases is becoming paramount to effective disease management, and growers need up-to-date information to help make important decisions on optimal use and timing of pesticides and other control options.

Preface

The world's favorite vegetables, the potato (*Solanum tuberosum L.*), are edible plant tubers first cultivated in the Andes region of South America. Potato is one of the important vegetable crops and are available year-round. Compared with grains, potato tubers are inherently more productive. Potato is the third most important food crop in the world after rice and wheat in terms of human consumption. More than a billion people worldwide eat potato, and global total crop production is approximately 400 million metric tons. According to International Potato Center (2016), one hectare of potato can yield two to four times the food quantity of grain crops. Potatoes produce more food per unit of water than any other major crop and are up to seven times more efficient in using water than cereals. They are produced in over 100 countries worldwide. Further, since the early 1960s, the growth in potato production area has rapidly overtaken all other food crops in developing countries.

Asia and Europe contribute to more than 80% of world production and are the world's major potato producing regions. The potato plays a strong role in developing countries with its ability to provide nutritious food for the poor and hungry hence the demand for potato is growing day by day for both fresh and processed potato food. The potatoes are good source of phytonutrients includes carotenoids, flavonoids and caffeic acid which are organic components of plants that are thought to promote human health. Though the potato protein content is fairly low but has an excellent biological value and potatoes are particularly high in vitamin C as well as good source of several B vitamins and potassium. The potatoes providing a low-cost source of critical nutrients, high-quality protein, and a satiating carbohydrate to the millions of poor population.

THE CHALLENGES

Potato is a critical crop in terms of food security in the face of population growth and increased human hunger rates. Potato can be grown on a wide range of soils, ranging from sandy loam, silt loam, loam and clay soils. It prefers friable, well aerated, fairly deep soils well supplied with organic matter. Well-drained sandy loam and medium loam soils, rich in organic carbon are most suitable for potato cultivation. Light soils are preferred because they tend to promote more uniform tubers and make harvesting of the crop easier. It is very sensitive to the temperature and frost. Potato is a temperate region crop and requires cool climate but adapted to wide range of climatic conditions. Higher day temperatures restrict some areas to less suitable for potato production due to lesser tuber yields and its quality. Tuber growth and yield can be severely reduced by temperature fluctuations outside 5-30 °C. Potato is cultivated at the marginal edges of temperatures and potato farming is an important economic activity in some parts of globe and provides livelihood for many poor and marginal farmers.

It is well known fact that agriculture production is dependent on set of climatic conditions. Each crop requires a particular climate for its growth, development and completion of its life cycle. This is the one of the reason that farmers can cultivate a specific crop in a particular region which is having suitable climatic condition to that crop (Londhe, 2016). Lobell and Gourdji (2012) noticed that climate trends over the past few decades have been fairly rapid in many agricultural regions around the world. Further, according to IPCC Assessment Report 5 (IPCC, 2013) the effects of climate change on crop and food production are evident in several regions of the world. Negative impacts of climate trends have been more common than positive ones. Further, IPCC also predicted that the average global temperature is likely to rise by another 2 to 8.6 degrees F by 2100.

Looking to these situations there are many uncertainties through which climate change may be impact potato availability, access, and utilization in various parts of the globe. This climate induced changes in potato farming will likely affect the incomes and livelihoods of potato farmers and potato prices. Potato being important food commodity and may have net effect on global food security. The changing climate is having important role in potato farming and individual climatic parameters and their interactions will play crucial role in future potato demand. There is an emerging consensus that changes in temperature precipitation and atmospheric CO_2 levels can have detrimental impacts on potato production. These impacts are difficult to quantify and depend on a range of assumptions and the land suitable for potato cultivation will greatly affected.

The impact of climate change will be different on different agro-ecological regions. Some of the regions will reduce land suitability for potato cultivation while others will have positive effect. There may be possibility, some of the newer areas which are presently not suitable for the cultivation of potato, may became suitable for its cultivation. Various chapters by the different authors viz. Ratneswar Poddar, V.K. Dua, Nawkor Ngozi has highlighted the potato crop and weather interactions in different part of world. The Probable effect of change in atmospheric gas concentration on potato crop is discussed in the chapter by author Bhagawan Bhillari. The existing package of practices, nutrient requirements, and management of potato diseases in changing climatic situation is also discussed in some of the chapters. The probable soil abiotic stress resulted from climate change is discussed in the chapter by John. The soil-site and climatic requirement for the potato crop is summarized by the author Ravindra Naitam.

There may be direct and indirect effects of climate change on potato cultivation and further will affect food security and livelihood. The direct impacts include changes in agro-ecological conditions of regions where as indirect impacts include changes in economic growth and distribution of incomes of farmers. Further, changes in climate patterns attached with population dynamics may result in higher complexity of potato demand and supply in food security on the globe. The complexity will further increase as quality and quantity of cropland available is projected to decrease. Some of the crop land areas will suffer from the various types of land degradation and increase in soil salinity level also.

SEARCHING FOR A SOLUTION

Projecting of climate change is complex process and these projections are inherently uncertain, due to the natural variability in the climate system, imperfect ability to model the atmosphere's response to any given emissions scenario. On the other hand demand for potato based food is increasing along with increasing population. According to Londhe (2016), though potato is sensitive to excess salt content of soil, the farmers' effort to cultivate potato on salt-affected soils shows the possibility of cultivation of this crop in such soils. Most of the presently available varieties show constraints for cultivation on abiotic stress-affected soils like alkaline, saline, sodic and saline sodic soils. Worldwide potato research is more focused on developing disease and pest-resistant, high-yielding and early-maturing varieties. Abiotic stress like soil salinity may be one of the impact of climate change and it may reduce the suitability of land for potato cultivation. But the demand is forcing the farmers to cultivate the potato on abiotic stress affected soils. There may be many other impacts of climate change such as rise in temperature, CO2, irregular behavior of monsoon etc. The combined interaction of these factors with potato crop will decide the potato farming in particular region. In order to overcome the situation the focused research is needed keeping eye on possible changes in climatic conditions and its impact on potato cultivation. The changing climate may also seriously interfere in the ecosystem services and its composition, function and production. The poor ecosystem performance may effect on the different ecosystem communities. This in turn may have consequences on cultivation of the potato crop. Hence, the holistic research may be more effective to address the problems arising from the changing climate for sustainable potato production.

Though there is ongoing debate about the impacts of carbon fertilization on plants and their yields. Carbon dioxide (CO2) is essential component of photosynthesis process. There are many evidences showing the potential for a 'CO2 fertilization' effect on various agricultural crops. It is also clear from Intergovernmental Panel on Climate Change (IPCC, 2007), that rise in atmospheric carbon dioxide (CO2) concentration from about 280 PPM before the industrial revolution to about 360 PPM at present. In principle, higher levels of CO2 should stimulate photosynthesis in certain plants. This is particularly true for C3 plants as potato is one of such plant and because increased Carbon Dioxide tends to suppress their photo-respiration. So the focused research on impact of increased atmospheric CO2 concentration and its impact on potato production is urgently required.

The impact of climate change is now visible in many parts of potato growing regions and in coming years may poses threats to the potato utilization. This may also affect the ability of individuals to use the potato as food in Effective way. The prices of this commodity will have significant impact of climatic changes. Being the important nutritional crop in poor population this may be having serious impacts on human health across the globe.

It is clear from the above discussions that there will be uncertainties about the earth's future climatic situation. The recommended package of practices for potato cultivation may not effective in changed climate. Hence, for sustainable potato production in projected climate changes will require modifications. This will call for the appropriate adaptations for its cultivation. Based on the area and prevailing climatic situations many options that are technologically, economically and socially feasible need to be employed in future potato production systems.

ORGANIZATION OF THE BOOK

The book is organized into 12 chapters. A brief description of each of the chapters follows:

Chapter 1 deals with the spatial and temporal variability in soils, agro climate, and the availability of water resources for the viable commercial potato production. The chapter focus on the probable reason for the gaps in potato productivity and its relationship with the specific soil-climatic requirement of the crop. The chapter also discusses the impact of climate change on the land resources requirement for potato crop with special reference to arid and semi-arid climates in Indian context.

Chapter 2 discusses about nutrient requirement of the potato crop and its behavior in changing climatic situation for better yield and quality of crop. The chapter elaborates about various macro and micro nutrient requirement of the potato crop and its effective time of application and the quantity and method of application. It also discuss on the impact of nutrient deficiency on the potato cultivation.

Chapter 3 reviews the brief account of potato production in West Bengal, one of the important potato growing region of India where potato is one of the main staple foods. It focus on the future potato production and potato crop interaction with various weather parameters. The chapter also discuss about the impact of various climatic parameters like temperature, rainfall, humidity etc. on future potato production in the region.

Chapter 4 discusses the impact and analysis of changing climatic parameter on potato production in India. The chapter elaborate the direct and indirect effect of climate change on potato yield and quality in the India. The chapter also suggests the Adaptation and mitigation measures for climate change for sustainable potato production in India.

Chapter 5 focuses on the various crop management practices such as improved nutrient application rate, mulching, raised beds and other improved technology to help farmers to raise the potato productivity in climate change situation. Various improved potato cultivation practices for climate change situation are also discussed in the chapter.

Chapter 6 discusses the impact on climate change on soil quality in terms of salinity and its consequences on potato productions. The present varieties are more suitable for non-saline environment and authors also discuss need of salt tolerant potato varieties for sustainable potato production in changing climatic situations.

Chapter 7 discusses the economic conditions of potato growers, present status of storage infrastructure facilities and improved technologies in the state of Rajasthan, India. The chapter also discuss about the bottleneck for sustainable potato production in changing climatic conditions. The authors tried to elaborate the constraints and possible suggestion for increasing cultivation of potato and requirement of various inputs for successful cash returns to the cultivators in short time span.

Chapter 8 is an experimental output of field trials and is discussed on increase of the atmospheric gases will have serious impact on the potato production in climate change situation. The authors tried to simulate aerosols: NH4Cl (reduced) and NaNO2 (oxidised) in different field trial for different varieties. The authors presented and discussed the results of various aerosol treatments on different potato varieties. The chapter gives future possible interactions of the atmospheric gases with the potato crop and possible positive and negative consequences.

Chapter 9 discusses the study conducted in Nigeria to assess the impact of Climate Change on root and tuber crop production among farmers in Nigeria. The authors discuss the climate change its impact on farmers involved in cultivation of root and tubers crops.

Chapter 10 discusses the host plants and pathogens interactions which are prone to various climatic factors like temperature, relative humidity, rainfall and CO2. These interactions will change and pathogens may behave in erratic manner. The authors discuss about some of the pathogens may adapted to higher temperature so there are chances to spread at a larger area. The authors also discussed about the viral diseases of potato which may be serious threat to potato industry.

Chapter 11 emphasizes the present state of potato production which is seriously compromised due to prevalence of a number of diseases. The present diseases of potato are major constraints in production resulting in significant yield reduction. The chapter discusses about the present practices of management of potato diseases in India. This chapter highlights various effective control measures of present diseases of potato crop for sustainable production.

Chapter 12 deals with the advent of technology and trends which provides tools to analyse predict about various consequences of potato cultivation in climate change scenario. The development in Information Technology (ICT) which includes Geoinformatics are playing a profound role in the agriculture sector and is helping to understand and assess the various factors affecting the growth of potato crops along with finding out the alternative suitability parameters for better production and distribution. This chapter focused on agro-geoinformatics as a linkage between technology usage and better potato production during adverse conditions.

Sunil Londhe
International Centre for Research in Agroforestry (ICRAF), India

REFERENCES

International Potato Center. (2016). *Potato Facts and Figures*. Retrieved July 09, 2016, from http://cipotato.org/potato/facts

IPCC. (2007). Summary for Policymakers. In *Climate Change 2007: The Physical Science Basis. Contribution of Working Group I to the Fourth Assessment Report of the Intergovernmental Panel on Climate Change*. Cambridge, UK: Cambridge University Press.

IPCC. (2013). Summary for Policymakers. In *Climate Change 2013: The Physical Science Basis. Contribution of Working Group I to the Fifth Assessment Report of the Intergovernmental Panel on Climate Change*. Cambridge, UK: Cambridge University Press.

Lobell, D. & Gourdji, S. (2012). The influence of climate change on global crop productivity. *Plant Physiology*.

Londhe, S. (2016). Climate change and agriculture: impacts, adoption and mitigation. In S. Dinda (Ed.), *Handbook of Research on Climate Change Impact on Health and Environmental Sustainability* (pp. 393–415). IGI Global. doi:10.4018/978-1-4666-8814-8.ch019

Acknowledgment

The editor would like to acknowledge the help of all the people involved in this project and, more specifically, to the authors and reviewers that took part in the review process. Without their support, this book would not have become a reality.

First, the editor would like to thank each one of the authors for their contributions. My sincere gratitude goes to the chapter's authors who contributed their time and expertise to this book.

Second, the editor wishes to acknowledge the valuable contributions of the reviewers regarding the improvement of quality, coherence, and content presentation of chapters. Most of the authors also served as referees; I highly appreciate their double task.

Sunil Londhe
International Centre for Research in Agroforestry (ICRAF), India

Introduction

Potato (*Solanum tuberosum L.*) is the most efficient crop and the third most important food crop in the world after rice and wheat. More than a billion people worldwide eat potato and this crop plays multiple and important roles in local food systems and is a critical crop in terms of food security in the face of population growth and increased hunger rates. Potato is wholesome food containing all important minerals and vitamins. Per unit production of potato is higher two to four times the food quantity than that of grain crops.

The present scenario of world potato production unfolds under conditions of climate change, which through rising temperatures, enhanced atmospheric CO_2, changing patterns of precipitation and more extreme weather events likely to affect the potato production. The potato is basically a crop of temperate climates but is grown in about 100 countries under temperate, subtropical and tropical conditions.

The Intergovernmental Panel On Climate Change (IPCC) in its AR 5 raised the concern about the globally averaged combined land and ocean surface temperature data as calculated by a linear trend, show a warming of 0.85°C (0.65 to 1.06), over the period 1880 to 2012, when multiple independently produced datasets exist. The total increase between the average of the 1850–1900 period and the 2003–2012 period is 0.78°C (0.72 to 0.85), based on the single longest dataset available. In addition to this increased levels of Carbon Dioxide will have more impact on C3 plants like potatoes. This increased temperature may have very serious impact on the cultivation of potato worldwide. There may be very severe impact on the cultivation of potato and millions of the farmers including small and marginal cultivators may face challenge of food security and livelihood.

The climatic situation for future potato cultivation is not favorable whereas the demand for potato is growing as both a fresh and processed food. Bridging the demand and supply gap is serious challenge for the scientific community and managers. Appropriate technology efforts may give some solutions. The some of the technologies like breeding biotic and abiotic stress tolerant varieties and Geoinformatics may provide options for increasing potato yields and improving yield stability in potato production. The successful generation and dissemination of technologies on potato production will brought economic prosperity to millions of small holder farming community in developing countries besides ensuring food and livelihood security.

The potato is important crop for global food security and farmers are increasingly dependent on the crop for their livelihoods. The crop is being cultivated on abiotic stressed soils at many places on the globe to full fill the demand. The climate change may bring many restrictions on cultivation of the profitable potato crop.

This book is the compressive analysis of the impact of climate change on future potato production. It is providing the detail analysis of various challenges like crop protection, need of biotic and abiotic stress tolerant varieties, livelihood and food security, improved agronomic practices and nutrient demand. The book also provides the focused review of interaction of the potato with various climatic parameters and emerging technologies for future potato cultivation.

Sunil Londhe
International Centre for Research in Agroforestry, India

Chapter 1
Climate Change and Land Suitability for Potato Cultivation in India

Ravindra Kashinath Naitam
National Bureau of Soil Survey and Land Use Planning, India

Preeti Deshmukt
Vasandada Sugar Institute, India

P. C. Moharana
National Bureau of Soil Survey and Land Use Planning, India

Indal K. Ramteke
Maharashtra Remote Sensing Applications Centre, India

R. S. Singh
National Bureau of Soil Survey and Land Use Planning, India

S. K. Singh
National Bureau of Soil Survey and Land Use Planning, India

ABSTRACT

Potato is one of the most important vegetable crops in India accounting for 20-25% of area under cultivation of vegetables and grown in a wide range of climatic conditions. It is grown in almost all states under diversified agro-climatic conditions. Nearly 80% of the crop is grown in Indo- Gangetic plains comprising Uttar Pradesh, West Bengal, Punjab, Haryana, Bihar and other parts of India like Gujarat and Karnataka. Moreover, within the country, there is a lot of heterogeneity in potato productivity depending upon mostly on management and climatic conditions. The viability of commercial potato production is influenced by spatial and temporal variability in soils, agro climate, and the availability of water resources. The inter and intra-regional variations in productivity within the country are attributed to the variations in bio-physical factors vis-a-vis specific soil-climatic requirements of the crop. The present chapter discusses the impact of climate change on the land resources requirement for potato crop with reference to Indian context.

DOI: 10.4018/978-1-5225-1715-3.ch001

INTRODUCTION

Potato (*Solanum tuberosum* L.) plays an important role in global food and nutritional security especially for the poor (Thiele et al., 2010). It is commonly known as 'The king of vegetables and it has emerged as fourth most important food crop in India after rice, wheat and maize. Indian vegetable basket is incomplete without Potato. It is an annual, herbaceous, dicotyledonous and vegetatively propagated plant. The dry matter, edible energy and edible protein content of potato make it nutritionally superior vegetable as well as staple food. Potato is a highly nutritious, easily digestible, wholesome food containing carbohydrates, proteins, minerals, vitamins and high quality dietary fiber. Fresh potato contains about 80 per cent water and 20 per cent dry matter of which 60- 80 per cent is starch. It has low fat content and high vitamin C. A single potato of 150 g can meet 100 mg of vitamin C requirement. Potato is also a good source of iron, vitamins B1, B3 and B6 and important minerals. It also contains dietary fibers, which benefit human health. (Jha, 2015). It produces more quantity of dry matter, edible energy and edible protein in lesser duration of time than cereals like rice and wheat. The crop has also shown better economic viability during the current trend of diversification from cereals to horticultural/vegetable crops. However impending global climate changes are set to alter the potato production systems in the country and hence it is imperative to examine its consequences.

Global warming also termed as greenhouse effect is the result of accelerated emission of greenhouse gases (GHGs) *viz.* carbon dioxide (CO_2), methane (CH_4) and nitrous oxide (N_2O) in to the atmosphere due to anthropogenic activities. Global warming also termed as greenhouse effect is the result of accelerated emission of greenhouse gases (GHGs) *viz.* carbon dioxide (CO_2), methane (CH_4) and nitrous oxide (N_2O) in to the atmosphere due to anthropogenic activities. Atmospheric CH_4 and $N_2O$are attributed to crop cultivation forestry and other land uses. Whereas rise in concentration of CO_2 is attributed to fossil fuel combustion. Agriculture has a minor role in contribution of CO_2 to the atmosphere. It is now established that the global atmospheric concentrations of CO_2, CH4and N_2O have increased markedly as a result of human activities since 1750. The increase in GHGs has resulted in warming of the climate system by 0.74°C between 1906 and 2005. The rate of warming has been much higher in recent decades (Dua et al., 2013).

Although increase in atmospheric CO_2 has a fertilization effect on crops with C_3 photosynthetic pathway and thus promotes their growth and productivity, on the other hand, it can reduce crop duration. Report of Working Group II of Inter-Governmental Panel on Climate Change (IPCC) and a few other global studies (Aggarwal, 2008) indicated a probability of 10–40% loss in crop production in India with increase in temperature by 2080–2100. The discussion in this chapter pertains to potato cultivation in the wake of climate changes.

Global Potato Production Scenario

The potato is the most popular food crop cultivated in the world and area under its cultivation is 4th largest following rice, wheat and maize. The major potato producing countries in the world are China, Russia, India, Poland, U.S.A., Germany and Spain. Presently, China is the biggest potato producer in the world and almost one third of all potatoes harvested globally come from China and India.

Potato Production in India

Potato is an important food crop of India and comes third after rice and wheat in terms of production and food security (Dua et al., 2015). India has made tremendous progress in potato production and the per capita availability of potato has increased from 4.37 kg in 1950 to 21.52 kg in 2012 (National Horticultural Board, 2012). India is the second largest producer of potato in the world after China and both the countries put together contribute nearly one third of the global potato production (Scott and Suarez, 2012). Potato is most important vegetable crop in India accounting for 20-25% of area under cultivation of vegetables. In India 46.609 million tons of potato production was obtained in the year 2013 - 2014 from an area of 2.032 million ha with an average productivity of 22.5 t/ha (Jha, 2015). In India, it is a major vegetable crop and is being grown in a wide range of climatic conditions (Pandit & Chandran, 2011). It is grown in almost all states under diversified agro-climatic conditions. Nearly 80% of the crop is grown in Indo- Gangetic plains comprising Uttar Pradesh, West Bengal, Punjab, Haryana, Bihar and Gujarat. The potato growing states and their distribution in different Agro-Ecological Regions (AER) and Agro-ecological sub regions (AESR) are presented in Table 1.

Though it is grown in all the three seasons with diversified soil-management practices, potato is grown largely in the *rabi* season in major production regions of Uttar Pradesh (UP) West Bengal (WB), Bihar, Gujarat and Punjab. The rainy season (*kharif*) potato production is taking place in Karnataka, Maharashtra and Chhattisgarh. However, the potato productivity in India is still very poor as compared to about 45 t/ha in many countries of Europe and U.S.A. (Bansal & Trehan, 2011). Moreover, within the country, there is a lot of heterogeneity in potato productivity depending upon mostly on nutritional management and climatic conditions (Table 2).

Table 1. Distribution of potato growing states in different AER and AESR in India

AER	AESR	States
5	5.2	Gujarat
9	9.2	Uttar Pradesh, Bihar
10	10.4	Madhya Pradesh, Maharashtra
11	-	Uttar Pradesh, Madhya Pradesh, Bihar
12	12.3	Bihar, West Bengal. Orissa
13	13.1	Uttar Pradesh, Bihar
14	14.3, 14.5	Himachal Pradesh, Uttar Pradesh
15	15.1, 15.2, 15.4	West Bengal. Assam
16	16.1, 16.2,16.3	West Bengal, Assam, Sikkim, Arunachal Pradesh
17	17.1, 17.2	Meghalaya, Assam. Nagaland, Arunchal Pradesh
18	18.2, 18.4, 18.5	Tamil Nadu, Andhra Pradesh, Orissa, West Bengal
19	19.2,19.3	Maharashtra, Goa, Karnataka, Kerala, Tamil Nadu

Source: Naidu et al., 2009

Table 2. State wise area, production and productivity of Potato India (2012)

State	Area (000 ha	Production (000 MT)	Productivity (MT/ha)
Uttar Pradesh	603.76	14430.28	23.9
West Bengal	386.61	11591.30	30.0
Bihar	322.46	6640.55	20.6
Gujarat	81.27	2499.73	30.8
Madhya Pradesh	108.87	2299.00	21.1
Punjab	85.25	2132.31	25.0
Assam	99.77	975.27	9.8
Karnataka	44.40	698.30	15.7
Haryana	29.47	676.02	22.9
Jharkhand	47.21	659.61	14.0

Source: NHB, Database, 2013

For example, Bihar is the 3rd largest potato growing and producing state of India accounting for nearly 9.8% of total potato area and 4.2% of the total potato production in the country, but with very poor productivity of 7.89 t/ha as compared to comparatively quite high productivity realized in the two neighboring states of UP (21.97 t/ha) and WB (24.7 t/ha). Similarly, productivity is very poor in Karnataka state as the climate is not very favorable for growing potato.

Potato can be grown on a wide range of soils, ranging from sandy loam to loam and best suited to acidic soils (Reddy & Shiva Prasad, 1999). It prefers friable, well aerated, fairly deep soils and well supplied with organic matter. Well-drained light soils are preferred for uniform tubers. The viability of commercial potato production is influenced by spatial and temporal variability in soils, agro climate, and the availability of water resources. Soil site characteristics and agro-climatic conditions greatly influence the choice of variety, land management practices and the economics of production. The inter and intra regional variations in productivity within the country are attributed to the variations in bio-physical factors *vis-a-vis* specific soil-climatic requirements of the crop. However, some researchers reported that the low use of fertilizers and severely imbalanced use of N, P and K fertilizers are some of the reasons responsible for low production of potato in many parts of the country (Bansal & Trehan, 2011). This study concentrates mainly on the land evaluation for growing potato crop in varied climatic conditions prevailing in India and simulation of likely impact because of climate change.

BACKGOUND

Indian agriculture has made tremendous progress in ensuring food security to its vast population with food grain production touching an on all-time record level of 263+ Mt in 2013-14 and estimated requirement in 2030 and 2050 is 345 and 494 Mt, respectively (Rattan, 2014). The country faces a daunting task of meeting future food demands and providing nutrition security to 17.5% of the global population and providing feed and fodder to 11% of the world's livestock population on 2.3% of the land and only 4% of the fresh water resources. The per capita arable land shall shrink from 0.34 ha in 1950-51 to 0.08 ha in 2025. Since, 1970, the net cultivated area has remained around 142 Mha and highly productive agricultural land has been lost due to urbanizaation, industrialization and other developmental activities which implies that out of total 120.8 Mha of degraded land, more marginal and degraded lands are being progressively brought under cultivation. Natural resources viz. soil, water and biodiversity have been over exploited and have been consequentially degraded due to loss of organic matter, depletion of nutrients, over exploitation of ground water and depletion of carrying capacity of pastures, grazing land etc. posing severe health hazards which need immediate attention.

LAND EVALUATION AND SOIL-SITE REQUIREMENT FOR POTATO PRODUCTION

The natural resources (soil, climatic features and water) profoundly influence the cropping patterns and control crop productivity in specified region. It is well known that the adaptability of crops in one or the other area is the interaction between edaphic properties and existing climatic features. It is also well known that each plant species requires specific soil-site conditions for its optimum growth. Although, the State -Agricultural Universities and Research Institutes have generated voluminous data for crop production through experimentation, it has not been correlated adequately to work out soil-site suitability models for optimizing agricultural land use in the country (Naidu et al., 2009).

Of late, the importance of land evaluation for better land use options has been receiving greater importance. An efficient land use maintains agricultural sustainability. Land evaluation has been defined by FAO (1976) as "the process of assessment of land performance when used for specified purpose. It involves the execution and interpretation of surveys and the studies on landforms, soils, vegetation, climate and other related aspects of land for a comparison between the promising land use and/or specific land use". Soil - site suitability studies provide information on the choice of crops to be grown on best suited soil unit for maximizing crop production per unit of land, labour and inputs. The land suitability for defined use and the impact of the use on environment is determined by land conditions and land qualities (Table 3). The sustainable land use depends on soil resilience. It is a balance between soil restorative and soil degradation processes. Ecologically every factor of environment exerts directly or indirectly a specific effect on growth and development of the plant. The factors such as, the water, temperature, sunlight, soil aeration and availability of plant nutrients directly control the crop growth. However, it varies from habitat to habitat and determines the suitability of a plant to any particular environment. For planning and effective utilization of soil resources, the information relating to the soil-site characteristics for cultivation of crops is necessary.

Each plant species requires specific soil-site conditions for its optimum growth. For rationalizing land use, the soil-site suitability for different crops needs to be determined. The suitability models provide guidelines to decide the policy of growing most suitable crops, depending on the suitability/capability of each soil unit.

The adaptability of crops in the area is the interaction between existing edaphic conditions and fitness of the cultivar under these conditions. Plant growth requires a reasonable range of moisture and nutrient supply linked to rooting depth, photosynthesis and in turn the biomass production. Each crop has specific growth requirements. Some of the crop requirements are not always directly measurable in the field. They are to be derived from other associated observations. Many workers attempted

Table 3. Important land qualities and related soil characteristics

Land quality	Land/Soil characteristics
Temperature and light energy for plant growth	Temperature (Max. and Min.), Sunshine hours and day length
Moisture availability in crop growing season	Rainfall, relative humidity frost. Potential evapotranspiration Soil depth and texture and length of growing period
Root development and anchorage	Soil depth, texture in root zone, structure and hard pans
Oxygen availability to roots	Drainage, depth of ground water table, frequency and period of flooding, moisture retention capacity of soils
Nutrient availability in root zone	Organic matter, CEC, base saturation, NPK status and pH
Sensitivity to Soil toxicity	pH, salinity, sodicity, $CaC0_3$, Al and heavy metals
Workability and management	Slope, surface stoniness/rockiness, moisture retention

(Sehgal et al., 1989; Sys et al., 1991; Davidson, 1992) to identify and specify the suitability criteria for different annual crops through land evaluation approach using multivariate regression models with the help of soil parameters and yield data.

The term Land Evaluation was first used in 1950 at the Amsterdam Congress of the International Society of Soil Science, where Visser (1950) presented a paper entitled, 'The trend of the development of land evaluation in the future'. Stewart (1968) then defined land evaluation as assessment of suitability of land for use in agriculture, forestry, engineering, hydrology, recreation, etc. According to Van Wambeke and Rossiter (1987) land evaluation is the ranking of the soil unit on the basis of its capabilities (under given management levels and socio-economic conditions) to provide optimum returns per unit area besides conserving the natural resources for future use. In other words, it is the process of estimating the potential of land for alternative kinds of use. The basic feature of it is the comparison of the requirements of a crop with the resources offered by the land (Dent & Young, 1981). The land evaluation thus is the assessment of land performance for a specific purpose. It involves the interpretation of basic data on the climate and the characteristics of land under the existing land use (Sehgal, 1991). The principal objective of land evaluation is to select the optimum land use for each defined land unit taking into account both physical and socio-economic considerations and the conservation of environmental resources for sustainable use (FAO, 1983).

The generally used methods for land evaluation are Storie index rating (Storie, 1933). Land and soil irrigability classification (USBR, 1953), USDA land capability classification (Klingebiel & Montgomery, 1961), Productivity Index (Requier et al. 1970) and land suitability classification (FAO, 1976; Sys 1985, 1993). These methods differ from each other in the original purpose for which they were proposed, in terms of terminology, in the number and kind of soil properties taken into account, and in the logic of the procedures followed to arrive at a suitability rating (Van Diepen et al. 1991). The Land Evaluation proposed by FAO (1976) defines the basic concepts and principles followed universally. The basic concepts include the land and its major use, utilization type, characteristics, qualities and diagnostic criteria. Land includes soil, vegetation, hydrology, landform and climate. The framework suggested the classification of land in to different categories viz., Orders, Classes, Subclasses and Units. There are two orders namely, 'S' for suitable lands, and 'N' for non- suitable lands; further, three classes (SI, S2, S3) within the order 'S' and two classes (N1, N2) under the order 'N' depending upon the degree of limitations with respect to specific land use. The appraisal of the classes, within the order is done according to the land limitations. The Subclasses reflect the kinds of limitations or major kinds of improvement measures required within these classes. They are indicated by the symbols using lower case letter following the Arabic numeral.

In the land evaluation, there are four steps namely:

1. Characterization of existing soil, climatic and land use conditions
2. Development of soil-site criteria or crop requirements
3. Matching of the crop requirements with the existing soil and climatic conditions and
4. Choosing of the best fit among the crops and selecting the same as alternative crop strategy.

Development of the Soil-Site Criteria

Among the above four steps given above the formulation of the soil-site criteria to meet the crop requirements forms a vital and important step. The land and soil-site characteristics considered in land evaluation are climate, topography and landscape, wetness conditions, physical and chemical soil conditions (texture, depth, CaC03, etc.) and soil fertility characteristics (CEC, BS, pH, EC and ESP). The soil-site characterization are expressed in terms of degree of limitations (0, I, 2, 3, or 4) as per their optimality with respect to specific land use (Table 4). Limitation '3' is considered critical at which the expected yields decline significantly (upto 50 per cent) and the cultivation is considered marginally economical. The degree of limitations as proposed by Sys (1985), ranging from 0 to 4 suggesting no limitation to severe limitation level was used for determining the criteria of the suitability classes. These soil-site criteria are to be matched with the land qualities of each mapping unit of a study area to arrive at land suitability class.

Table 4. Climatic and soil-site limitations

Limitations	Symbol	Limitations	Symbol
Climatic hazard	c	Erosion hazard	e
Flood hazard	f	Gravelliness	g
Workability	k	Topography	t
Moisture availability	m	Nutrient availability	1
Crusting	p	Rooting condition	r
Texture	s	Drainage	w
Excess of salt / calcareousness	m		

Matching Crop Growth Requirements with Environmental Data

It is an exercise of comparing existing climatic, soil and physiographic conditions with the soil-site criteria with respect to individual crop. The matching of land qualities with crop requirements exercise includes two steps viz. climatic and a soil-physiographic evaluation. Initially, the climate of the soil unit is compared with that of the crop requirements. For this the specific soil and crop-linked growing season is calculated, using rainfall data, in particular, consumptive use of water of the plant as obtained from the calculated potential evapotranspiration (PET), the crop factor and the soil moisture storage capacity.

The second step refers to the comparison of the individual soil and physiographic properties with the crop requirements in terms of nutrient supply, rooting depth, susceptibility to toxic elements and workability. On the basis of the degree and number of limitations identified the suitability class is established, viz. highly suitable (SI), moderately suitable (S2), marginally suitable (S3) and unsuitable lands (N 1 and N2) for specific kind of land use. S 1 classes correspond to areas which have a yield potential above 80% of the maximal attainable harvest within the climatic area. This figure drops to 60% and 40% for classes S2, and S3 respectively. The soil-site requirement for optimum potato production under Indian conditions worked out by Reddy and Shiva Prasad (1999) is given in Table 5.

CLIMATE CHANGE AND ITS IMPACT ON LAND SUITABILITY AND POTATO CULTIVATION IN INDIA

Climate change is the largest threat ever faced by the world as it affects the earth's natural resources widely from tropical to arctic and from sea to land and atmosphere

Table 5. Soil-site suitability criteria (crop requirements) for potato

Soil-site characteristics			Rating			
		Unit	Highly suitable S1	Moderately suitable S2	Marginally suitable S3	Not suitable N
Climatic regime	Mean temperature in growing season	°C	16-25	26-30 13-15	31-32 10-12	>32 <10
Land quality	**Land characteristics**					
Oxygen availability to roots	Soil drainage	class	Well drained	Moderately /imperfectly drained	Poorly drained	Very poorly drained
Nutrient availability	Texture-surface	Class	sl, I, ls	s, scl	sil, cl	Heavy c
	Sub-surface texture	class	scl, sil	s, sil	s	Heavy c
	pH	1:2.5	5.5-6.5	6.6-8.2 5.0-5.4	>8 <5	
	CEC	C mol (p+)/ kg	>16	<16	<5	
	OC	%	High	Medium	Low	
Rooting conditions	Effective soil depth	cm	75-100	50-75	25-50	<25
	Stoniness	%	0-10	10-15	15-35	>35
Soil toxicity	Salinity (Ece)	dSm^{-1}	>16	<16		
	Sodicity (ESP)	%	Non sodic	10-15	>15	
Erosion hazard	Slope	Hills %	<5	5-10	10-15	>15
		Plains %	<3	3-5	5-8	>8

Source: Reddy and Shiva Prasad, 1999

(Pant, 2009). According to IPCC (2007) reports, warming of the climate system is unequivocal and scientists are more than 90% certain that it is primarily caused by increasing concentrations of greenhouse gases (CO_2, CH_4, O_3, CFCs and N_2O) produced by human activities such as the burning of fossil fuels and deforestation.

The effects of climate change on crop production can be complex. Depending on the temperature regime and the crop, high temperatures can lead to low yields due to increased development rates and higher respiration. However, a short growth cycle can also be beneficial, e.g., to escape drought or frost, and the use of late-maturing cultivars could offset the effect of high development rates.

In environments where low temperatures now limit production, global warming could lead to a beneficial lengthening of the growing season and temperatures close to optimal for assimilation. Moreover, global warming is related to the increase of atmospheric CO_2, con- centration, which is likely to increase crop yields, particularly when water limits crop production (Nonhebel, 1993). Increase in atmospheric CO_2 has a fertilization effect on crops with C_3 photosynthetic pathway as Rubisco, the primary enzyme in leaf photosynthesis of C_3 plants, competes better with dissolved O_2 for binding sites on the Rubisco protein when CO_2 concentration is higher and leads to increased photosynthesis and promotes growth and productivity (Allen & Prasad, 2004). However, on the other hand, it can reduce crop duration.

Potato is grown in many different environments, but it is best adapted to temperate climates. Increase in temperature above optimal is reported to decrease in crop yields, particularly due to increased respiration rate (Takashi et al., 2007). At high temperatures (above 17^{0}C; Stol et al. 1991) tuberization diminishes (Reynolds and Ewing, 1989a). Potato is also frost sensitive and severe damage may occur when temperature drops below 0^{0}C (Hijmmans et al. 2003). In this regard various reports on climate change have given their views on rise in temperature in the coming years. It is likely that the currently observed trend of global warming, which has been 0.6 °C ± 0.2 since 1900, will continue and that the average global temperature will increase by between 1.4 and 5.8 °C over the period 1990 to 2100 (Houghton et al. 2001). For the Indian region (south Asia), the IPCC, (2007) has projected 0.5-1.2^{o}C rise in temperature by 2020, 0.88-3.16 ^{o}C by 2050 and 1.56-5.44 ^{o}C by 2080. Report of Working Group II of IPCC and a few other global studies (Aggarwal, 2008) indicate a probability of 10–40% loss in crop production in India with increase in temperature by 2080–2100. This would lead to more frequent heat extremes, floods, droughts, cyclones and gradual recession of glaciers, which in turn would result greater instability in food production (Aggarwal, 2008). Having look on these data on changes in temperature over the years, it seems that the classes of land suitability in respect of climatic region with special reference to mean temperature during growing season will go on changing and the lands categorized as highly suitable S1 class may become moderately suitable (S2), while moderately suitable S2 will become marginally suitable (S3) class and marginally suitable S3 class will change to unsuitable class up to 2080. As far as the rise in temperature during growing season is concerned in the coming years there will be great loss in potato production in the arid and semi arid regions of India compared to northern India. Increase in temperature and atmospheric CO_2, both are interlinked and occur simultaneously and the CO_2 enrichment does not appear to compensate for the detrimental effects of higher temperature on tuber yield (Singh *et al.*, 2010).

The climate model projections for the Fifth Assessment Report of the IPCC made using the newly developed representative concentration pathways (RCPs) under the Coupled Model Inter-comparison Project 5 (CMIP5) (Chaturvedi et al. 2012) reveals that surface air temperature including night time temperatures are expected to increase further. The all India rainfall and extreme rainfall events are also expected to increase in future. Under the business-as-usual scenario, mean warming over India is likely to be in the range of 1.7-2.0 °C by 2030s and 3.3-4.8 °C by 2080s relative to pre-industrial time. Likewise, all India precipitation is projected to increase by 4-5% by 2030s and by 6-14% by 2080s compared to the 1961-1990 baselines. There is consistent positive trend in frequency of extreme precipitation days (e.g. > 40 mm/day) for decades 2060s and beyond.

Climate change impacts are variable in different parts of the country. Western Rajasthan, southern Gujarat, Madhya Pradesh, Maharashtra, northern Karnataka, northern Andhra Pradesh, and southern Bihar are likely to be more vulnerable in terms of extreme events (Mall et al. 2006). For every 1°C rise in temperature, yield of wheat, soybean, mustard, groundnut and potato are expected to decline by 3-7 percent (Agrawal, 2009).

Potato can be grown on a wide range of soils, ranging from sandy loam, silt loam, loam and clay soils. It prefers friable, well aerated, fairly deep soils well supplied with organic matter. Well-drained sandy loam and medium loam soils, rich in organic carbon are most suitable for potato. Light soils are preferred because they tend to promote more uniform tubers and make harvesting of the crop easier. Studies on climate change impact on soils of the US Great Plains (Follet et al. 2012) shows that soil carbon and nitrogen stocks are strongly negatively related mean annual temperature and positively related to the ratio of mean annual precipitation to potential evapo-transpiration, suggesting that they are equally vulnerable to increased temperature and decreasing water availability. Crop is well suited to acidic soils (pH 5.0-6.5).

Rising temperature and decease in precipitation would also lead to deterioration of water quality. In north east India under the business-as-usual scenario, soil acidification may further intensify under the influence of rising atmospheric CO_2 concentration. This possibility stems from a frequent experimental observation of increased CO_2 production in soil due to increased root and soil microbial respiration under elevated CO_2 atmosphere. As the CO_2 produced this way forms carbonic acid (H_2CO_3) in soil waters, which removes base cations from the soil systems after leaching and produce soil acidity, enhanced CO_2 production in soil under elevated CO_2 can increase carbonic acid leaching and therefore intensify the already existing acute problem of soil acidity in north east India (Kumar, 2011).

Majority of the Indian soils are unhealthy in terms of very low in organic matter content, nutrient status and degradation. About 120.8 Mha constituting 36.5% of total geographical area of the country are degraded due to soil erosion, salinity/alkalinity, soil acidity, waterlogging and other problems (NBSS & LUP, 2008). There is no doubt about climate change as it is happening very rapidly resulting in drastic effects on the environment, particularly on soil system. Soils are the outcome of different soil forming factors and processes which are directly linked to the atmospheric-climate system through exchange of various gases, solar radiation, temperature, hydrologic cycles (Brevik, 2012). Soil erosion rates, driven by the climatic variables like precipitation and number of rainy days and raining intensity, are sensitive to global climate changes. Soil erosion, major factor causing a decline in agricultural productivity and impairing environmental quality, reduces soil fertility and water availability. Direct impacts on soil erosion include changes in the erosive power of rainfall due to changes in rainfall amounts and intensities (Nearing, 2001); Indirect impacts suggest changes in soil erodibility and soil surface covers.

Increased precipitation will increase the surface runoff in hilly lands; increase infiltration and water storage within the soil in the flat lands; enhance ground water recharge in the highly permeable and well drained soils; and increase the evaporation on soils having low infiltration and transpiration in the case of well-developed canopies (Varallyay, 2007). Soil erosion due to water is the major cause of soil degradation (82.6 Mha) followed by chemical degradation (24.7 Mha). About 5.34 billion tonnes (Gt) of soil is eroded in India at an average rate of 16.3 t ha^{-1} yr^{-1} (Dhruvanarayana & Ram Babu, 1983). While 61% of eroded sediments get redistributed on land, nearly 29% are lost permanently to the sea. The remaining 10% are deposited in reservoirs reducing their holding capacity by 1 to 20% annually (Sharma & Singh, 2012).

Rise in temperature will increase the potential evapo-transpiration and decrease the surface runoff, infiltration, water-storage and ground water recharge, especially if accompanied by low precipitation. The influence of global warming on soil processes is more complex and needs elaborate discussion.

FUTURE RESEARCH DIRECTIONS

There is uncertainty about the future climate on different parts of the globe which may have positive and negative impact on the agricultural production and food security. The crops like potato which are more sensitive to soil and climatic conditions may affect badly due to climate change. Better understanding of past, recent and possible future changes in the climate is an important element for developing suitable soil climatic criteria based suitability of cultivation. Further research is needed ideally be targeted at locally relevant climate impacts and risks. The information provided in the chapter is of value to a wide range of different stakeholders in potato production system eager to learn more about future weather and climate of the Earth and its consequences on potato cultivation. Nevertheless, there is broad agreement that the climate is set to shift significantly, and that new cultivation techniques are needed in order to meet the growing demand of potato. Further, research is needed on the improved varieties of potatoes suitable for wide range of soil and climatic conditions.

CONCLUSION

The world's climate is changing, and the changes will have an enormous impact on the globe. There is threat that climate change will impact potato availability, access, and utilization due to change in the climatic parameters like temperature, precipitation, CO_2.etc. It is also expected to increase soil salinity in many part of the globe which may reduce the suitability of potato cultivation on such soils. In addition, the change in environmental parameters may change availability of water for cultivation of this important crop. However, the demand and changing food demand of growing population is crucial and will exaggerate the problem to greater extent. The impact of non-availability of favorable conditions for potato cultivation will lead to more poverty and degradation of land resources. Though the aggregate impact of climate change on future potato cultivation is not fully understood. The changing climate is playing central role in availability of sufficient supply of potato and individual climatic parameters and their interactions are equally important. There is an emerging consensus that systematic soil site analysis will help farmers in decision making on potato cultivation.

REFERENCES

Aggarwal, P. K. (2008). Climate change: implications for Indian agriculture. In S. S. Lal, P. M. Govindakrishnan, V. K. Dua, J. F. Singh, & K. Pandey (Eds.), *Impact Assessment of Climate Change for Research Priority Planning in Horticultural Crops* (pp. 1–10). Central Potato Research Institute, Shimla. Retrieved August 17, 2016, from http://krishikosh.egranth.ac.in/bitstream/1/2054278/1/CPRI034.pdf

Aggarwal, P. K. (2009). Global climate change and Indian agriculture: Case studies from ICAR network project. Indian Council of Agricultural Research.

Allen, L. H., Jr., & Prasad, P. V. (2004). Crop responses to elevated carbon dioxide. Encyclopedia of Plant and Crop Sci., 346-348.

Brevik, E. C. (2012). Soil and Climate Change: Gas Fluxes and Soil Processes. *Soil Horizons, 53*. Retrieved August 04, 2016, from https://www.researchgate.net/publication/262260059_Soils_and_Climate_Change_Gas_Fluxes_and_Soil_Processes

Chaturvedi, R. K., Joshi, J., Jayaraman, M., Bala, G., & Ravindranath, N. H. (2012). Multi-model climate change projections for India under representative concentration pathways. *Current Science, 103*, 791–802.

Dua, V. K., Singh, B. P., Govindakrishnan, P. M., Kumar, S., & Lal, S. S. (2013). Impact of climate change on potato productivity in Punjab – a simulation study. *Current Science, 105*(6), 787–794.

Haverkort, A. J. (1990). Ecology of potato cropping systems in relation to latitude and altitude. *Agricultural Systems, 32*(3), 251–272. doi:10.1016/0308-521X(90)90004-A

Houghton, J. T., Ding, J., Griggs, D. J., Noguer, M., van der Linden, P. J., & Xiaosu, D. et al. (2001). *The Scientific Basis. Contribution of Working Group I to the Third Assessment Report of the Intergovernmental Panel on Climate Change.* Cambridge University.

IPCC. (2007). *The Physical Science Basis Contribution of Working Group I to the Fourth Assessment Report of the Intergovernmental Panel on Climate Change.* Cambridge, UK: Cambridge University Press.

Kumar, M. (2011). North East India: Soil and water management imperatives for food security in a changing climate? *Current Science, 101*, 1119.

NBSS & LUP. (2008). *Harmonization of Degraded lands/Wasteland Dataset of India- GIS- based Approach.* Nagpur: NBSS&LUP.

Nearing, M. A. (2001). Potential changes in rainfall erosivity in the U.S. with climate change during the 21st century. *Journal of Soil and Water Conservation, 56*, 229–232.

Nonhebel, S. (1993). *The importance of weather data in crop growth sim- ulation models and assessment of climatic change effects* (Ph.D. thesis). Wageningen Agricultural University, Wageningen, The Netherlands.

Pant, K. P. (2009). Effects of Agriculture on Climate Change: A Cross Country Study of Factors Affecting Carbon Emissions. *J. Agric. Envi., 10*, 72–88.

Rattan, R. K. (2014). Soil processes and climate change. *Journal of the Indian Society of Soil Science, 62*(Supplement), s5–s24.

Reynolds, M. P., & Ewing, E. E. (1989a). Effects of High Air and Soil Temperature Stress on Growth and Tuberization in *Solanum tuberosum*. *Annals of Botany*, *64*(3), 241–247. Retrieved from http://aob.oxfordjournals.org/content/64/3/241.short

Sharma, P. D., & Singh, M. V. (2012). State of health of Indian soils. In N. N. Goswami, B. Singh, D. K. Pal, R. K. Rattan, & D. L. N. Rao (Eds.), *Soil Science in the Service of Nation* (pp. 191–213). New Delhi: Indian Society of Soil Science.

Singh, J. P., Lal, S. S., Govindakrishnan, P. M., Dua, V. K., & Pandey, S. K. (2008). Impact of climate change on potato. In S. S. Lal, P. M. Govindakrishnan, V. K. Dua, J. F., Singh, & K. Pandey (Eds.), *Impact Assessment of Climate Change for Research Priority Planning in Horticultural Crops* (pp. 125–136). Central Potato Research Institute, Shimla. Retrieved August 17, 2016, from http://krishikosh.egranth.ac.in/bitstream/1/2054278/1/CPRI034.pdf

Stol, W, de Koning, G. H. J., Haverkort, A. J., Kooman, P.L., van Keulen, H. & Penning de Vries, F. W. T. (1991). *Agro-ecological characterization for potato production. A simulation study at the request of the International Potato Center (CIP), Lima, Peru.* CABO-DLO, Report 155.

Varallyay, G. (2007). Potential impacts of climate change on agro-ecosystems. *Review Agriculturae Conspectus Scientificus*, *72*, 1–8.

Chapter 2
Nutrient Management for Sustainable Potato Production in India:
New Initiative

R. P. Sharma
National Bureau of Soil Survey and Land Use Planning, India

V. K. Dua
Central Potato Research Institute, India

M. K. Jatav
Central Institute for Arid Horticulture, India

Manoj Kumar
Central Potato Research Institute, India

ABSTRACT

Potato is important crop for solving food and nutritional security problem of growing population of India. Application of N in two split dose i.e. half at planting time and rest at time of earthing up produce higher yields and higher N recovery. At the time of planting, calcium ammonium nitrate or ammonium sulphate should be preferred by furrow application. Selection of suitable variety may play major role beside time and method of application in improving nutrient use efficiency. Balanced use of major and micronutrients plays an important role in improving quality of produce besides good yield. Potato based cropping system mostly shows build up of P and negative balance of N and K which may be overcame by organic residues recycling. Intensive cropping system has resulted in wide spread deficiency of secondary and micro nutrients particularly Zn and these must be applied on soil test basis. Integrated nutrient management is a must for an exhaustive and responsive crop like potato.

DOI: 10.4018/978-1-5225-1715-3.ch002

INTRODUCTION

The book chapter "Nutrient Management for sustainable potato production" deals with the problems and potentials of natural resources of India in relation to Potato production. The results presented in this chapter are summarized findings of the research conducted in India; its agricultural universities and Indian council of Agricultural Research (ICAR). Nutrient management (major, secondary and micronutrients) for potato production in tropical and subtropical climatic are discussed with the findings of multi-location trials of All India Coordinated Research Projects. The potato production, area and productivity dynamics are also discussed with time series of Indian continent. Knowledge of improved nutrient management for potato growing area can increase production potential of land resources, improve soil and fruit quality and reduce economic burden of low income group of farmers.

BACKGROUND

Potato, a carbohydrate rich food highly popular worldwide, can be important crop in solving food and nutritional security problem of India. This crop is third in production after rice and wheat with only 0.8% of gross cropped area. Production of rice, wheat and maize is 99.15, 80.58 and 19.29 m tonnes from 45.35, 27.88 and 8.19 m ha of land, whereas, potato produces 28.47 m tonnes from 1.55 m ha of land area. In addition, potato produces more dry matter (47.6 kg/ha/day) and edible protein (3 kg/ha/day) than the major cereal food crops and therefore, requires higher amount of nutrients on per day basis. Freshly harvested potato tubers contain about 80% water and 20% dry matter of which 70% is starch. On the dry weight basis, the protein content of potato is similar to that of cereals and is very high in comparison with other root and tuber crops. The quality of protein in potato is very high with its biological value similar to egg. In addition, the potato is low in fat (0.1%) and energy (80 k cal/100g edible portion) and is rich in several nutritional components, especially vitamin C (17 mg/100 g edible portion). The potato is a moderate source of Iron, and the non-haem form of Iron is more readily available for absorption by intestines in the presence of ascorbic acid. It is a good source of vitamins like thiamine, riboflavin and niacin and minerals such as potassium, Phosphorus, Magnesium, Calcium, Sodium, Iron and Zinc. It can supply at least part of the daily requirement of trace elements like Copper, Manganese, Molybdenum and chromium. Dietary antioxidants and fibre in potato tubers take part in preventing diseases related to ageing and benefit health (Ezekiel *et al.,* 1999).

Potato has emerged as one of the most important cash crop due to its fitting into many cropping systems and resulting into expansion of area of the crop. The short duration nature of this crop makes it convenient to fit it into diverse type of cropping systems in various agro-ecoregions. Potato is a unique crop in the sense that it can be harvested flexibly early or late depending upon market price and requirement of field for any other subsequent crop. Being line planted crop, it has also been identified as one of the most suitable crop for inter and relay cropping with other prevalent crops of different agro-ecoregion. No other field crop gives an opportunity to double the variable cost invested in form of gross return in just 70 to 80 days. However, this requires very intelligent farming approach and adoption of all appropriate technologies. These facts have led to development of a number of highly profitable cropping systems suitable for different zones. The diversification of conventional rice-wheat system into potato based systems like rice-potato-wheat or rice-potato-onion almost doubles the total return from the farm.

This crop is very labour intensive as most of the operations starting from planting to harvesting and storage and marketing require lot of man power thus cultivation of this crop generates very high farm employment. Moreover, to this for a marginal and small farmer this crop become a livelihood crop as from a very small piece of land in very short period using his household labour, more produce is obtained. For a country like India, where human labour is surplus and food and nutritional security is still a problem this crop has both advantage of producing more of quality food and adding to the farm employment.

Area, Production, and Productivity Trends

Potato is among few agricultural commodities which have maintained positive growth of production in last decade. Major food crops showed either negligible increase or a decrease in the overall production and yield after the year 1997 with annual compound growth rate between -2.83 to 1.14. Whereas, production of potato jumped from 1.54 mt in year 1949-50 to 24.22 mt in 1996-97 and 28.47 mt in 2007-08 and reached 41.5 mt in year 2013-14 (Figure 2). The area under potato cultivation increased from 1135,000 ha in 1991 to 1973,000 ha in year 2013-14 (Figure 1). Productivity at national level increased by 27 times from 1949-50 to 2013-14. (Figure 3) (Indian Horticulture Database, 2014).

Potato is cultivated almost throughout the country; however, its acreage distribution varies. The northern plain, also called Indo-Gangetic plain comprising of north western, west central and north eastern plains accounts for almost 86% of the potato area of the country. The plateau region of Karnataka, Gujarat and Maharashtra accounts for about 8% of the potato area. The hill zone including north western, north eastern and southern hills accounts for about 6% of the potato area of the country. Among different states maximum area and production comes from Uttar Pradesh followed by West Bengal and Bihar (Table 1 & Figures 4 & 5). Productivity wise,

Table 1. Top Ten Potato Producing States: 2013-14

State	Area ('000 ha)	Production ('000 t)	Productivity (mt/ha)
Uttar Pradesh	564.3	13808.8	24.5
West Bengal	412.3	9030	21.9
Bihar	318.5	6535	20.5
Madya Pradesh	110	2322.4	21.1
Gujarat	73.6	2267.4	30.8
Punjab	87.2	2189.2	25.1
Assam	98	700.1	7.1
Haryana	30	696.5	23.2
Jharkhand	49.1	653.1	13.3
Karnataka	40.7	539.7	13.3
Others	189.7	2813.2	14.8
Total	1973.4	41555.4	215.6

Source: Indian Horticulture Database, 2014

Figure 1. Area dynamics of potato in India during last three years

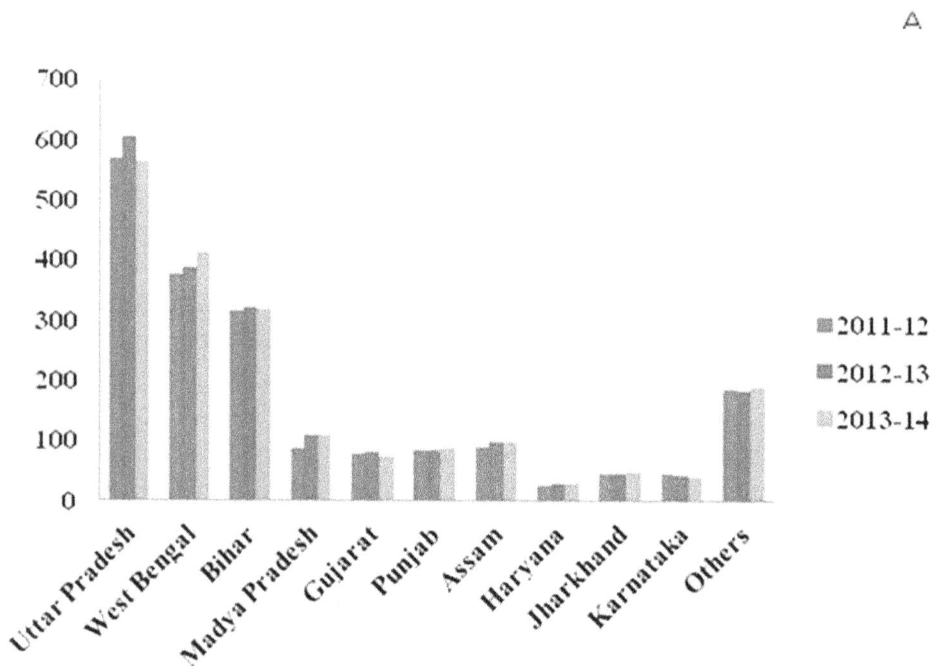

Figure 2. Production dynamics of potato in India during last three years

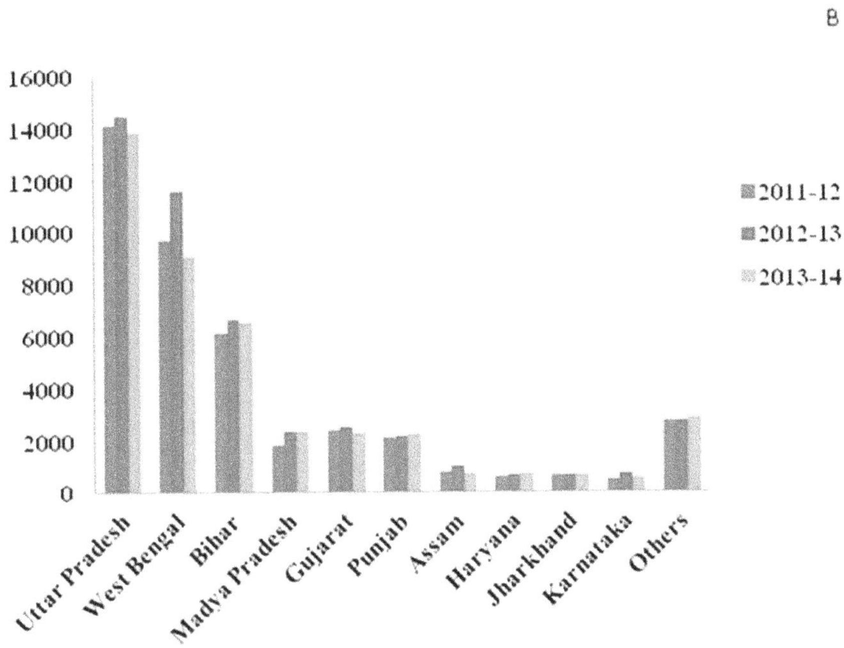

Figure 3. Productivity dynamics of potato in India during last three years

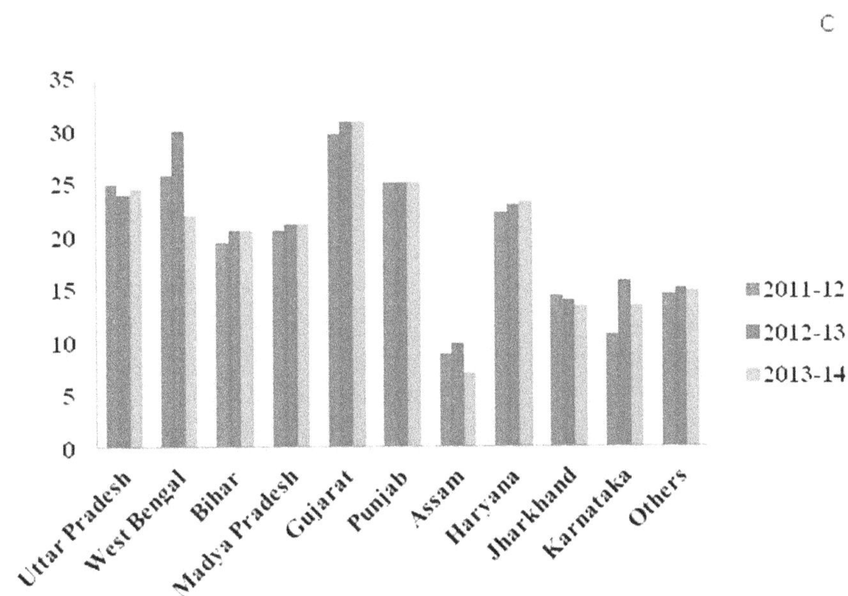

Figure 4. Top 10 potato growing states of India (2012-13)

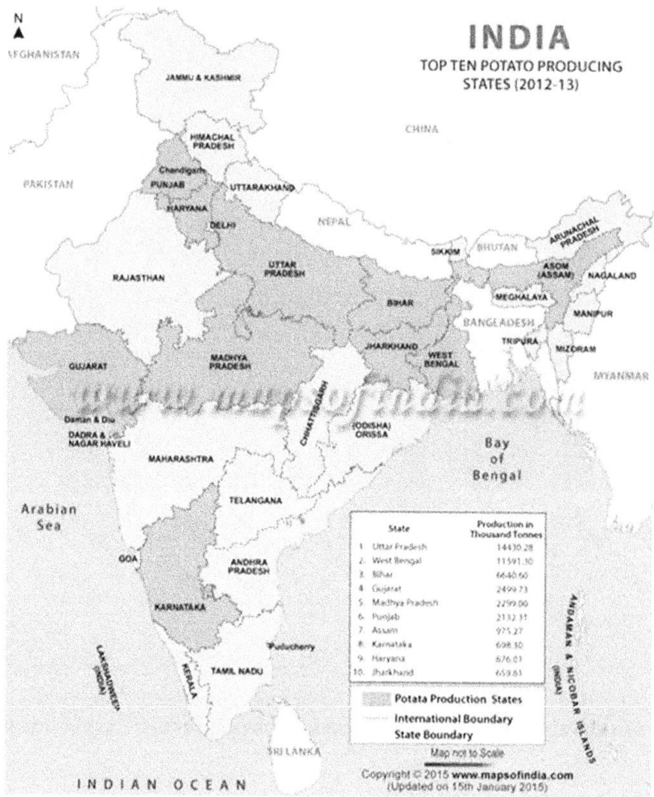

however, Gujarat tops the list followed by Punjab, Uttar Pradesh and West Bengal. Increase in area and production during 1980-2007 was maximum in West Bengal (193 & 258%) followed by Utter Pradesh (81 & 199%) and Punjab (73 & 77%).

Nutrient Management

The basic philosophy of nutrient management should be to apply fertilizers at rates to ensure high fertilizer use efficiency so that the amount of unutilized fertilizers is reduced to environmentally acceptable levels. Used properly, fertilizers can, in fact, lead to better environment because greater vegetative growth and higher crop yields leave large quantities of residues in the soil, thus reducing erosion; absorb more carbon dioxide, a major source of green house effect and lessen the need to bring marginal and forest land into cultivation. The declining or stagnating yield trend in India has been attributed to multiple nutrient deficiencies and imbalance of nutrients.

Figure 5. Leading potato growing states of India

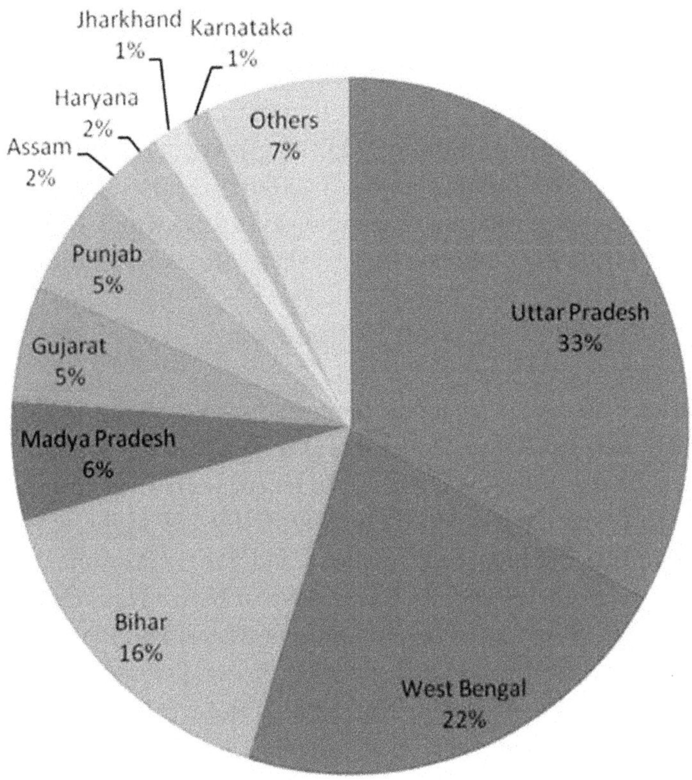

Nutrients need of potato depends upon the tuber yield, response to applied nutrients from fertilizers, agro-climatic conditions, variety and nutrient availability from the soil pool. The nutrient status of the soil further depends upon the quantity of total and available nutrients present in the soil pool.

MAJOR NUTRIENTS

Nitrogen

The Indian soils are generally deficient in organic matter thus application of Nitrogen in form of fertilizers and manures become indispensable to meet N needs of the crop. Significant response to N application has been reported in different agro-ecological regions of the country, however, magnitude of response and optimum dose differed with soil type, cultivar and climatic conditions. In plains, as the crop is taken under short day and irrigated conditions, optimum temperature, sunshine hours and moisture plays an important role in getting high responses. While in north-western and eastern hills where the crop is grown under rain fed and long day conditions, the soils are, although, rich in total N yet require high N dose because of slow mineralization rate of organic matter owing to low temperature prevailing at initial growth stage and luxury vegetative growth during tuber bulking phase. The fertilizer use efficiency for N by potato is 30-50 per cent (Sharma & Upadhayay, 1993) which can be increased considerably if following components are taken into consideration.

Varietal Differences in Nitrogen Requirement

Potato varieties differ widely among themselves with regard to their N needs (Trehan, 2003, Trehan, 2006). Thus there is a need to keep in view the variety grown while applying Nitrogen fertilizer. In recent years with newer cultivar the optimum dose of Nitrogen was in the range of 189-239 kg/ha at Patna (Singh et al., 2008a). Kufri Pukhraj, Kufri Ashoka and Kufri Chipsona-1 yielded 23, 13 and 13% more at 15, 34 and 4% lower dose of N, respectively as compared to standard cultivar Kufri Jyoti at economic optimum and therefore, it can be concluded that newly released cultivars were better yielder at lower doses of N (Singh et al., 2008b). At Patna, Kufri Pukhraj gave response of 1.84 q/kg of Nitrogen at optimum Nitrogen level, whereas, Kufri Jyoti and Kufri Giriraj responded 1.03 and 1.14 q/kg Nitrogen applied, respectively (Kumar et al., 2009a). At Ooty in the Nilgiri hills, Kufri Swarna was found more N efficient than Kufri Giriraj and Kufri Jyoti (Anonymous, 2006).

Among ten cultivars tested at Jalandhar, Punjab (Trehan, 2006), the most N efficient cv. Kufri Pukhraj, which give particular fixed yield with the lower dose of N as compared to other cultivars, had the highest agronomic efficiency (kg tubers produced/kg N applied) and the least N efficient cv. Kufri Jyoti had the lowest agronomic efficiency (Figure 6). The cultivar Kufri Pukhraj had the highest physiological efficiency both with inorganic fertilizer alone and with combined application of fertilizer and green manure. However, the N uptake efficiency of this cultivar was similar to that of Kufri Jyoti in the presence of inorganic N alone but was higher with combined application of fertilizer and green manure, showing that Kufri Pukhraj had the higher capacity to use/take up organic N efficiently than least N efficient cultivar Kufri Jyoti. It was also observed that cultivars Kufri Sutlej and Kufri Sindhuri had the highest total (soil+fertilizer/manure) N uptake/use efficiency and highest fertilizers/manure N recovery among the ten cultivars tested.

Figure 6. Agronomic efficiency of different potato cultivars for nitrogen (kg tuber/kg N applied)
Source: Trehan, 2006

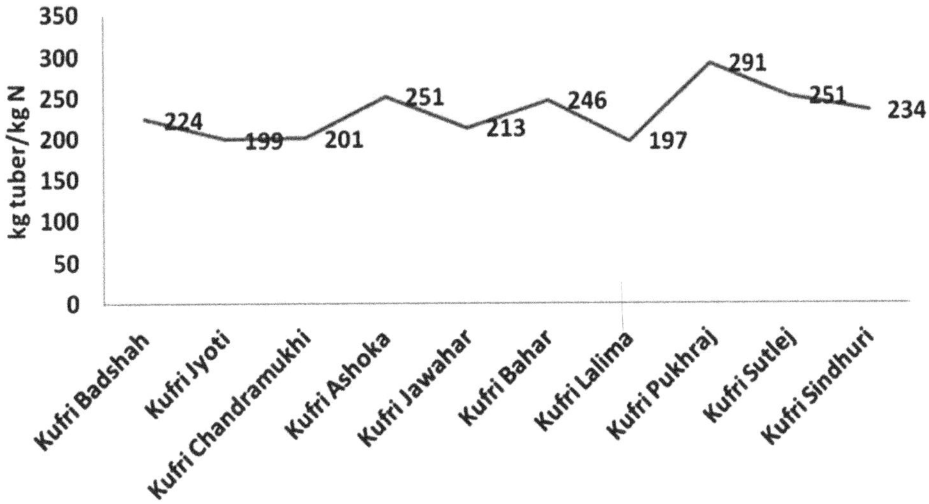

Sources of N Fertilizer

The performance of different sources of N varies with the agro-climatic zone and soil type. Among the N fertilizers available, ammonium sulphate has been found to be the best Nitrogenous carrier for potato. The better performance of ammonium sulphate might be attributed to the S supplied by it as it contains 25 per cent S (Sud & Grewal, 1994). However, because of its high cost it may not be within reach of small and marginal farmers but can be of great use in S deficient soils. Moreover in long use, it causes acidity in soil. The Calcium ammonium nitrate and urea are widely used in potato production. The poor performance of urea has been found to be positively related to its dose because of its adverse effect on plant emergence particularly at higher doses. Moreover the adverse affect is observed more in alluvial and calcareous soils having high pH as compared to acidic soils with low pH. Urea can safely be used by applying it 2-3 days before planting of potato.

Time and Method of N Application

In potato crop, sufficient N is needed in early stages to build up crop canopy and to enhance leaf area thus giving longer period for tuber development during tuber bulking phase which is 65-85 days in hills and 40-60 days in plains. Hence the time of application of N should have to be scheduled in such a way that sufficient N is made available at the aforesaid critical period.

Split application of N has proved to be more effective than its full application at planting (Sud *et al.,* 1991). This not only reduces leaching losses of N and better utilization of applied N but at the same time also improves N use efficiency. In higher reaches of Himalayas where the growing season is of 4-5 months and crop is grown under irrigated conditions, three split of N is better than its two splits (Chadha *et al.,* 2006). However, where late blight is a regular phenomenon and variety grown is susceptible to late blight, application of N 2/3rd at planting and 1/3rd at earthing time is more beneficial than its two splits in equal half. Results from different agro-ecological zones have revealed that application of mixture of urea and Calcium ammonium nitrate in equal ratio at planting and urea at the time of earthing up gives not only higher yields but results in higher N recovery and nutrient use efficiency in potato crop. Studies indicate that application of urea by broadcast and mixed with moist soil before 2-3 days of planting is as effective as Calcium ammonium nitrate or ammonium sulphate in alluvial soils where crop is taken under assured irrigation. Broadcasting does not ensure proper utilization of N under moisture stress particularly in rainfed area due to potato having sparse root system. Therefore, it is important to place the fertilizer where interception by active roots is more. The placement of N fertilizer *i.e.* urea in side bands at planting has been found to be beneficial to broadcasting and furrow placement with respect to tuber yield and nutrient recovery (Trehan *et al.,* 2008).

Phosphorus and Potassium

Being a shallow rooted crop the fertilizer use efficiency for P is 10-15% while that for K it ranges between 50-60% thus the residual effect of P and K applied to potato affects the fertilizer needs of succeeding crops. Requirement of Potato for P and K largely depends upon:

1. Soil pH
2. Soil capacity to supply these nutrients in water soluble forms
3. Parent minerals present in the soil and
4. Climatic factors, therefore, varies with agro-climatic regions.

The optimum dose varies from 60 to 100 kg P_2O_5 depending upon soil P status with response raging from 25 to 40 q/ha (Grewal & Sud, 1990). The potato response to P is higher in spring crop as compared to winter crop in north western plains. The responses of K are high in alluvial and hill soils followed by red and black soils. The low response to K in black soils can be due to high K supplying capacity of these

Figure 7. Agronomic efficiency of K (kg tuber produced/kg K supplied) of potato cultivars with and without green manure (GM); (The CD at 0.05 was 6.6 without GM and 6.0 with GM for cultivar mean) Source: Trehan, 2009

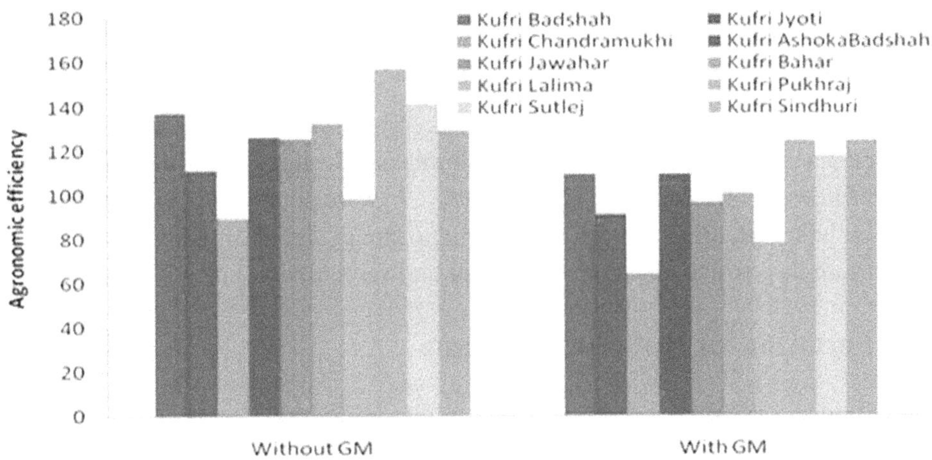

**For a more accurate representation of this figure, please see the electronic version.*

soils showing that a smaller dose of K is necessary for potato. The results shows that the optimum doses for K in alluvial, hill, red and black soils are 100-150, 100, 110 and 70 kg K_2O/ha, respectively (Trehan *et al.,* 2008).

Similar to N response K response also varied with cultivar (Figure 7). It can thus be concluded that varieties behave differently in their response to K application so that selection of the rate of K application should be based on variety in addition to the soil test value.

Sources of P and K

The most common P fertilizer used in potato cultivation is a single super phosphate. The other P carriers tried are di ammonium phosphate, pyrophosphate and rock phosphate. The later can be used in acid soils by mixing it with the single super phosphate in ratio 3:1 (Grewal & Sud, 1990, Sud *et al.,* 1996). Farmyard manure is another important source of P, which over the long run increases the available P status of these soils (Grewal *et al.,* 1981, Grewal & Trehan, 1984, Grewal & Trehan, 1988).

Potassium sulphate (SOP) has been found to be best in term of its beneficial effect on tuber quality *viz.,* dry matter, ascorbic acid and sugar content particularly in sulphur deficient soils but due to its high cost, it has not found much use in potato crop (Sharma *et al.,* 1976). Potassium chloride (MOP) is most commonly used and constitutes 97% of Potassium fertilizers consumption in potato crop. Another

Potassium fertilizer, Potassium schoenite an indigenous source having double salt of Potassium and Magnesium sulphate has also been found to be equally good for potato in acidic and alluvial soils (Trehan *et al.,* 2008).

Time and Method of P and K Application

Furrow placement of P and K fertilizers has been more economical than broadcasting or band placement in acidic hill soils of Shimla and Shillong (Grewal & Sud, 1990, Sud & Grewal, 1992, Verma & Grewal, 1978). In ware potato crop, split application of K *i.e.* half at planting and rest at earthing up along with N is highly remunerative in terms of its positive effect on the yield of large size tubers (Sud & Grewal, 1991).

Bio Fertilizer for Nutrient Economy

Bio-fertilizers have shown a good promise and have emerged as an important component of integrated plant nutrients supply (IPNS). Bio-fertilizers are cheaper, pollution free and based on renewable energy source and improve soil physical properties, tilth and soil health in long run.

- **Plains and Plateau Regions:** *Azotobacter* proved to be superior over *Azospirillum* with 4% to 22% higher yield in tuberlet and sprouts crop, respectively, in alluvial soils and the beneficial effect was more pronounced at lower dose of N *i.e.* 100 kg N/ha. Similarly, the effect of Mycorrhizae (VAM), a P solubilizing fungi on potato was more conspicuous in crop raised from multisprouts and tuberlets than seed tuber crop. Integrated application of FYM (either 25% or 37.5% N through FYM), inorganic (75% or 62.5% N through urea) and bio-fertilizers proved to be effective in obtaining higher tuber bulking rate, tuber yield and the percentage of bigger tubers than that of 100% Nitrogen application through inorganic source (Singh *et al.,* 2007). Application of *Rhizobium* to green manure crop of *dhaincha* and *Azotobacter* and Phosphorus solubilizing bacteria (PSB) to the succeeding potato crop improved the productivity of GM-potato-okra sequence. Likewise, combined application of bio-fertilizers, green manuring with *dhaincha* and FYM (30 t/ha) was as effective as 100% NPK treatment without GM (Upadhyay *et al.,* 1999). Combined inoculation of both *Azotobacter chroococcum* and *Pseudomonas striata* and 50 per cent NPK through fertilizer was found best in obtaining economically higher potato yield in Hasan (Nandekar et al., 2006). The growth attributes and tuber yield of potato increased significantly due to inoculation of plant growth promoting bacteria (*viz. Bacillus cereaus* and *Bacillus subtilis*) at Hooghly area of West Bengal (Chettri et al., 2003).

- **North Western and Eastern Hills:** *Azotobacter* inoculation has been found beneficial and increased the yield of potato under rainfed condition in absence of N. However, as the N doses increased the effect of this non symbiotic N fixer decreased and it became non significant at 180 kg N/ha (Anonymous 2008). Seed inoculation with PSB in conjunction with 50, 75 and 100% recommended dose of P application not only resulted in higher yields but also better nutrient utilization as was evident by its positive effect on nutrients uptake and in higher nutrient recovery of NPK (Sud & Jatav, 2007). Use of plant growth promoting bacteria *Bacillus cerus* economized on NPK fertilizer dose by 25%. *Bacillus subtilis* and *Bacillus cereus* separately increased the tuber yield of potato (Sud & Sharma, 2001). At Shillong, application of different biofertilizers in combination with chemical fertilizers increased potato tuber size and yield and economized N and P upto 25%. (Singh, 2000; Singh, 2001; Singh, 2002)

Secondary and Micronutrients

Secondary Nutrients

In potato, sulphur is required in many metabolic activities in the plant. Sulphur deficiency is similar to N in many ways, however, on contrary to N deficiency in case of S deficiency upper leaves show chlorosis. In alluvial and red loam soils, the optimum dose varies from 30-40 kg S depending upon the S status of the soil with response ranging from 26 to 78 q/ha under assured irrigation. In acidic soils of Shimla hills, application of S at 36 kg/ha is needed to meet S needs of potato under rainfed conditions in soils containing less than 10 ppm available sulphur (Trehan *et al.,* 2008). The major sulphur carriers, ammonium sulphate (23.7% S), super phosphate (12.0% S), Potassium sulphate (18.0% S) and gypsum (13-15% S) have active S-source in form of sulphate which is readily available to the plant. Sources which initially do not contain sulphate-S are Pyrites having 20-50% S and elemental S (85-100% S) and they have to undergo transformation by chemical or biological means in the soil so as to convert the S to sulphate form. In plains, where the crop is taken in winter under assured irrigation, the use of gypsum has been found to be best while in hills, where the crop duration is more and soils contains appreciable quantity of organic matter, the element sulphur has been found to be superior than ammonium sulphate (Trehan *et al.,* 2008).

Failure of development of terminal buds of the apical tips is the first sign of Ca deficiency. Leaves do not develop normally and have wrinkled appearance. On the other hand Magnesium deficiency appears at lower leaves. Plants become slightly pale, older leaves develop central necrosis, turn yellow or brown, interlineal areas become bronzed and wither prematurely. In severe deficiency of Mg, leaflets become thick and brittle and show a definite bulging with leaves rolling upwards. Leaves have higher Ca content than tubers and *vice versa* in case of Mg where tubers contain more Magnesium than leaves. Soil application of $CaCl_2$ at 20 kg/ha was more effective than foliar application of $CaCl_2$ (0.01%) in increasing the potato yield mainly due to increase in yield of large and medium sized tubers (Sud et al., 2008).

MICRONUTRIENTS

About 3 billion people in the world are affected with micronutrient malnutrition. Emerging deficiencies of multi- micronutrients in Indian soils is due to depletion in fertility with exhaustive cropping systems during last decades. It is becoming evident that without the use of some of the micronutrients, it is not possible to get the maximum benefit of other inputs. Response of potato to micronutrients differed with soil group. Zinc is the most deficient micronutrient in almost all potato growing soils. Alluvial soils are more responsive to Iron, Manganese, Boron and Copper as compared to other soils. Marked differences in potato cultivars regarding their sensitivity to micronutrients reaction exist (Singh et al., 1986, Trehan et al., 2008). Cultivation of crop varieties less susceptible to a particular nutrient stress can assist in economizing on the cost of alleviation of its deficiency. A fairly wide differential response of cultivars of potato to Zn, Fe, Mn and B has been demonstrated under field conditions (Trehan & Grewal, 1991).

Nutrient Management Through Fertigation

The sprinkler fertigation with 100% N produced significantly (28%) higher yield of potatoes (336 q/ha) and higher net return of Rs. 43,474/ha along with higher benefit: cost ratio of Rs. 2.07 as against traditional furrow method of irrigation which gave tuber yield 261q/ha and net return of Rs. 26,168/ha along with B:C ratio of Rs. 1.67. When yield target was fixed, it economized Nitrogen dose by 25% along with 40% water saving in comparison to conventional furrow method of irrigation (Anonymous, 2006, Anonymous, 2008, and Anonymous, 2009). The fertigation improved the efficiency of Nitrogen fertilizer and produced 9.4 and 16.5% higher yields with 75 and 100% Nitrogen fertigation, respectively as compared to sprinkler irrigation alone.

Soil Related Constraints

Potato is cultivated in wide range of soils but the best potential of any cultivar may be harvested in well-drained soil having pH 6.5 to 7.5, medium to light in texture and high in organic carbon. Saline or alkali soils, heavy textured soils with poor drainage limits the productivity of potato crop. The main soil related constraints is low organic carbon in potato growing soils of Indo-Gangatic plains, particularly in the western and central part due to intensive cultivation in light textured soils (Anonymous, 2010). Another problem of potato growing pockets in plains is nutrient imbalances. It has been reported that excessive buildup of P in soil is making certain micronutrient unavailable particularly Zn to potato. In a recent study in the states of Punjab more than 90% soil samples from different potato growing pockets were low in organic carbon (Anonymous, 2010) and more than 80% soil samples were having Olsons' P more than 20 ppm (Anonymous, 2009) creating nutritional imbalances particularly to Zinc. In a study in potato growing pocket of Bihar also 68% of soil samples were found high in available P (Kumar et al., 2009b). Intensive cropping systems and decline in organic manuring in potato based cropping systems has led to wide spread deficiency with Zn and other micronutrients Fe, Mn, Cu, B and Mo in soils to lesser extent (Trehan & Sharma, 1999). According to a report, different parts of plains of India which contributes more than 80% of potato area has 40 to 55% soils deficient in Zn and about 10, 3 and 2% samples deficient in Fe, Mn and Cu, respectively (Singh, 2006). A large potato growing area in the states of Bihar and other states suffers from B deficiency. Iron and Boron deficiency was reported in sandy loam soils of Nagaur district (Sharma et al., 2003) and Tonk district (Meena et. al., 2006) of Rajasthan.

In hill region of Shimla more than 50% soils were deficient in available P and about 16% soils were deficient in available Zn (Jatav et al., 2007). This region, however, suffers most due to the acidic nature soil which is very favorable for P fixation. This is reason we get high response to applied P and poor response to N in these soils.

Nutrient Removal and Addition, Balance Sheet

Nutrient removal by potato (Table 2) and potato based cropping system is higher than other field crops due to its higher dry matter production per unit area and time.

Due to the sparse root system of this crop nutrients which are not mobile in soil are poorly utilized and remain in soil resulting into its buildup. In general, this crop shows positive balance for P but negative for N and K. Several studies have highlighted the use of organic sources like crop residue, FYM, *etc.* to overcome the negative balance of N and K. Singh *et al.,* (1994) studied the effect of residue incorporation on the nutrients balance of two intensive potato based cropping systems

Table 2. Average nutrient removal by a healthy potato crop yielding tuber in the range of 25 to 30 t/ha

Primary Nutrient	Kg/ha	Micronutrient	g/ha
N	130-150	Zn	200-300
P_2O_5	25-30	Fe	1200-1600
K_2O	170-230	Mn	700-900
		Cu	50-70
		Mo	8-12

Source: Trehan et al., 2008

Table 3. Nutrients balance (kg/ha) in soil under intensive potato based cropping systems

Fertilization Schedule	Residue incorporation	Rice-Potato-Sunflower					Maize-Potato-Sunflower				
		N	P	K	S	Zn	N	P	K	S	Zn
F1	0	-86	-15	-331	+16	-0.84	-53	-06	-285	+21	-0.73
F2	0	-86	+11	-264	+61	-0.84	-53	+20	-235	+74	-0.73
F1	R	+71	+09	-29	+42	-0.38	+88	+14	-27	+42	-0.37
F2	R	+71	+35	+38	+87	-0.38	+88	+40	+23	+87	-0.37

Source: Singh et al., 1994

F1 = PK fertilization to potato only and F2= potato and cereals both; R= in situ residues incorporation of all the crops.

in north western plains. Residue incorporation not only reduced the depletion of P, K and Zn but also resulted in net gain of soil N besides enhancing the yields of all crops in these two systems (Table 3).

In the western hill region there was positive balance of N, P and K in the potato-garlic (1: 1) intercrop when crop residues of wheat and paddy straw and FYM was applied (Sud et al., 2007). Thus application of paddy straw along with FYM not only economizes on PK dose by 50% but also resulted in the higher tuber yields. In a recent study with a prominent cropping system it was found that most of the time P balance was positive but K and N was negative when crop residue was not incorporated (Table 4). The results emphasized the value of crop residues along with fertilizers for maintaining better nutrient balance (Trehan et al., 2008). Replacement of 50% of inorganic P and K fertilizers from FYM on K basis along with recommended dose of N was best for sustainable yields of potato and improved soil fertility in mid hills of Shimla (Jatav et al., 2010).

Table 4. Effect of treatment sequences on nutrient balance (kg/ha) after 3 crop cycles in potato-onion-groundnut rotation

Fertilizers to Potato-Onion-Groundnut*		Balance		
Onion	Groundnut	N	P	K
100% NPK	100% NPK	88	136	-42
0	100% NPK	-249	75	-125
50% NPK	50% NPK	-30	67	-138
100% NPK	0	92	60	-236
50%N	100%NPK	-95	80	-213
50%N+CR**	50%N+CR*	773	105	299
50%NPK+CR**	50%NPK+CR*	773	183	488

* Recommended rate of NPK to potato - 180, 35, 100 kg/ha, respectively, was applied in all the treatments; Recommended rate of NPK to onion - 140, 26, 50 kg/ha, respectively.

**CR denotes application of crop residue of each crop to the succeeding crop. Mean addition of crop residue per year on dry weight basis was 13.7 q/ha potato haulms, 6.1 q/ha onion top and 42.2 q/ha groundnut top.

Source: Singh et al., 2008a

The decline in K status and build up of available P in potato based cropping systems compared to initial soil test values after three crop cycles was also reported by Singh *et al.,* (2008a) in rice-potato-wheat and rice/groundnut/rice-potato-wheat systems (Figure 8).

Many long-term experiments have illustrated this situation more clearly. In cropping systems with two to three crops per year, removal of K from soil was far in excess of applied K, resulting in a severe negative balance of K in the soil to the detriment of crop yields (Roy et al., 2001, Sharma & Singh, 1989, Singh et al., 2001). The rice-potato-wheat cropping system suffered the most from this negative balance (Figure 9).

A negative balance of K in the soil was shown to be responsible to a large extent for the decline in the yield of crops and has rendered potato based cropping systems unsustainable in the long run (Singh et al., 2001).

Kumar and Lal (2004) while studying with seven potato based cropping systems in the eastern Indo-Gangatic plains found that most of the systems decreased the organic C, available K and Zn when green manuring was not done. The maize-potato-green gram system improved the soil physical conditions and gave better potato yield. Residue incorporation of leguminous crop had more beneficial effect on subsequent maize crop. Soil physical conditions improved in the system as indicated by decreased bulk density. This system also increased organic carbon marginally in sub soil (Kumar et al., 2009c).

Figure 8. Status of available P and K in different cropping systems compared to initial soil test values after three crop cycles. Legends are cropping systems CS1 (rice-wheat), CS2 (rice-potato-wheat), CS3 (rice/groundnut/rice-potato- wheat) and CS4 (rice-vegetable peas-wheat)
Source: Singh et al., 2008a

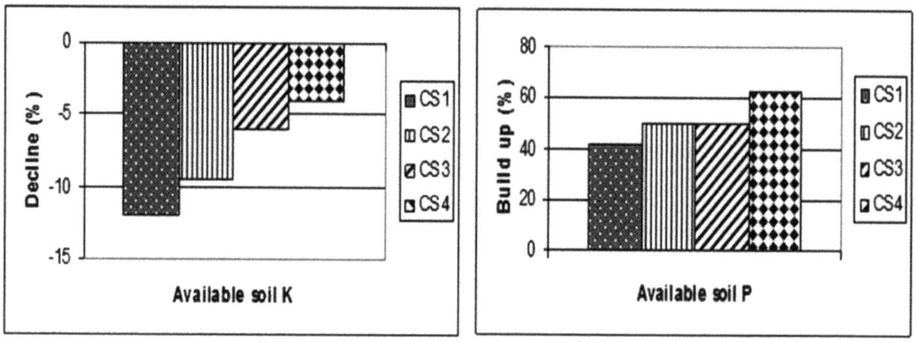

Figure 9. Mean balance of Potassium in soil after 17 years (1970-1987) in different cropping systems
Source: Singh et al., 2001

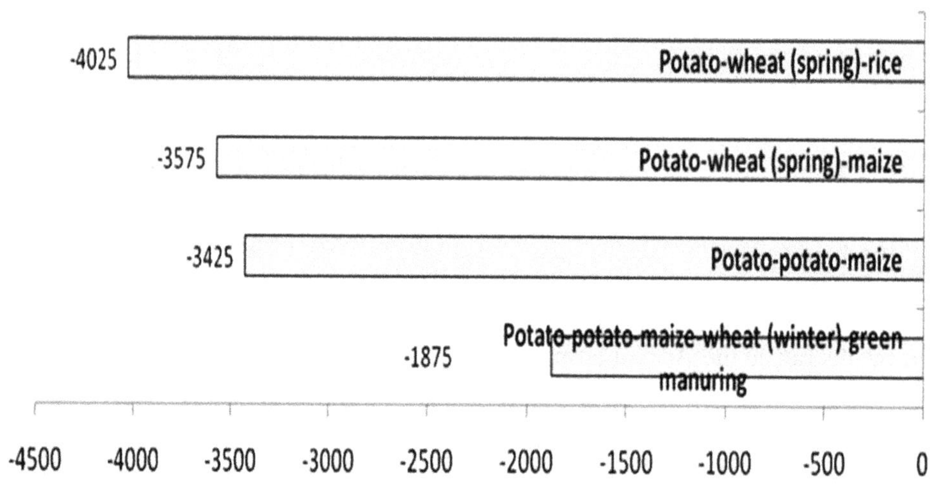

Soil and Plant Tests for Nutrient Management in Potato

Various soil and plant tests and their deficiency/sufficiency (critical) limits for potato is given in Table 5 & 6. These tests are indicative of severity and urgency of fertilization. Moderation in rate of fertilization is possible on the basis of soil test

Table 5. Critical levels of soil available P and K for potato

Soil Nutrient Status	Soil Test Values (ppm)			
	Hill Soils		Alluvial Soils	
	Bray P	Olsen's P	Olsen's P	Ammonium Acetate-K
Low	< 32	< 16	< 10	< 105
Medium	32 - 90	16 - 40	10 - 20	105 - 150
High	> 90	> 40	> 20	> 150

Source: Sud et al., 2008

Table 6. Critical deficiency limits of micronutrients for determining nutrient status in soil and plant for potato

Micronutrient	Critical Concentrations of DTPA-Extractable Micronutrients in Soil and in 4th Leaves of Plant				
	In Soil (ppm)	4th Leaves (Days After Planting)			
		30	40	50	60
Zn	0.75	27.5	25	18	16
Cu	0.32	8.5	8.5	6.0	4.5
Fe	6.6	-	80	83	73

Source: Sud et al., 2008

value after calibration depending upon the soil, climate, season and variety, because of the effect of these factors on response to fertilizers (Trehan *et al.,* 2008). Plant tests made during the crop season enables only a corrective measures as set back already incurred on growth and yield is often not possible to recover due to short duration of the potato crop. Plant tests also suffer from diurnal and temporal variations in field making them less reliable and less efficient than soil tests.

Linear models for prescription of the optimal fertilizer dose taking into account the soil test values have been developed for different part of Indo-Gangetic plains and hills (Table 7 & 8) and may be used for more precise nutrient management.

At Jaladhar, this has helped in increasing fertilizer use efficiency and benefit: cost: ratio by 109 and 3.3%, respectively over the recommended fertilization without affecting the tuber yield (Trehan *et al.,* 2008). Upadhayay and Sharma (2002) developed variety specific fertilizer prescription equations for judicious use of N, P and K for the western Indo-Gangetic plains and subsequently verified them on the cultivators' fields in Meerut area of Utter Pradesh for four commercial potato cultivars (Table 8).

Table 7. Target yield equations for potato production

Location	Soil Type	Variety	Prescription Equation			Source
			N	P	K	
Ludhiana (Punjab)	Alluvial	Kufri Chandramukhi	FN = 0.88T-0.33SN	FP_2O_5 = 0.71-0.875SP	FK_2O = 0.76T-0.32SK	81
Jaladhar (Punjab)	Alluvial	-	FN = 1.21T-0.76SN	FP_2O_5 = 0.44T-2.93SP	FK_2O + 1.36T-1.50SK	78
Pantnagar (Uttrkhand)	Mollisol	Kufri Chandramukhi	F = 1.19T-0.98SN	FP_2O_5 = 0.50T-0.3.33SP	FK_2O = 0.47T-0.60SK	2
		Kufri Dewa	FN = 1.09 T-0.90SN	FP_2O_5 = 0.70T-4.47SP	FK_2O = 0.38T-0.41SK	1
Pusa (Bihar)	Calcareous	Kufri Sindhuri	FN = 1.50T-0.24SN	FP_2O_5 = 1.35T-4.26SP	FK_2O = 0.81T—0.57SK	28
Patna (Bihar)	Alluvial	Kufri Ashoka	FN = 0.47T-0.17SN-0.34CN	FP_2O_5 = 0.14T-0.43SP_2O_5-0.09C P_2O_5	FK_2O = 0.30T-0.15SK_2O-0.08CK_2O	29

T= Targeted yield(q/ha), SN, SP/SP2O5and SK /SK2O are the soil test values in kg/ha and F is fertilizer dose

Table 8. Targeted yield equations for predicting N, P and K dose for potato varieties

Variety	Nitrogen	Phosphorus	Potassium
Kufri Bahar	FN= 2.67T-2.265SN	FP_2O_5= 2.685T-13.285SP	FK_2O=3.249T-4.70SK
Kufri Badshah	FN=2.48T-0.784SN	FP_2O_5=2.634T-12.316SP	FK_2O=2.429T-3.12SK
Kufri Jyoti	FN=2.86T-2.161SN	FP_2O_5=0.892T-4.212SP	FK_2O=2.910T-1.02SK
Kufri Sutlej	FN=1.45T-2.150SN	FP_2O_5=1.725T-10.414SP	FK_2O=2.595T-3.42SK

T= Targeted yield(q/ha), SN, SP and SK are the soil test values in kg/ha and F is fertilizer dose;
Source: Upadhayay & Sharma, 2002

Nutrient Interactions

Nutrient interaction is a common feature in agricultural crops as they play a vital role in modifying the nutrients needs of many crops including potato. Studies show that P does not interact with N nutrition in potato, whereas, K has a significant interaction with N indicating that the doses of N and K are interdependent. Studies conducted in cold desert soils of Lahaul Valley (Chadha et al., 2006) and in Shimla hills (Sud et al., 2008) showed significant interaction between N and K. According to them, application of N induces K deficiency in plant thereby necessitating the balanced K application. Application of K on the other hand helps in increasing N utilization in drought years. Singh and Ragav (2000) attributed the significant interaction between N and K in tarai belt of Uttrakhand to the better utilization of

Figure 10. Effect of K application on per cent increase in agronomic use efficiency of N by six potato cultivars/hybrids
Source: Trehan, 2010

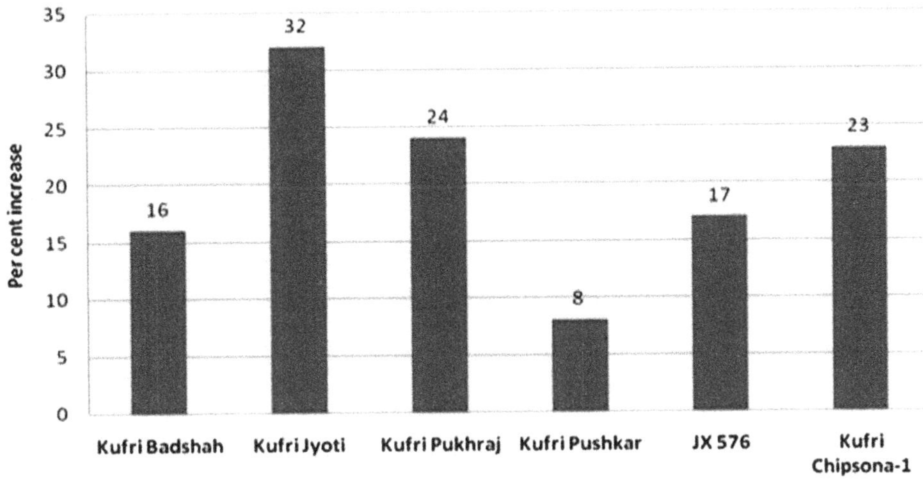

N in presence of K. Application of K increased agronomic use efficiency of N by 8 to 32% in 6 cultivars/hybrid (Figure 10) with the highest increase of 32% in least N efficient cv. Kufri Jyoti (Trehan, 2010). Higher levels of K decrease Ca concentration in potato leaf petioles and haulms. The synergistic relationship between P and S at lower levels and antagonistic at higher levels have been reported in potato in alluvial, red and acidic soils. This antagonistic behavior of P on S and *vice versa* is mainly due to the competition between these anions on the absorbing surface of plant root. Similarly P and Zn & P and Mg interaction have been observed in potato crop (Sud et al., 2008).

Balance Use of Nutrients to Improve Yield and Quality of Tuber and Disease Resistance

Balance application of nutrients including that of secondary and micronutrient improves the yield and quality of tubers and also imparts resistance to different biotic and abiotic stresses to potato crop. Highest yield (Figure 11) was obtained with balance application of N, P & K (Singh, 2001) in all the six potato cultivars/hybrid at Jalandhar (Trehan, 2009).

Balanced application of N and S has been found to enhance nutrient recovery from applied fertilizers to potato crop both in hills and plains. The optimum N: S ratio in potato leaf at stolon formation and tuber initiation has been found to be 28.4:1 and 17.9:1, respectively and is positively correlated with yield and N & S

Figure 11. Effect of balance application of N, P and K on potato tuber yield (q/ha) in different potato cultivars/hybrid
Source: Trehan, 2010

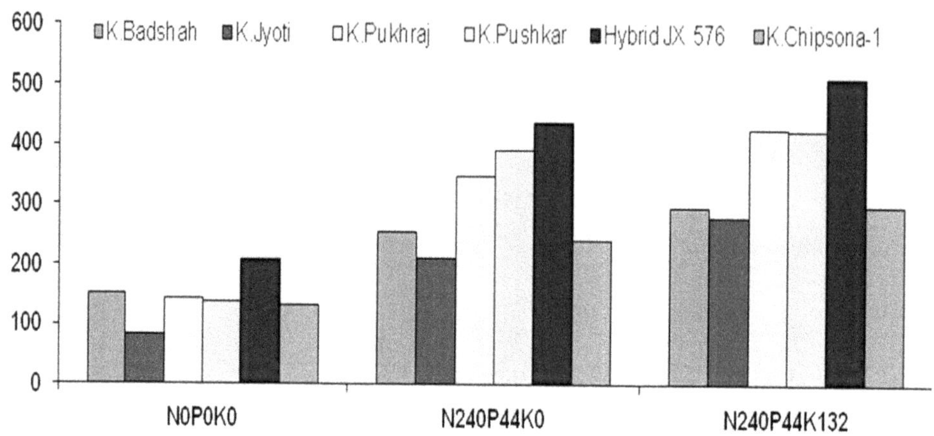

uptake by potato (Sud, 1996). On the other hand, higher levels of S or P have an antagonistic relationship in potato crop as indicated by the adverse association of P: S ratio in leaf with potato tuber yield showing the importance of balance application of these two bio-anions (Sud, 2006). Optimum N:P:S ratio at stolon formation and tuber initiation stage should be 15.8:1.37:1.0 and 12.9:0.65:1.0, respectively for optimum potato yields.

Balanced Nutrition and Tuber Quality

Apart from their effect on tuber size and number, application of P and K has a significant influence on nutrient composition of tubers and tuber quality parameters *viz.*, dry matter content, sugar, vitamin C content. Application of P at optimum rates increases tuber starch and vitamin C content but at higher levels adversely affects protein content. Likewise, application of K has a direct influence on tuber quality. It increases total and reducing sugar content but decreases tuber dry matter content, the decrease being more with muriate of potash than sulphate of potash. The negative effect of MOP on dry matter is mainly due to the accompanying chloride ion in MOP. Application of SOP produces tubers with higher dry matter, starch and Vitamin C content. Application of K through MOP has been reported to decrease enzymatic discoloration and phenol content thereby reducing the browning of potato chips (Joshi et al., 1982).

Zinc, Copper, Manganese, Boron and Molybdenum have been shown to increase ascorbic acid content of tubers (Grewal & Trehan, 1979, Grewal et al., 1979). Zinc fertilization reduced the content of tyrosine, orthodihydroxy and total phenols in tubers (Marwaha, 1989). The potato used for processing should contain minimum quantities of tyrosine and phenolic compounds as they are implicated in enzymatic discoloration which occurs in raw peeled potatoes due to oxidation of tyrisine and chlorogenic acid formation of ferric dihydric phenolic complexes after cooking.

Nutrient and Tolerance of Potato Crop to Stresses

The high K concentration in leaves helps in lowering the freezing point of cell sap thereby helping the plant to escape frost. A decrease in intensity of frost with increase in K dose in alluvial soils has been reported in western plains. Potato response to K application is positively correlated with frost damage. In general, varieties which are more susceptible to frost injury are highly responsive to K. Potassium application although, do not economize on water needs of potato crop but increases water use efficiency in terms of tubers yield/mm water. In addition adequate K also imparts resistance to late blight, a common feature in hills.

Zinc sulphate used alone had some fungicidal value (Ram, 1987). Dipping of tubers in ZnSO4 solution could control the black scurf disease. Likewise, boric acid treatment of potato tubers could greatly help in controlling common scab and black scurf (Somani, 1986, Somani, 1988). At Jalandhar, leaf spots observed in the early planted crop were controlled by foliar sprays of 0.2% Manganese sulphate (Trehan & Grewal, 1994). Spray application of Boron also reduced leaf spot intensity significantly. Application of ferrous sulphate (50 kg/ha) significantly reduced the incidence of pseudo common scab from 81% to 58% in cultivar Kufri Sindhuri. The incidence of pseudo common scab was negatively correlated with the Iron concentration of potato tubers.

The effectiveness of micronutrients in controlling late blight was more at low level of disease incidence at Jalandhar (Trehan & Grewal, 1995, Trehan et al., 1995). Foliar sprays of micronutrients (Zn, Cu, Zn + Cu + Mo) proved effective in reducing late blight by 65.4% to 92.6% and increased tuber yield by 36.3 to 45.9%. However, in blight epidemic year the micronutrient (Zn + Cu + Mo + Mn) and mancozeb resulted in a significant control of late blight and increased tuber yield significantly, but were less effective than systemic fungicide metalaxyl + mancozeb (95.7%). Potato leaves showing high concentration of either Cu, Zn or Mn showed minimum late blight intensity and apparent infection rate. The result thus show that in plains, where late blight appears in mild to moderate intensities in most of the years (Bhattacharaya et al., 1990), prophylactic sprays of micronutrients Cu, Zn and Cu Zn + Mo may be helpful in reducing late blight and increasing yield. Besides, micronutrients are easily available, have low cost and pose little problem of resistance to pathogen (Dowley, 1981). However, under the high disease pressure the micronutrients, sprays should be supplemented with fungicides for effective control of late blight.

Integrated Nutrient Management for Higher Yield and Profit

Potato being very resource intensive crop, integration of all possible sources of nutrients including organic sources like green manuring, FYM, vermicompost, crop residue and bio fertilizers and chemical fertilizer pays dividends. Several workers have reported yield enhancement and fertilizer savings through the use of organic sources. Therefore, the integration of all possible sources depending upon region and situation is very important to have most economic production. In general region wise recommendation is given in Table 9 which may further be modified considering soil test values using yield adjustment equations (Table 7 & 8). Application of secondary and micronutrients should also be taken up if found deficient in soil.

The utilization efficiency of absorbed K from native soil sources in tuber production is 104% more than from fertilizer sources (Singh & Grewal, 1995). High utilization efficiency of native soil K indicates the desirability of maintenance fertilization of K commensurate with crop removal to avoid possible decline in yield of crops over a period.

In intensive potato based cropping systems, incorporation of crop residues into the soil improved soil physical condition, arrested the depletion of macro and micro nutrients by recycling from lower soil layers to surface layer in the profile, maintained

Table 9. Fertilizer requirement of potatoes grown in different zones

Zone	Soil Type	Dose in kg/ha		
		N	P_2O_5	K_2O
North western hill zone	Acidic hill soil	120-150	100-150	110-130
North-eastern hill zone	Acidic hill soil	100-120	120-150	50-70
North-western plain zone	Alluvial	180-240	80-100	100-150
North-central plain zone	Alluvial	180-240	80-100	100-150
North-eastern plain zone	Alluvial	180-240	80-100	100-150
Plateau zone	Black	100-120	50-70	50-70
Nilgiri zone	Acidic hill soil	90-120	130-150	80-100

Source: Sharma et al., 1999

Table 10. Doses of micronutrient application for correction of their deficiency in potato

Micronutrient	Soil Application (kg/ ha)	Spray Application (g/100 lit. water)	Tuber Soaking Treatment (g/100 lit. water)
Zinc sulphate	25	200	50
Ferrous sulphate	50	300	75
Manganese sulphate	25	200	50
Copper sulphate	25	200	50
Ammonium molybdate	2	100	20
Sodium borate	2	100	20

Source: Trehan and Sharma, 1999

soil fertility and sustained systems productivity economically (Singh et al., 1997, Trehan et al. 2008). In presence of organic sources like green manure, FYM and rice straw there was decrease in N requirement of cultivars Kufri Jyoti and Kufri Pukhraj to produce a fixed yield. Incorporation of green manure crop (Dhaincha) showed maximum saving of fertilizer N (Anonymous, 2009).

Tuber inoculation with biofertilizers (Azotobacter, PSB and growth promoting bacteria) is beneficial specially in north western and eastern hills saving upto 25% of recommended dose of N and P (Singh, 2000, Singh, 2001, Singh, 2002).

Severe deficiency of micronutrients may requires soil application, tuber soaking or foliar spray at 40-50 days after planting as mentioned in Table 10 (Trehan & Sharma, 1999).

CONCLUSION AND FUTURE RESEARCH NEEDS

For an exhaustive and resource intensive crop like potato integrated nutrient management is inevitable. Soil inherent fertility, organic sources like *in situ* green manuring with *dhaincha*, FYM, vermicompost and biofertilizer along with inorganic fertilizers must be integrated depending upon available resources, cultivar and cropping system to get optimum yield and return while maintaining soil health and minimum environmental damage. Use of high analysis fertilizers in the intensive cropping system has resulted in wide spread deficiency of secondary nutrients, particularly of sulphur and micronutrients most importantly Zn need to be addressed on soil test basis. Nutrient interactions as well as balanced application of nutrients are key to sustained productivity under different stresses and to improve quality of produce.

However, following aspects need to be further strengthening for developing efficient nutrient management strategy in potato crop:

1. Development of nutrient management schedules through diagnostic tools on regional scale by mapping and delineating nutrient deficiency and imbalances in major potato growing areas
2. Identification and quantification of nutrient efficiency traits in different cultivars to develop strategy for development of highly efficient cultivars.
3. Development of sound and convenient method of delivery system for nutrients in prominent potato based cropping systems using recent technologies like Nano dispersed nutrients, mineral based micronutrients (e.g. Zeolite *etc.*), organic chelates, and other techniques such as tuber soaking *etc.*
4. Investigating further the role of balance nutrition including macro and micronutrients and trace elements on plant metabolism and in overcoming physiological and pathological disorders of potato crop.
5. Study the role of balanced nutrition on nutritional quality of produce and examining further the possibility of bio-fortification of potato tubers of suitable cultivars by manipulating the nutrient supply system.
6. In depth study of the soil rhizosphere including microbial ecology and enzymatic reaction and exploiting them for improving efficiency and yield with emphasis on making available fixed P in soil to the crop.
7. Adaptation and modification of Site Specific Nutrient Management methodology to potato including development of leaf colour chart and omission plot techniques.
8. Developing, standardizing and evaluating suitable technology for fertigation using modern pressurise system of irrigation and liquid fertilizer formulations for saving of nutrient and water.

REFERENCES

Anonymous, . (2006). *Annual Scientific Report, 2005-06*. Shimla: Central Potato Research Institute.

Anonymous. (2008). Annual Scientific Report, 2007-08. Central Potato Research Institute.

Anonymous. (2009). Annual Scientific Report, 2008-09. Central Potato Research Institute.

Anonymous, . (2010). *Annual Scientific Report, 2009-10*. Shimla: Central Potato Research Institute.

Bhattacharaya, S. K., Shekhawat, G., & Singh, B. P. (1990). *Potato Late Blight. Technical Bulletin No. 27*. Shimla: Central Potato Research Institute.

Chadha, S., Rana, S. S., Rameshwar, D., & Chaudhary, D. R. (2006). Effect of split application of N, K and FYM on the productivity of seed potatoes in cold desert region of H.P. *Potato Journal*, *33*, 94–96.

Chettri, M., Basu, A., & Konar, A. (2003). Effect of plant growth promoting bacteria on the growth and yield of potato in the plains of West Bengal. *J. Indian Potato Assoc.*, *30*(1-2), 83–84.

Dowley, L. J., & Sullivan, E. O. (1981). Metalaxyl-resistant strains of *Phytophthora infestons* (Mont.) de Bary in Ireland. *Potato Research*, *24*(4), 417–421. doi:10.1007/BF02357324

Ezekiel, R., Sukumaran, N. P., & Shekhawat, G. S. (1999). *Potato: A Wholesome Food*. Tech. Bull. No. 49. CPRI.

Grewal, J. S., Sharma, R. C., & Sud, K. C. (1981). Effect of continuous application of FYM and P K fertilizers on potato yield and some soil properties. *Journal of the Indian Society of Soil Science*, *28*, 29–31.

Grewal, J. S., & Sud, K. C. (1990). Potato response to phosphorus in northern India. *Fertilizer News*, *35*, 25–30.

Grewal, J. S., & Trehan, S. P. (1979). Direct, residue and cumulative effects of phosphorus and potassium applied through organic and inorganic sources on potato (Solanum tuberosum L.) in acidie brown bill soils. *J. Indian Potato Assoc.*, *6*, 41–51.

Grewal, J. S., & Trehan, S. P. (1984). Long-term effects of phosphorus and potassium fertilizers, farmyard manure and their combinations on potato yield, nutrient uptake and soil fertility. *Journal of the Indian Society of Soil Science*, *32*, 384–386.

Grewal, J. S., & Trehan, S. P. (1988). Results of use of P and K fertilizers and farmyard manure on potato yield, nutrient uptake and P and K status of soil. *Journal of Potassium Research*, *4*, 24–30.

Grewal, J. S., Verma, R. S., & Bist, B. S. (1979). Method, time and level of nitrogen application on growth, yield and quality of potato grown on acidic brown hill soils of Shimla. *Indian Journal of Agricultural Sciences*, *49*, 683–688.

Indian Horticulture Database. (2014). *National Horticulture Board, Ministry of Agriculture*. Govt. of India.

Jatav, M. K., Sud, K. C., & Dua, V. K. (2007). Nutrient status of soils from high hills of potato growing areas of Shimla district in Himachal Pradesh. *Potato Journal, 34*(3-4), 216–220.

Jatav, M. K., Sud, K. C., & Trehan, S. P. (2010). Effect of organic and inorganic sources of P and K on their different fraction in potato-radish sequence in Brown Hill Soils. *Journal of the Indian Society of Soil Science, 58*(4), 388–393.

Joshi, K. C., Grewal, S. S., Mishra, J. B., & Verma, S. C. (1982). Discoloration of potato tubers in relation to K fertilization. In B. B. Nagaich, G. S. Shekhawat., P. C. Gaur, & S. C. Verma (Eds.), Potato in Developing Countries (pp. 265-69). Indian Potato Association.

Kumar, D., & Lal, S. S. (2004). Evaluation of potato based cropping systems in Bihar plains. *Potato Journal, 31*(1-2), 49–53.

Kumar, M., Jatav, M. K., Trehan, S. P., & Lal, S. S. (2009a). Integrated nutrient management in potato based cropping systems for eastern Indo-Gangetic plains of India. *Potato Journal, 36*, 136–142.

Kumar, M., Rawat, S., Jatav, M. K., Dua, V. K., & Lal, S. S. (2009c). GIS mapping of spatial variability of available nutrients for site specific nutrient management in potato. In P. M. Govindakrishnan, J. P. Singh, S. S. Lal, V. K. Dua, S. Rawat, & S. K. Pandey (Eds.), Information Technology Application in Horticultural Crops (pp. 99–105). Academic Press.

Kumar, M., Trehan, S. P., Jatav, M. K., & Lal, S. S. (2009b). Efficacy of potato (Solenum tuberosum) cultivars under varying levels of nitrogen and growth duration in eastern Indo-Gangetic plains. *Indian Journal of Agronomy, 54*, 63–68.

Marwaha, R. S. (1989). Changes in chemical constituents of potato tuber, with the application of Zinc. *Proceedings of the National Academy of Sciences of the United States of America, 59*, 229–233.

Meena, H. B., Sharma, R. P., & Rawat, U. S. (2006). Status of macro and micronutrients in some soils of Tonk district of Rajasthan. *Journal of the Indian Society of Soil Science, 54*, 508–512.

Nandekar, D. N., Sawarkar, S. D., & Naidu, A. K. (2006). Effect of bio fertilizers and NPK on the growth and yield of potato in Satpura plateau. *Potato Journal, 33*(3-4), 168–169.

Ram, S. (1987). Performance of zinc sulphate and lime combination for controlling late blight in the north-east hills. *Journal of Hill Research, 1*, 96–101.

Roy, S. K., Sharma, R. C., & Trehan, S. P. (2001). Integrated nutrient management by using farmyard manure and fertilizers in potato-sunflower–paddy rice rotation in the Punjab. *The Journal of Agricultural Science, 137*(03), 271–278. doi:10.1017/S0021859601001472

Sharma, R. C., Grewal, J. S., & Sud, K. C. (1976). Relative suitability of different phosphatic sources for potato on brown hill soils of Shimla. *Journal of the Indian Society of Soil Science, 24*, 95–97.

Sharma, R. C., Trehan, S. P., Roy, S. K., & Kumar, D. (1999). Nutrient management in potato. *Indian Farming, 49*(9), 52–54.

Sharma, R. C., & Upadhayay, N. C. (1993). Nitrogen nutrition of potato. In K. L. Chadha & J. S. Grewal (Eds.), Advances in Horticultural (vol. 7, pp. 231-59). Malhotra Publishing House.

Sharma, R. P., Singh, M., & Sharma, J. P. (2003). Correlation studies on DTPA – extractable micronutrients vis-à-vis soil properties in some soils of Nagaur district in semi-arid region of Rajasthan. *Journal of the Indian Society of Soil Science, 51*, 522–527.

Sharma, U. C., & Singh, K. (1989). Nutrient balance sheet and economics of three potato-based cropping systems in acidic soils. *Journal of the Indian Society of Soil Science, 37*, 754–758.

Singh, J. P., & Grewal, J. S. (1995). Requirement of potassium to potato crop and effect of some agrotechniques on yield and fertilizer use efficiency. *J. Potassium Res., 11*, 160–165.

Singh, J. P., Jaiswal, V. P., & Trehan, S. P. (1994). Potato: Present and Future. GS Indian Potato Association.

Singh, J. P., Lal, S. S., & Sharma, R. C. (2001). Productivity and sustainability of potato based cropping systems with reference to fertilizer management in north-western plains of India. In *Proceedings of Global Conference on Potato*. Indian Potato Assoc.

Singh, J. P., Salaria, A., Singh, K., & Gangwar, B. (2008a). Efficiency of diversified rice-wheat cropping systems including potato, vegetable peas and groundnut crops in trans-gangetic plains. *Potato Journal, 35*(1-2), 53–60.

Singh, J. P., Trehan, S. P., & Sharma, R. C. (1997). Crop residue management for sustaining the soil fertility and productivity of potato based cropping systems in Punjab. *Journal of Indian Potato Association*, *24*, 85–89.

Singh, J. P., Trehan, S. P., Upadhayay, N. C., & Lal, S. S. (2008b). Potato based cropping system. In S. K. Pandey & S. K. Chakrabarti (Eds.), *Twenty steps towards hidden treasure* (pp. 105–125). Shimla, India: Central Potato Research Institute.

Singh, K. (2000). Effect of inoculation with Azotobacter and Phosphobacteria on potato (Solanum tuberosum) in north- eastern hills. *Indian Journal of Agricultural Sciences*, *70*(6), 385–386.

Singh, K. (2001). Effect of phosphobacteria, nitrogen and phosphorus on the tuber yield of potato (Solanum tuberosum) under East Khasi hill conditions of Meghalaya. *Indian Journal of Agronomy*, *46*(2), 375–379.

Singh, M. V. (2006). Micro- and secondary nutrients and pollutant elements research in India. Coordinator Report-AICRP Micro and secondary nutrients and pollutant elements in soil and plants. *IISS, Bhopal.*, *30*, 1–110.

Singh, N. P., Ojha, A. P., & Kumar, V. (1986). Response of potato (Solanum tuberosum L.) varieties on various fertility levels. *Progressive Horticulture*, *18*, 114–117.

Singh, N. P., & Raghav, M. (2000). Response of potato to nitrogen and potassium fertilization under U.P. tarai conditions. *Journal of Indian Potato Association*, *27*, 47–48.

Singh, S. K. (2002). Effect of inoculation with Azotobacter and Phosphobacteria on potato (Solanum tuberosum) in north- eastern hills. *Indian Journal of Agronomy*, *47*(2), 273–277.

Singh, S. N., Singh, B. P., Singh, O. P., Singh, R., & Singh, R. K. (2007). Effect of nitrogen application in conjunction with bio-inoculants on the growth, yield and quality of potato under indo-gangetic plain region. *Potato Journal*, *34*(1-2), 103–104.

Singh, S. N., Trehan, S. P., & Manoj, K. (2008c). Response of potato cultivars to nitrogen in the plains of Bihar. *Potato Journal*, *35*(3-4), 158–160.

Somani, A. K. (1986). Non-hazardous chemical control of black scurf of potato. *Indian Journal of Agricultural Sciences*, *56*, 366–369.

Somani, A. K. (1988). Control of black scurf (*Rhizoctonia solani*) and common scab (*Streptomyces scabies*) of potato (*Solanum tuberosurn* L.) with boric acid. Control of black scurf (*Rhizoctonia solani*) and common scab (*Streptomyces scabies*) of potato (*Solanum tuberosurn* L.) with boric acid. *Indian Journal of Agricultural Sciences*, *58*, 693–698.

Sood, M. C., & Sharma, R. C. (2001). Value of growth promoting bacteria, vermicompost and Azotobacter on potato production in Shimla hills. *Journal of Indian Potato Association*, *28*(1), 52–53.

Sud, K. C. (1996). Effect of phosphorus and sulphur on potato nutrition in acid soils of Shimla. *Journal of the Indian Society of Soil Science*, *44*, 440–444.

Sud, K. C. (2006). Anionic balance of nitrogen, phosphorus and sulphur in potato leaves under rainfed conditions in Shimla hills. *Journal of the Indian Potato Association*, *33*, 68–73.

Sud, K. C., & Grewal, J. S. (1991). Effect of time of K application on potato nutrition in Shimla hills. *Journal of Potassium Research*, *7*, 266–276.

Sud, K. C., & Grewal, J. S. (1992). Evaluation of methods of potassium application in potato production on acidic soils of Shimla. *Journal of Potassium Research*, *8*, 137–143.

Sud, K. C., & Grewal, J. S. (1994). Evaluation of Nitrogen sources and levels on potato nutrition and their effect on soil fertility under potato-fallow-potato crop sequence in Shimla hills. In Potato: Present and Future. Indian Potato Association.

Sud, K. C., Grewal, J. S., & Sharma, R. C. (1991). Efficient use of urea by potato (*Solanum tuberosum* L.) in Shimla hills. *Indian Journal of Agricultural Sciences*, *61*, 389–394.

Sud, K. C., & Jatav, M. K. (2007). Response of potato to phosphorus and phosphorus solubilizing bacteria in brown hill soils of Shimla. *Potato Journal*, *34*, 109–110.

Sud, K. C., Sharma, R. C., & Verma, B. C. (1996). Evaluation of levels and sources of sulphur on potato nutrition in Shimla hills. *Journal of the Indian Potato Association*, *23*, 134–139.

Sud, K. C., Trehan, S. P., & Lal, S. S. (2007). Performance of organic and inorganic in potato –garlic sequence under zero tillage in the mid hills of Himachal Pradesh. *Potato Journal*, *34*, 203–208.

Sud, K. C., Upadhayay, N. C., Trehan, S. P., Jatav, M. K., Kumar, M., & Lal, S. S. (2008). *Soil and Plant Tests for Judicious Use of Fertilizer in Potato*. Technical Bulletin No. 91. Central Potato Research Institute.

Trehan, S. P. (2003). Evaluation of nitrogen efficiency of different potato cultivars. *Journal of the Indian Potato Association*, *30*, 63–64.

Trehan, S. P. (2006). Genetic control of different potato cultivars in the manipulation of nitrogen uptake from green manured soil. *Advances in Horticultural Science*, *20*, 199–207.

Trehan, S. P. (2009). Improving nutrient use efficiency by exploiting genetic diversity of potato. *Potato Journal*, *36*, 121–135.

Trehan, S. P. (2010). Nitrogen and Potassium use efficiency vis-à-vis N & K interaction in different potato cultivars/hybrid. Presented in Round Table Discussion on "Enhancing Nutrient Use Efficiency through Potassium fertilization" being held by Fertilizer Association of India and International Potash Institute Switzerland on 9th Feb., 2010 at the Fertilizer Association of India House, New Delhi, India.

Trehan, S. P., & Grewal, J. S. (1991). *Micronutrients management in potato. In Annual Scientific Report* (p. 65). Jalandhar, India: Central Potato Research Station.

Trehan, S. P., & Grewal, J. S. (1994). *Micronutrients management in potato. In Annual Scientific Report* (p. 100). Jalandhar, India: Central Potato Research Station.

Trehan, S. P., & Grewal, J. S. (1995). Response of zinc, copper and iron and their critical levels for potato. *Journal of the Indian Society of Soil Science*, *43*, 89–91.

Trehan, S. P., Sharma, K. K., & Grewal, J. S. (1995). Effect of soil and foliar application of micronutrients and fungicides on late blight (*Phytophthora infestans*) and tuber yield of potato (*Solanum tuberosum* L.). *Indian Journal of Agricultural Sciences*, *65*, 836–841.

Trehan, S. P., & Sharma, R. C. (1999). *Micronutrients Requirement of potato*. Technical Bulletin 23 (revised). CPRI.

Trehan, S. P., Upadhayay, N. C., Sud, K. C., Kumar, M., Jatav, M. K., & Lal, S. S. (2008). *Nutrient Management in Potato*. Technical Bulletin No. 90. Central Potato Research Institute.

Upadhayay, N. C., & Sharma, R. C. (2002). Fertilizer prescription equations for N, P and K in potato. In S. M. Paul Khurana, G. S. Shakawat, S. K. Pandey & B. P. Singh (Eds.), Potato, Global Research & Development (vol. 2, pp. 838-842). Indian Potato Association.

Upadhyay, N. C., Sharma, R. C., & Singh, N. (1999). *Effect of organic and biofertilizer economy in green manure-potato-okra system.Global Conference on Potato*, New Delhi, India.

Verma, R. S., & Grewal, J. S. (1978). Evaluation of levels and methods of phosphatic fertilizer application to potatoes in acidic brown hill soil. *Journal of the Indian Potato Association*, *5*, 76–82.

Chapter 3
Crop Weather Interaction in Potato in South Bengal Plains

S. Maji
Bidhan Chandra Krishi Viswavidyalaya, India

Ratneswar Poddar
Bidhan Chandra Krishi Viswavidyalaya, India

Pramiti K. Chakraborty
Bidhan Chandra Krishi Viswavidyalaya, India

R. Nath
Bidhan Chandra Krishi Viswavidyalaya, India

S. Basu
Siksha 'O' Anusandhan University, India

P. Bandopadhyay
Bidhan Chandra Krishi Viswavidyalaya, India

Sarika Jena
Odisha University of Agriculture and Technology, India

P. K. Chakraborty
Bidhan Chandra Krishi Viswavidyalaya, India

ABSTRACT

Potato is one of the main staple foods in West Bengal, where it ranks second in production after Uttar Pradesh. There is lots of variation in productivity of the crop. It is due to climatic variability which causes widespread disease infection in potato crop. The shifting of onset and withdrawal of monsoons has also proved to be a barrier in the productivity of the crop. The farmers are habituated to plant the crop within 15th of November; however this is being disrupted because of the shifting of withdrawal of monsoons. Potato is a thermo sensitive crop. The crop growth rate of potato is significantly affected by cumulative maximum and minimum temperatures.

DOI: 10.4018/978-1-5225-1715-3.ch003

Leaf area index significantly decreased with the increase in cumulative maximum and minimum temperatures. Rainfall and relative humidity are two crucial factors that determine the incidence of late blight in potato, the most devastating disease in Bengal. Rainfall increased productivity by lowering soil temperature and reducing hydrolysis of starch respiratory losses from tubers.

INTRODUCTION

In the Gangetic plains of Bengal, appreciable contributions in research have been made on a few crop plants which include rice, jute, wheat, mustard and few pulses. The assessment of weather on these crops remains incomplete. Potato is a new entrant in this arena of study. The growth of potato plant and its tubers depends largely on weather factors particularly temperature, solar radiation and day length. The potato is surface feeder. The root system is usually confined to the top layers of the soil (up to 30 cm). Therefore, nutrient content within the top layers of the soil principally regulates the growth of haulm as well as tuber. Soil temperature influences the uptake of nutrients and water from the top soil layer. Growth of the buds starts after planting of the tuber, which is largely dependent on temperature. At 15 °C or above, the apical sprout grows faster to restrict the other buds at a small size (Goodwin, 1967). Under non favourable temperatures, many buds continue to grow for a longer period. The dominant sprout is maintained by storing the tuber under favourable and sub-favourable temperatures. Therefore, it is possible to have pre-determined population by manipulating the storage temperature (Moorby & Milthorpe, 1975). The rate of dry weight gain of sprout in single-sprout tubers remains constant over a prolonged period irrespective of temperature.

The crop growth period may be divided into three distinct phases:

1. Planting to establishment of sprouts
2. The first phase of autotrophic growth when biomass accumulation in haulm is prominent
3. Tuber growth period comprising of the translocation of photosynthates with the gradual senescence of haulms

In the Gangetic plains of West Bengal, potato cultivation is generally confined to several districts scattered on both the banks of the Ganges. The winter here is mild and short. The winter creeps in with the withdrawal of the south-west monsoons. This plain welcomes the cool north east continental wind from the end of October. The growing season of potato is thus restricted to November to February. The potato must be planted within 15th of February for better yield. The present scenario is different due to weather variability. The late onset and concomitant delayed withdrawal of monsoons impede land preparation for potato. The temperature doesn't drop even during December. Under this changing scenario, the present two year field experiment has been performed to understand the crop weather interaction in potato.

The experiments were carried out during the winter seasons of 2010-11 and 2011-12 at Kalyani, B.C.K.V. (Lat: 22°58' N and Long: 88°32' E). The experimental site is located in the eastern bank of the river Ganges of Nadia district. The district is famous for vegetable and potato cultivation. The experimental site falls under tropical sub-humid climate and experiences three distinct seasons. The annual rainfall is approximately 1500 mm, 80% of which occurs during monsoons.

In the present chapter, impact of temperature humidity and rainfall on the growth and the yield of the crop were investigated. The crop was exposed to variable weather condition by planting it on five different dates viz. D1- 15th November, D2- 22nd November, D3- 29th November, D4- 6th December and D5- 13th December. Macro climatic analysis of maximum and minimum temperatures, relative humidity and rainfall were done. The weather data were collected from the meteorological observatory adjacent to the experimental site. The leaf area index (LAI) of the crop was measured using LAI 220 Plant Canopy Analyzer (LI-COR). Crop Growth Rate (CGR) indicates the rate at which the crop is growing and is expressed as g of dry matter produced per day.

$$CGR = \frac{W_2 - W_1}{T_2 - T_1}$$

where, W1 and W2 are dry weights of plants in g m-2 at times T1 and T2

MAXIMUM TEMPERATURE DURING THE GROWTH OF POTATO

The crop was first planted on 15th November, 2010. In 2010, the maximum temperature was 31.2 °C, marginally higher than the normal value. In 2011, the same was 28.7 °C, marginally lower than it's normal. The observed and normal maximum temperatures during the growth period of potato are given in (Figure 1). The weather variability leads to the temperature variation and this variation may be negative or positive when compared to the normal temperature. The duration of the exposure of the crop to this deviated temperature from normal may either reduce or increase the duration of a particular phenophase. The dry matter accumulation and yield primarily depend on this phenological duration (Parya et al., 2010). The deviation of this maximum temperature from normal during the growth period is shown in Table

Figure 1. Observed maximum and normal maximum temperatures (^{0} C) during the growth of potato

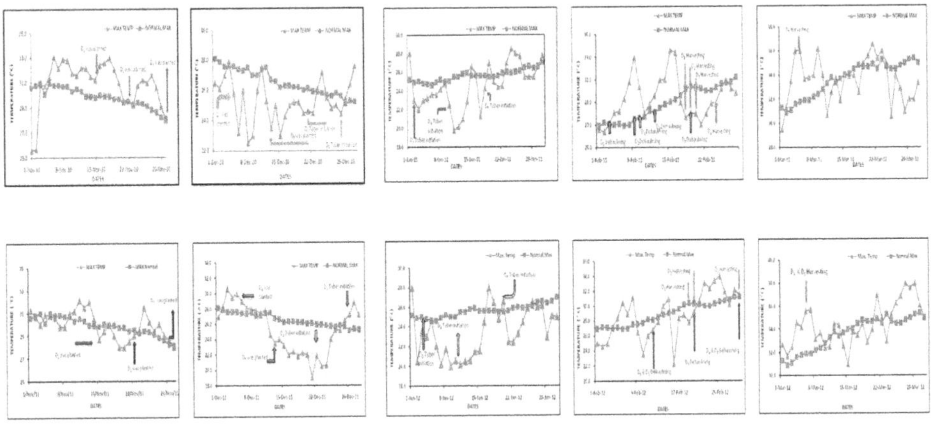

Table 1. Deviation of maximum temperature from normal from November to March (2010-11 and 2011-12)

Date	T_{max}-T_{normal}	Duration Exposed to Max. T(days)		Date	T_{max}-T_{normal}	Duration Exposed to Max. T(days)	
		> Normal	< Normal			> Normal	< Normal
01-Nov-10	-5		1	01-Dec-10	-1.6		1
02-Nov-10	-5.1		1	02-Dec-10	-1.8		1
03-Nov-10	-0.1		1	03-Dec-10	-0.6		1
04-Nov-10	0.7	1		04-Dec-10	0.4	1	
05-Nov-10	0.5	1		05-Dec-10	-1.8		1
06-Nov-10	2.3	1		06-Dec-10	-4.2		1
07-Nov-10	1.4	1		07-Dec-10	-1.3		1
08-Nov-10	2.2	1		08-Dec-10	-5		1
09-Nov-10	2.1	1		09-Dec-10	-4		1
10-Nov-10	1.5	1		10-Dec-10	-1		1
11-Nov-10	1.1	1		11-Dec-10	0.1	1	
12-Nov-10	1.9	1		12-Dec-10	-2.2		1
13-Nov-10	2.3	1		13-Dec-10	-3.2		1
14-Nov-10	1.6	1		14-Dec-10	-1.6		1
15-Nov-10	1.4	1		15-Dec-10	-3.2		1
16-Nov-10	2.4	1		16-Dec-10	-2		1
17-Nov-10	2.7	1		17-Dec-10	-1.3		1
18-Nov-10	3.1	1		18-Dec-10	-1		1
19-Nov-10	2.4	1		19-Dec-10	-1		1

continued on following page

Table 1. Continued

Date	T_{max}-T_{normal}	Duration Exposed to Max. T(days)		Date	T_{max}-T_{normal}	Duration Exposed to Max. T(days)	
		> Normal	< Normal			> Normal	< Normal
20-Nov-10	1.3	1		20-Dec-10	-1.5		1
21-Nov-10	0.1	1		21-Dec-10	-1.5		1
22-Nov-10	0.1	1		22-Dec-10	-1.4		1
23-Nov-10	0	0		23-Dec-10	0.2	1	
24-Nov-10	1.4	1		24-Dec-10	1.4	1	
25-Nov-10	1.9	1		25-Dec-10	-0.2		1
26-Nov-10	1.9	1		26-Dec-10	-0.8		1
27-Nov-10	2.8	1		27-Dec-10	-0.8		1
28-Nov-10	2	1		28-Dec-10	-1.2		1
29-Nov-10	0.9	1		29-Dec-10	0.4	1	
30-Nov-10	-0.1		1	30-Dec-10	1.1	1	
TOTAL		**25**	**4**	31-Dec-10	2.4	1	
				TOTAL		**7**	**24**
01-Jan-11	2.8	1		01-Feb-11	-0.3		1
02-Jan-11	-1.9		1	02-Feb-11	-0.8		1
03-Jan-11	-3.0		1	03-Feb-11	0.0	0	
04-Jan-11	-1.8		1	04-Feb-11	1.1	1	
05-Jan-11	-1.5		1	05-Feb-11	1.1	1	
06-Jan-11	-1.0		1	06-Feb-11	2.4	1	
07-Jan-11	-1.3		1	07-Feb-11	3.8	1	
08-Jan-11	-1.0		1	08-Feb-11	5.5	1	
09-Jan-11	-0.2		1	09-Feb-11	2.8	1	
10-Jan-11	-2.5		1	10-Feb-11	0.3	1	
11-Jan-11	-5.7		1	11-Feb-11	0.6	1	
12-Jan-11	-5.3		1	12-Feb-11	1.5	1	
13-Jan-11	-4.8		1	13-Feb-11	2.5	1	
14-Jan-11	-3.0		1	14-Feb-11	2.2	1	
15-Jan-11	1.0	1		15-Feb-11	4.6	1	
16-Jan-11	0.2	1		16-Feb-11	4.2	1	
17-Jan-11	-4.3		1	17-Feb-11	-0.6		1
18-Jan-11	-1.3		1	18-Feb-11	-2.5		1
19-Jan-11	1.5	1		19-Feb-11	-2.4		1
20-Jan-11	-0.5		1	20-Feb-11	-2.3		1
21-Jan-11	-1.2		1	21-Feb-11	-3.2		1

continued on following page

Table 1. Continued

Date	T_{max}-T_{normal}	Duration Exposed to Max. T(days)		Date	T_{max}-T_{normal}	Duration Exposed to Max. T(days)	
		> Normal	< Normal			> Normal	< Normal
22-Jan-11	-1.4		1	22-Feb-11	-2.0		1
23-Jan-11	1.0	1		23-Feb-11	-1.1		1
24-Jan-11	2.6	1		24-Feb-11	-1.3		1
25-Jan-11	2.0	1		25-Feb-11	-0.4		1
26-Jan-11	1.6	1		26-Feb-11	0.3	1	
27-Jan-11	-1.0		1	27-Feb-11	-0.6		1
28-Jan-11	-1.1		1	28-Feb-11	-1.4		1
29-Jan-11	-0.9		1	**TOTAL**		**14**	**13**
30-Jan-11	0.0	0					
31-Jan-11	0.8	1					
TOTAL		**9**	**21**				
01-Mar-11	-1.8		1	01-Nov-11	0.7	1	
02-Mar-11	0.4	1		02-Nov-11	-0.1		1
03-Mar-11	2.4	1		03-Nov-11	-1		1
04-Mar-11	4.4	1		04-Nov-11	0.8	1	
05-Mar-11	4.2	1		05-Nov-11	-0.2		1
06-Mar-11	1.8	1		06-Nov-11	-0.2		1
07-Mar-11	1.9	1		07-Nov-11	-1		1
08-Mar-11	1.8	1		08-Nov-11	-0.9		1
09-Mar-11	3.8	1		09-Nov-11	0.3	1	
10-Mar-11	1.2	1		10-Nov-11	0.9	1	
11-Mar-11	-2.6		1	11-Nov-11	1.7	1	
12-Mar-11	-1.2		1	12-Nov-11	1.2	1	
13-Mar-11	0.4		1	13-Nov-11	2.1	1	
14-Mar-11	-1.8		1	14-Nov-11	0.6	1	
15-Mar-11	0	1		15-Nov-11	-1.1		1
16-Mar-11	-0.2		1	16-Nov-11	-0.6		1
17-Mar-11	-0.5		1	17-Nov-11	0.1	1	
18-Mar-11	-2.2		1	18-Nov-11	-0.9		1
19-Mar-11	0.3	1		19-Nov-11	-1.7		1
20-Mar-11	0.9	1		20-Nov-11	-1.7		1
21-Mar-11	1.7	1		21-Nov-11	-0.9		1
22-Mar-11	0.9	1		22-Nov-11	-0.5		1
23-Mar-11	0.7	1		23-Nov-11	0.3	1	

continued on following page

Table 1. Continued

Date	T_{max}-T_{normal}	Duration Exposed to Max. T(days)		Date	T_{max}-T_{normal}	Duration Exposed to Max. T(days)	
		> Normal	< Normal			> Normal	< Normal
24-Mar-11	0.3	1		24-Nov-11	2.1	1	
25-Mar-11	-3.9		1	25-Nov-11	0.9	1	
26-Mar-11	-0.5		1	26-Nov-11	0.3	1	
27-Mar-11	-1.9		1	27-Nov-11	1.2	1	
28-Mar-11	-4		1	28-Nov-11	0.4	1	
29-Mar-11	-3.3		1	29-Nov-11	0.4	1	
30-Mar-11	-3.4		1	30-Nov-11	-0.1		1
31-Mar-11	-1.6		1	**TOTAL**		**16**	**14**
TOTAL		**16**	**15**				
01-Mar-11	-1.8		1	01-Nov-11	0.7	1	
02-Mar-11	0.4	1		02-Nov-11	-0.1		1
03-Mar-11	2.4	1		03-Nov-11	-1		1
04-Mar-11	4.4	1		04-Nov-11	0.8	1	
05-Mar-11	4.2	1		05-Nov-11	-0.2		1
06-Mar-11	1.8	1		06-Nov-11	-0.2		1
07-Mar-11	1.9	1		07-Nov-11	-1		1
08-Mar-11	1.8	1		08-Nov-11	-0.9		1
09-Mar-11	3.8	1		09-Nov-11	0.3	1	
10-Mar-11	1.2	1		10-Nov-11	0.9	1	
11-Mar-11	-2.6		1	11-Nov-11	1.7	1	
12-Mar-11	-1.2		1	12-Nov-11	1.2	1	
13-Mar-11	0.4		1	13-Nov-11	2.1	1	
14-Mar-11	-1.8		1	14-Nov-11	0.6	1	
15-Mar-11	0	1		15-Nov-11	-1.1		1
16-Mar-11	-0.2		1	16-Nov-11	-0.6		1
17-Mar-11	-0.5		1	17-Nov-11	0.1	1	
18-Mar-11	-2.2		1	18-Nov-11	-0.9		1
19-Mar-11	0.3	1		19-Nov-11	-1.7		1
20-Mar-11	0.9	1		20-Nov-11	-1.7		1
21-Mar-11	1.7	1		21-Nov-11	-0.9		1
22-Mar-11	0.9	1		22-Nov-11	-0.5		1
23-Mar-11	0.7	1		23-Nov-11	0.3	1	
24-Mar-11	0.3	1		24-Nov-11	2.1	1	
25-Mar-11	-3.9		1	25-Nov-11	0.9	1	

continued on following page

Table 1. Continued

Date	T_{max}-T_{normal}	Duration Exposed to Max. T(days)		Date	T_{max}-T_{normal}	Duration Exposed to Max. T(days)	
		> Normal	< Normal			> Normal	< Normal
26-Mar-11	-0.5		1	26-Nov-11	0.3	1	
27-Mar-11	-1.9		1	27-Nov-11	1.2	1	
28-Mar-11	-4		1	28-Nov-11	0.4	1	
29-Mar-11	-3.3		1	29-Nov-11	0.4	1	
30-Mar-11	-3.4		1	30-Nov-11	-0.1		1
31-Mar-11	-1.6		1	**TOTAL**		**16**	**14**
TOTAL		**16**	**15**				
01-Feb-12	-1.9		1	01-Mar-12	1.4	1	
02-Feb-12	-2.6		1	02-Mar-12	0.8	1	
03-Feb-12	-2.1		1	03-Mar-12	1.4	1	
04-Feb-12	0.2	1		04-Mar-12	3	1	
05-Feb-12	1.3	1		05-Mar-12	2.4	1	
06-Feb-12	3.6	1		06-Mar-12	3.7	1	
07-Feb-12	1.9	1		07-Mar-12	3.8	1	
08-Feb-12	3.7	1		08-Mar-12	1	1	
09-Feb-12	-1.1		1	09-Mar-12	1.3	1	
10-Feb-12	-4.2		1	10-Mar-12	-0.4		1
11-Feb-12	-2.2		1	11-Mar-12	-0.8		1
12-Feb-12	-1.3		1	12-Mar-12	0.7	1	
13-Feb-12	-0.6		1	13-Mar-12	1	1	
14-Feb-12	1.2	1		14-Mar-12	-1.1		1
15-Feb-12	2	1		15-Mar-12	-3		1
16-Feb-12	-7		1	16-Mar-12	-0.3		1
17-Feb-12	-1.2		1	17-Mar-12	-1		1
18-Feb-12	-1.4		1	18-Mar-12	-0.3		1
19-Feb-12	-2.6		1	19-Mar-12	-1.8		1
20-Feb-12	-1.1		1	20-Mar-12	-0.7		1
21-Feb-12	0.5	1		21-Mar-12	2.1	1	
22-Feb-12	3.1	1		22-Mar-12	0.5	1	
23-Feb-12	2.4	1		23-Mar-12	-2.3		1
24-Feb-12	3.1	1		24-Mar-12	0.6	1	
25-Feb-12	3.4	1		25-Mar-12	1.8	1	
26-Feb-12	1.3	1		26-Mar-12	2.4	1	
27-Feb-12	-0.6		1	27-Mar-12	3.2	1	

continued on following page

Table 1. Continued

Date	T_{max}-T_{normal}	Duration Exposed to Max. T(days)		Date	T_{max}-T_{normal}	Duration Exposed to Max. T(days)	
		> Normal	< Normal			> Normal	< Normal
28-Feb-12	1	1		28-Mar-12	2.6	1	
29-Feb-12	0.3	1		29-Mar-12	2.7	1	
TOTAL		15	14	30-Mar-12	0.7	1	
				31-Mar-12	0	0	
				TOTAL		20	10

Table 2. The observed and normal maximum temperatures (^{0} C) on different DOP

DOP	Maximum Temperature(°C)		Normal Maximum(°C)
	2010-11	2011-12	
D_1	31.2	28.7	29.8
D_2	29.6	29.0	29.5
D_3	29.1	28.6	28.2
D_4	23.2	29.0	27.4
D_5	23.5	23.5	26.7

1. In 2010-11, 25 days in the month of November experienced higher maximum temperature than the normal. The extent of rise was as high as 3.1 °C. In the second year, maximum temperature was higher than the normal for 16 days. In the month of December, the first and second year recorded higher maximum temperature than the normal for 7 and 14 days respectively.

The maximum temperature during the time of plantings recorded a wide variation in the two year of study (Table 2). The maximum temperature in 2011-12 up to D3 was lower as compared to 2010-11. The maximum temperature has an important bearing in the process of sprouting of potato. During planting, the maximum temperatures under different DOP were lower in the second year than the normal value (except D4). This low maximum temperature adversely affected the sprouting process. The duration of planting to emergence recorded wide variation in two different years (Table 3). In 2010-11, the duration was maximum under D1 planting whereas in 2011-12, the duration ranged from 9-14 days. The above analysis showed that a wide variation in the durations of emergence existed due to the DOP. In 2010-11, D2 to D5 plantings recorded a lower duration than D1 although the maximum temperature on 15th November was 31.2 0 C, higher than the normal. During 2011-12, the durations ranged from 9-11 days for D1 to D4 planting (Table 4). However,

Table 3. Cumulative maximum temperature (day 0C) and duration (days) from planting to emergence under different DOP

DOP	Date of emergence		Duration (days)		Cumulative Maximum Temperature(day ^{0}C)	
	2010-11	2011-12	2010-11	2011-12	2010-11	2011-12
D_1	27-11-10	26-11-11	12	11	373.6	321.8
D_2	29-11-10	02-12-11	7	10	214.0	292.2
D_3	08-12-10	08-12-11	9	9	239.7	259.2
D_4	13-12-10	15-12-11	7	9	173.5	192.9
D_5	21-12-10	27-12-11	8	14	195.7	181.1

Table 4. The cumulative maximum temperatures (day 0C) and duration (days) during emergence to tuber initiation under different DOP

DOP	Date of tuber initiation		Duration (days)		Cumulative Maximum Temperature(day ^{0}C)	
	2010-11	2011-12	2010-11	2011-12	2010-11	2011-12
D_1	21-12-10	22-12-11	24	26	621.9	683.4
D_2	28-12-10	29-12-11	29	27	737.4	675.2
D_3	02-01-11	04-01-12	25	27	629.6	654.8
D_4	10-01-11	11-01-12	28	27	692.7	624.0
D_5	18-01-11	20-01-12	28	24	677.1	567.5

the duration was maximum when potato was planted on 13th December, when an enormous dip in maximum temperature was observed. The discrepancy observed in the durations of germination might be attributed to the cumulative maximum temperature. The cumulative maximum temperature shown in the above table clearly indicated the lower values in D4 and D5 planting in both the years.

Tuber initiation began from 21st December and continued to 18th January in the first year, whereas in the second year, it ranged from 22nd December to 20th January. In the first year, the maximum temperature during this phase ranged 19.8 to 28.0 °C whereas in the second year it was in the range of 20.0 to 28.8 °C. When the crop was planted on 15th November, the cumulative maximum temperature from emergence to tuber initiation was 621.9 day 0 C and the duration of this phenophase was 24 days in 2010-11; whereas, in 2011-12, it was of 26 days duration with 683.4 day 0 C (Table 4). In case of D2 planting the cumulative maximum temperature increased with the increment of the duration of this phenophase. In both the years, the D2 planted crop recorded the tuber initiation during the last phase of December when the maximum temperature was ranging from 24.4 to 27.2 0 C. The duration of this phenophase was found to be maximum under D2 planting, in both the years. The tuber initiation was not delayed with the delay in planting. The cumulative maximum temperatures in two experimental years varied remarkably because of the difference in the maximum temperatures observed during these two years. In 2010-11, the D4 and D5 plantings recorded similar duration whereas in 2011-12, the duration was found to be minimum under D5 planting. In potato, tuber initiation is highly sensitive to both radiation and temperature (Yuan & Bland, 2004). Potato growth in well managed field followed an exponential growth pattern until the onset of senescence (Yuan & Bland, 2004). Generally the potato has different temperature requirement for shoot and tuber growth. High temperature favours vegetative growth. Ezekiel and Bhargava (2000) observed that 25 or 30 0C day temperatures were conducive for potato top growth under short day conditions. However, these authors failed to identify the nature of maximum temperature variation for tuber initiation under different DOP. In the first year experiment under D1, the duration from emergence to tuber initiation was 24 days with the minimum amount of cumulative maximum temperature (621.9 day 0C). Under D5 planting, the duration was 28 days. In 2011-12, the duration of this phenophase was 26 days under D1. The delayed tuber initiation under D5 accelerated the vegetative growth which enhances the photosynthetic efficiency of the crop which ultimately recorded high yield under this planting in the first year. In 2011-12, the D1 planting recorded highest yield. If the cumulative maximum temperatures from emergence to tuber initiation were critically analyzed under D5 and D1 plantings, it would be seen that a marginal difference did exist in these two plantings. In 2010-11, the mean maximum temperature during this phenophase under D5 planting was 24.20C; whereas in 2011-12, the mean maximum temperature under D1 planting was 26.30C. The biological response in terms of DM production and tuber yield was found to be maximum under D5 and D1 plantings in two experimental years respectively. The close similarity in mean maximum temperature during the emergence to tuber initiation phase might be the reason for this contradictory result obtained in two years. On the basis of this observation, it may be concluded that the delayed planting of potato will not show the reduction in tuber yield always and it will depend on the mean maximum temperature prevailing during the vegetative phase. If the mean maximum temperature lies in between 24.2 to 26.3 0C it will definitely increase the vegetative growth under W.B situations. Ezekiel and Bhargava (2000) observed that the optimum day temperature for leaf growth should lie in between 25 and 300 C.

Tuber initiation to dehaulming comprises of three sub phases: Early bulking phase (40-50 DAP), mid bulking (50-60 DAP) and maximum bulking (60-70 DAP). This phase ends with the sensescence of leaves and tuber maturity. In both the year, the duration reduced with the delay in planting (Table 5). In general, duration was less in

Table 5. The impact of maximum temperatures (day time in 0C) during the tuber initiation to dehaulming

DOP	Date of dehaulming		Duration (days)		Cumulative maximum temperature (day °C)		Mean maximum temperature (°C)	
	2010-11	2011-12	2010-11	2011-12	2010-11	2011-12	2010-11	2011-12
D_1	03-02-11	12-02-12	44	52	1097.4	1273.4	24.9	24.5
D_2	08-02-11	12-02-12	42	45	1063.2	1109.6	25.3	24.7
D_3	09-02-11	20-02-12	38	47	964.3	1177.3	25.4	25.0
D_4	12-02-11	29-02-12	33	49	864.1	1312.1	26.2	26.8
D_5	19-02-11	29-02-12	32	40	899.3	1103.9	28.1	27.6

2010-11 than in 2011-12. In the first year the duration was minimum when the crop was planted on 13th December. In the first year, the duration ranged from 32 to 44 days whereas in the second year, it ranged from 40 to 52 days under different DOP.

MINIMUM TEMPERATURE DURING THE GROWTH OF POTATO

The minimum temperature has a significant effect on the growth processes of potato. One of the principal perspectives in climate change is the increase in minimum temperature. This adversely affects the winter crops by altering the duration of phenophases (Parya et al., 2010). The minimum temperature during the planting of potato is presented in Table 6.

Table 6. The observed and normal minimum temperatures (0 C) on different DOP

DOP	Minimum Temperature(°C)		Normal Minimum(°C)
	2010-11	2011-12	
D_1	20.7	21.0	17.2
D_2	14.5	18.0	16.7
D_3	21.0	14.4	14.8
D_4	19.0	16.0	13.0
D_5	15.6	15.0	12.7

The minimum temperatures observed on different DOP were higher than their normal values in most of the cases. The minimum temperature during planting was lowest under D2 in 2010-11 but in 2011-12, it was found to be the least under D3. Low minimum temperature during planting may affect the process of sprouting in potato. As the minimum temperature during planting was higher than the normal value, the sprouting accelerated in all DOP except D2 in 2010-11 and D3 in 2011-12. In growth processes, impact of minimum temperature during winter is more pronounced than the maximum temperature. Increasing minimum temperature in winter hastens the biological process also shortens the duration of a particular phenophase (Parya et al., 2010).

PLANTING TO EMERGENCE

Minimum temperature influences the process of sprouting in potato. There are three principal reactions which influence the starch sugar ratio in potato such as breakdown of starch to sugar, conversion of sugar to starch and oxidation of sugar during respiration (Emilsson & Lindblom, 1963). Low soil temperature often causes the accumulation of sugar in seed tuber which activates the enzymes hydrolyzing starch to sucrose. The relative proportions between sucrose and hexoses are also subjected to change due to change in temperature. Fischnich (1959) observed that at high temperature, only sucrose is formed while at low temperature glucose and fructose as well as sucrose accumulate. During sprouting the hydrolysis of starch is accelerated and this acceleration is due to rise in minimum temperature.

Table 7 shows the duration from planting to emergence, cumulative minimum temperature requirement, and the mean minimum temperature during this phenophase. The results revealed the following facts-

1. The duration was maximum under D_1 planting in 2010-11; the durations under D_2, D_4 and D_5 plantings were almost similar. The cumulative minimum temperature requirement was lowest in D_5.In 2011-12, the duration of this phenophase ranged from 9 to 14 days, the D_5, required the longest duration. The mean minimum temperature in 2011-12 under D_5 was 10.7^{0}C, the lowest among all the DOP.
2. The D_1 planted crop required 11to12 days for emergence, when the mean minimum temperature ranged from 18.1 to 19.2 ^{0}C.
3. The germination was hastened in D_2, D_3, D_4 plantings when the mean minimum temperature ranged from 15.4 to 17.9^{0}C.

Table 7. Cumulative and mean minimum temperatures and duration (days) from planting to emergence under different DOP

DOP	Date of emergence		Duration (days)		Cumulative Minimum temperature (day ^{0} C)		Mean Minimum temperature (^{0} C)	
	2010-11	2011-12	2010-11	2011-12	2010-11	2011-12	2010-11	2011-12
D_1	27-11-10	26-11-11	12	11	229.9	156.6	19.2	18.1
D_2	29-11-10	02-12-11	7	10	121.4	139.9	17.3	15.8
D_3	08-12-10	08-12-11	9	9	138.7	143.2	15.4	15.9
D_4	13-12-10	15-12-11	7	9	125.1	144.4	17.9	16.0
D_5	21-12-10	27-12-11	8	14	84.2	149.3	11.8	10.7

It may be concluded that the emergence of potato will be hastened if the average minimum temperature during the planting to emergence lies in between 15.4 to 17.90C. This is an intermediate minimum temperature range observed in between D1 and D5. At intermediate temperatures, hydrolysis of starch produces all monosaccharides and there is no presence of sucrose in the system. This domination of monosaccharide facilitates the sprouting process with the concomitant emergence of the haulm.

EMERGENCE TO TUBER INITIATION

Tuber initiation takes place after 25 days of vegetative growth, when photosynthates are accumulated in the modified stem of the plant. It is cognate with the understanding of the mechanisms which lead to the formation of organs that plant did not produce earlier in the ontogeny-whether these are flowers, as in most species, or tubers as in the potato. The development of tuber in potato is largely controlled by the factors of the environment, particularly the temperature and photoperiod. Borah and Milthorpe (1959), Milthorpe (1963) and Headford (1961) say that the increased concentration of carbohydrate in stolon tips favours the tuber formation which is affected by radiation, temperature, soil moisture condition and nutrient supply. Among these physical factors, temperature is the pivotal.

In the two years of experimentation, duration from emergence to tuber initiation ranged from 24 to 29 days in 2010-11 and 24 to 27 days in 2011-12 (Table 8). In the first year, D1 planted crop required 24 days while in the second year, the tuber initiation was observed after 26 days of the vegetative growth. The cumulative minimum temperatures for this DOP were 361.7 and 373.3 day 0C respectively

Table 8. Cumulative and mean minimum temperatures and the duration (days) from emergence to tuber initiation, under different DOP

	Date of tuber initiation		Duration (days)		Cumulative Minimum temperature (day ^{0} C)		Mean Minimum temperature (^{0} C)	
DOP	2010-11	2011-12	2010-11	2011-12	2010-11	2011-12	2010-11	2011-12
D_1	21-12-10	22-12-11	24	26	361.7	373.3	15.1	14.4
D_2	28-12-10	29-12-11	29	27	379.8	356.2	13.1	13.2
D_3	02-01-11	04-01-12	25	27	288.2	351.7	11.5	13.0
D_4	10-01-11	11-01-12	28	27	277.4	356.8	9.9	13.2
D_5	18-01-11	20-01-12	28	24	248.6	337.7	8.9	14.1

for first and second years of experimentation respectively. The delay in planting increased the duration marginally; the cumulative minimum temperature however decreased remarkably in the first year. In the second year, the duration of this phenophase was almost similar upto D4, while the minimum duration was observed in D5. The cumulative minimum temperature requirement ranged from 337.7 to 373.3 day 0C for different DOP, the minimum being observed under D5. The mean minimum temperature during this phenophase ranged from 8.9 to 15.1 0C in the first year, whereas in the second year it ranged from 13.0 to 14.4 0C. In the first year experiment, the D4 and D5 recorded a very low minimum temperature as compared to other three DOP. In the second year, the variation in mean minimum temperature under different DOP was very nominal. From this result no definite effect of minimum temperature on the process of tuber initiation could be identified. The variation in minimum temperature affected the temperature of the soil which ultimately determined the tuber growth. However, it could safely be said that the late planted potato will initiate tuber at lower minimum temperature condition than the early planted crop.

TUBER INITIATION TO DEHAULMING

Tuber initiation to dehaulming is the active phase of tuber growth which is highly sensitive to temperature variation. The transport of carbohydrate from leaf to tuber is dependent on the interaction of climatic conditions, cultural practices as well as genotypic variation (Kooman & Rabbinge, 1996). The variation in assimilate allocation is determined by the time of initiation as well as temperature condition of the soil (Kooman & Rabbinge, 1996). The citations already mentioned clearly show

Table 9. Cumulative and mean minimum temperatures and the duration (days) from tuber initiation to dehaulming, under different DOP

	Date of dehaulming		Duration (days)		Cumulative Minimum Temperature (day °C)		Mean Minimum temperature (°C)	
DOP	2010-11	2011-12	2010-11	2011-12	2010-11	2011-12	2010-11	2011-12
D_1	03-02-11	12-02-12	44	52	417.7	658.6	9.5	12.7
D_2	08-02-11	12-02-12	42	45	419.5	585.3	10.0	13.0
D_3	09-02-11	20-02-12	38	47	386.2	611.8	10.2	13.0
D_4	12-02-11	29-02-12	33	49	350.8	630.9	10.6	12.9
D_5	19-02-11	29-02-12	32	40	395.3	530.2	12.4	13.3

the importance of temperature in the tuber growth. However, the authors could not identify the impact of minimum temperature which is so important in winter crops. The minimum temperature condition during this phenophase is given in Table 9.

The duration of this phenophase reduced due to delay in planting in 2010-11; however in 2011-12, the trend was not regular. This discrepancy emanated from the variation in the amount as well as the distribution of rainfall during the second year experiment. In second year, the durations were found to be higher than the first year under all DOP which was also due to the rainfall receipt. The cumulative minimum temperature requirement did not show a definite pattern under different DOP. The mean minimum temperature during this phenophase ranged from 9.5 to 12.40C in 2010-11 and in 2011-12, it ranged from12.7 to 13.30C. The minimum temperature (night temperature) has a remarkable influence on the tuber growth. In the first year experiment, the D5 planted crop recorded the maximum yield where the mean minimum temperature was 12.40C; in the second year the D1 planted crop recorded maximum tuber yield when the mean minimum temperature was 12.70C. The unique feature is that the mean minimum temperature during this phenophase should be within the range of 12.4 to 12.70C for better tuberization process, in spite of the delay in planting. Ezekiel and Bhargava (2000) observed that the optimum night temperature for leaf growth in potato was 15 to 20 0C under controlled condition. In the present experiment, it is evident that the mean minimum temperature during the tuberization stage should be within 12 to 130C for better productivity under W.B. condition.

The Figure 2 describes the observed and normal minimum temperatures during the potato growing season of both the years. The result shows that in the month of November when three plantings were done the observed minimum temperature was higher than its normal value for a major duration. Out of 30 days, 23 days experi-

Figure 2. Observed minimum and normal minimum temperatures (^{0} C) during the growth of potato

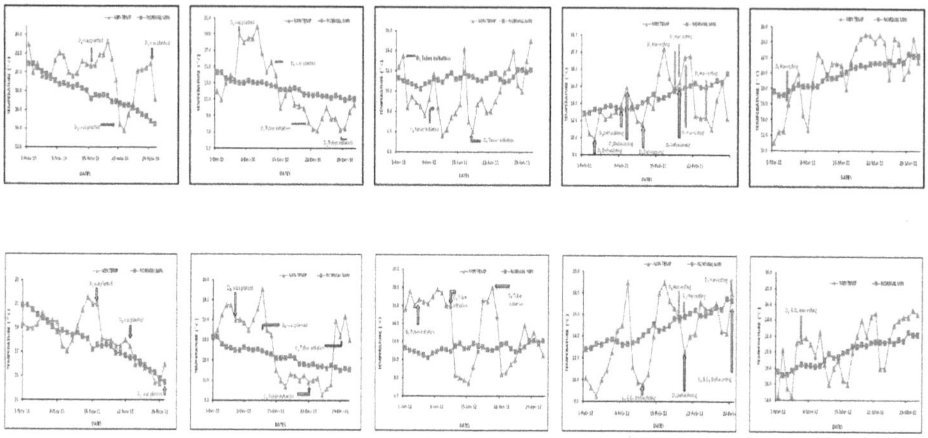

enced higher minimum temperature than its normal (Table 10). During this period, the magnitude of minimum temperature increase ranged from 0.4 to 6.20C. In the month of December, the first fortnight experienced higher minimum temperature than the normal; however in the second fortnight, the observed minimum temperature was lower than its normal value when the tuber initiation occurred in D1 and D2 plantings. The month December, has 12 days with higher minimum temperature and 19 days with lower minimum temperature than their normal values. In this month the observed minimum temperature exceeded the normal value by 7.0 0C on one occasion (Table 11). The January experienced lower minimum temperature than the normal for 23 days. The extent of reduction ranged from 1.2 to 5.40C. Therefore, it was observed that the tuber initiation occurred under lower minimum temperature condition (except D3) than the normal. The February has 17 days with lower minimum temperature than the normal.

In the second year, D1 and D2 crops were planted when the minimum temperature was above normal. The D3 crop was planted when the minimum temperature was above normal in the first year; whereas in the second year, contradictory situation was noted. In the month of November 2011, 16 days experienced higher minimum temperature than their normal values. The magnitude of rise in the minimum temperature ranged from 0.4 to 3.80C. The 13 days of this month recorded lower minimum temperature than their normal values. The month December had 17 days higher minimum temperature and 13 days lower minimum temperature than their respective normal values. Tuber initiation of D1 planted crop occurred when the observed minimum temperature was lower than the normal value but the reverse was observed in case of D2 planted crop. In the first year experiment, the

Table 10. Deviation of minimum temperature from normal from November to March (2010-11 and 2011-12)

Date	T_{min}-T_{normal}	Duration Exposed to Min. T(days)		Date	T_{min}-T_{normal}	Duration Exposed to Min. T(days)	
		> Normal	< Normal			> normal	< normal
01-Nov-10	2.1	1		01-Dec-10	-2.3		1
02-Nov-10	-0.9		1	02-Dec-10	-3.3		1
03-Nov-10	0.5	1		03-Dec-10	-0.6		1
04-Nov-10	-0.5		1	04-Dec-10	0.7	1	
05-Nov-10	-0.3		1	05-Dec-10	0.3	1	
06-Nov-10	-0.2		1	06-Dec-10	6.0	1	
07-Nov-10	2.2	1		07-Dec-10	5.0	1	
08-Nov-10	3.4	1		08-Dec-10	5.2	1	
09-Nov-10	2.9	1		09-Dec-10	5.4	1	
10-Nov-10	1.4	1		10-Dec-10	7.0	1	
11-Nov-10	1.2	1		11-Dec-10	3.6	1	
12-Nov-10	1.3	1		12-Dec-10	1.5	1	
13-Nov-10	2.9	1		13-Dec-10	2.9	1	
14-Nov-10	2.8	1		14-Dec-10	1.4	1	
15-Nov-10	3.5	1		15-Dec-10	-2.2		1
16-Nov-10	3.2	1		16-Dec-10	-0.7		1
17-Nov-10	4.4	1		17-Dec-10	0.3	1	
18-Nov-10	4.3	1		18-Dec-10	-2.0		1
19-Nov-10	5.8	1		19-Dec-10	-2.0		1
20-Nov-10	4.6	1		20-Dec-10	-1.7		1
21-Nov-10	2.3	1		21-Dec-10	-3.4		1
22-Nov-10	-2.2		1	22-Dec-10	-4.1		1
23-Nov-10	-2.7		1	23-Dec-10	-4.4		1
24-Nov-10	-1.0		1	24-Dec-10	-2.7		1
25-Nov-10	0.4	1		25-Dec-10	-1.4		1
26-Nov-10	4.3	1		26-Dec-10	-2.5		1
27-Nov-10	4.8	1		27-Dec-10	-2.8		1
28-Nov-10	5.2	1		28-Dec-10	-4.0		1
29-Nov-10	6.2	1		29-Dec-10	-3.5		1
30-Nov-10	2.7	1		30-Dec-10	-1.7		1
TOTAL		**23**	**7**	31-Dec-10	-0.7		1
01-Jan-11	1.1	1		**TOTAL**		**12**	**19**

continued on following page

Table 10. Continued

Date	T_{min}-T_{normal}	Duration Exposed to Min. T(days)		Date	T_{min}-T_{normal}	Duration Exposed to Min. T(days)	
		> Normal	< Normal			> normal	< normal
02-Jan-11	2.5	1		01-Feb-11	0.1	1	
03-Jan-11	-2.5		1	02-Feb-11	-2.4		1
04-Jan-11	-1.2		1	03-Feb-11	-3.5		1
05-Jan-11	-1.6		1	04-Feb-11	-0.1		1
06-Jan-11	-1.7		1	05-Feb-11	-1.4		1
07-Jan-11	-3.1		1	06-Feb-11	-1.1		1
08-Jan-11	-1.7		1	07-Feb-11	-0.5		1
09-Jan-11	0.5	1		08-Feb-11	0.6	1	
10-Jan-11	-2.7		1	09-Feb-11	2.6	1	
11-Jan-11	-5.3		1	10-Feb-11	0.6	1	
12-Jan-11	-5.3		1	11-Feb-11	-1.6		1
13-Jan-11	-4.4		1	12-Feb-11	-2.5		1
14-Jan-11	-3.4		1	13-Feb-11	-0.2		1
15-Jan-11	-2.6		1	14-Feb-11	-1.6		1
16-Jan-11	2.4	1	1	15-Feb-11	3.0	1	
17-Jan-11	-4.5		1	16-Feb-11	5.7	1	
18-Jan-11	-5.4		1	17-Feb-11	2.0	1	
19-Jan-11	-2.4		1	18-Feb-11	-0.1		1
20-Jan-11	-1.7		1	19-Feb-11	0.8	1	
21-Jan-11	-3.4		1	20-Feb-11	3.6	1	
22-Jan-11	-3.7		1	21-Feb-11	3.5	1	
23-Jan-11	-2.8		1	22-Feb-11	-3.8		1
24-Jan-11	-1.2		1	23-Feb-11	-3.6		1
25-Jan-11	1.0	1		24-Feb-11	-3.5		1
26-Jan-11	1.3	1		25-Feb-11	-5.2		1
27-Jan-11	2.6	1		26-Feb-11	-2.3		1
28-Jan-11	-1.5		1	27-Feb-11	0.0	1	
29-Jan-11	-2.1		1	28-Feb-11	-5.2		1
30-Jan-11	1.1	1		**TOTAL**		**10**	**17**
31-Jan-11	2.9	1		01-Nov-11	-1.5		1
TOTAL		**9**	**23**	02-Nov-11	-1.9		1
01-Mar-11	-6.6		1	03-Nov-11	-1.5		1

continued on following page

Table 10. Continued

Date	T_{min}-T_{normal}	Duration Exposed to Min. T(days)		Date	T_{min}-T_{normal}	Duration Exposed to Min. T(days)	
		> Normal	< Normal			> normal	< normal
02-Mar-11	-4.7		1	04-Nov-11	-0.8		1
03-Mar-11	-4.7		1	05-Nov-11	0.5	1	
04-Mar-11	-1.7		1	06-Nov-11	-0.1		1
05-Mar-11	0.3	1		07-Nov-11	0.5	1	
06-Mar-11	1.5	1		08-Nov-11	0.8	1	
07-Mar-11	-3.7		1	09-Nov-11	-1.3		1
08-Mar-11	-5.6		1	10-Nov-11	-1.5		1
09-Mar-11	0.8	1		11-Nov-11	-0.4		1
10-Mar-11	4.2	1		12-Nov-11	0.6	1	
11-Mar-11	2.6	1		13-Nov-11	2.2	1	
12-Mar-11	-0.4		1	14-Nov-11	3.5	1	
13-Mar-11	1.9	1		15-Nov-11	3.8	1	
14-Mar-11	1.9	1		16-Nov-11	3.4	1	
15-Mar-11	1.1	1		17-Nov-11	0.6	1	
16-Mar-11	-2.1		1	18-Nov-11	0.4	1	
17-Mar-11	0.8	1		19-Nov-11	0.5	1	
18-Mar-11	2.6	1		20-Nov-11	0.6	1	
19-Mar-11	3.5	1		21-Nov-11	0.6	1	
20-Mar-11	4.2	1		22-Nov-11	1.3	1	
21-Mar-11	3.7	1		23-Nov-11	1.0	1	
22-Mar-11	2.7	1		24-Nov-11	-0.5		1
23-Mar-11	3.7	1		25-Nov-11	-0.2		1
24-Mar-11	2.8	1		26-Nov-11	-0.2		1
25-Mar-11	-2.2		1	27-Nov-11	0.0	0	
26-Mar-11	3.4	1		28-Nov-11	-0.8		1
27-Mar-11	2.3	1		29-Nov-11	-0.4		1
28-Mar-11	-2.0		1	30-Nov-11	1.5	1	
29-Mar-11	-2.1		1	**TOTAL**		**16**	**13**
30-Mar-11	2.3	1		01-Jan-12	4.3	1	
31-Mar-11	-0.7		1	02-Jan-12	6.6	1	
TOTAL		**19**	**12**	03-Jan-12	5.1	1	
01-Dec-11	0.0	0		04-Jan-12	6.0	1	

continued on following page

Table 10. Continued

Date	T_{min}-T_{normal}	Duration Exposed to Min. T(days)		Date	T_{min}-T_{normal}	Duration Exposed to Min. T(days)	
		> Normal	< Normal			> normal	< normal
02-Dec-11	0.7	1		05-Jan-12	5.9	1	
03-Dec-11	2.9	1		06-Jan-12	6.0	1	
04-Dec-11	4.2	1		07-Jan-12	6.3	1	
05-Dec-11	4.3	1		08-Jan-12	6.9	1	
06-Dec-11	3.0	1		09-Jan-12	6.2	1	
07-Dec-11	3.0	1		10-Jan-12	4.9	1	
08-Dec-11	2.4	1		11-Jan-12	5.0	1	
09-Dec-11	1.9	1		12-Jan-12	-3.2		1
10-Dec-11	3.0	1		13-Jan-12	-3.8		1
11-Dec-11	4.0	1		14-Jan-12	-3.4		1
12-Dec-11	6.2	1		15-Jan-12	-3.9		1
13-Dec-11	2.3	1		16-Jan-12	-2.6		1
14-Dec-11	1.9	1		17-Jan-12	-0.5		1
15-Dec-11	-1.2		1	18-Jan-12	5.2	1	
16-Dec-11	-2.2		1	19-Jan-12	5.4	1	
17-Dec-11	-2.8		1	20-Jan-12	7.0	1	
18-Dec-11	-1.8		1	21-Jan-12	3.5	1	
19-Dec-11	-1.7		1	22-Jan-12	-3.3		1
20-Dec-11	-1.6		1	23-Jan-12	-3.2		1
21-Dec-11	-1.1		1	24-Jan-12	-1.7		1
22-Dec-11	-1.7		1	25-Jan-12	-0.9		1
23-Dec-11	-1.6		1	26-Jan-12	0.8	1	
24-Dec-11	-1.2		1	27-Jan-12	2.4	1	
25-Dec-11	-2.8		1	28-Jan-12	-0.1		1
26-Dec-11	-2.0		1	29-Jan-12	0.9	1	
27-Dec-11	-1.9		1	30-Jan-12	-0.1		1
28-Dec-11	4.6	1		31-Jan-12	-1.6		1
29-Dec-11	3.4	1		**TOTAL**		**18**	**13**
30-Dec-11	5.2	1		01-Mar-12	-3.1		1
31-Dec-11	2.9	1		02-Mar-12	3.2	1	
TOTAL		**17**	**13**	03-Mar-12	-1.7		1
01-Feb-12	-2.6		1	04-Mar-12	-3.2		1

continued on following page

Table 10. Continued

Date	T_{min}-T_{normal}	Duration Exposed to Min. T(days)		Date	T_{min}-T_{normal}	Duration Exposed to Min. T(days)	
		> Normal	< Normal			> normal	< normal
02-Feb-12	-3.6		1	05-Mar-12	3.1	1	
03-Feb-12	-4.8		1	06-Mar-12	2.9	1	
04-Feb-12	-3.2		1	07-Mar-12	3.5	1	
05-Feb-12	-2.7		1	08-Mar-12	2.7	1	
06-Feb-12	-1.2		1	09-Mar-12	1.3	1	
07-Feb-12	1.0	1		10-Mar-12	4.2	1	
08-Feb-12	1.8	1		11-Mar-12	-0.5		1
09-Feb-12	5.7	1		12-Mar-12	-3.8		1
10-Feb-12	-2.2		1	13-Mar-12	-1.3		1
11-Feb-12	-3.8		1	14-Mar-12	-0.4		1
12-Feb-12	-4.5		1	15-Mar-12	-3.4		1
13-Feb-12	-4.0		1	16-Mar-12	-4.1		1
14-Feb-12	-1.1		1	17-Mar-12	-0.7		1
15-Feb-12	3.5	1		18-Mar-12	1.3	1	
16-Feb-12	4.3	1		19-Mar-12	3.0	1	
17-Feb-12	2.3	1		20-Mar-12	1.2	1	
18-Feb-12	1.0	1		21-Mar-12	3.3	1	
19-Feb-12	-0.2		1	22-Mar-12	3.6	1	
20-Feb-12	-3.3		1	23-Mar-12	-3.3		1
21-Feb-12	-2.1		1	24-Mar-12	-3.2		1
22-Feb-12	-2.2		1	25-Mar-12	0.0	1	
23-Feb-12	1.2	1		26-Mar-12	2.3	1	
24-Feb-12	0.6	1		27-Mar-12	2.4	1	
25-Feb-12	0.3	1		28-Mar-12	3.0	1	
26-Feb-12	0.3	1		29-Mar-12	1.9	1	
27-Feb-12	-2.0		1	30-Mar-12	3.1	1	
28-Feb-12	-3.2		1	31-Mar-12	2.6	1	
29-Feb-12	0.7	1		**TOTAL**		**19**	**12**
TOTAL		**12**	**17**				

Table 11. Average maximum, minimum and mean R. H. s (%) from planting to emergence under different DOP

DOP	2010-11			2011-12		
	Max R. H.	Min R. H.	Mean R. H.	Max R. H.	Min R. H.	Mean R. H.
D_1	92.1	53.7	72.9	94.3	59.5	76.9
D_2	90.9	50.3	70.6	91.6	50.0	70.8
D_3	94.4	56.7	75.6	91.2	53.3	72.3
D_4	96.0	74.6	85.3	94.4	59.4	76.9
D_5	94.1	50.9	72.5	97.9	60.0	78.9

tuber initiation under D1 and D2 plantings was recorded when the minimum temperature was lower than the normal. The month of January recorded 18 days with higher minimum temperature and 13 days with lower minimum temperature than their corresponding normal values. The extent of rise in the lower minimum temperature ranged from 0.8 to 7.00C. The tuber initiation under D3, D4 and D5 occurred under higher minimum temperature regimes than their normal values. This was contradictory to the first year observation. The month February in 2012, had 12 days with higher minimum temperature and 17 days with lower minimum temperature than their respective normal values. In this month, the dehaulming was done under low minimum temperature regimes. In the first year experiment, dehaulming for D2 and D3 plantings was done under higher minimum temperature regimes. In the second year experiment, the tuber initiation occurred under lower minimum temperature regime only in case of D1 planting; whereas in the first year experiment, tuber initiation occurred under lower minimum temperature regimes for all DOP except D3. The tuber initiation under D5 planting was noted when the reduction of minimum temperature from the normal value was 5.40C. The lower minimum temperature (night temperature) reduces the soil temperature which increases the rate of bulking under low temperature regime. The dissimilarities obtained in the tuber yield in two year experiment under different DOP were due to different minimum temperature levels observed during tuberization phase.

RELATIVE HUMIDITY DURING THE GROWTH PERIOD

The R. H. is the ratio of actual vapour pressure to the saturated vapour pressure at a specified temperature. This is the measure of vapour pressure around the plant canopy. The vapour pressure surrounding the leaf regulates the stomatal movement,

Figure 3. Observed maximum, minimum and mean R. H. s (%) during the growth of potato

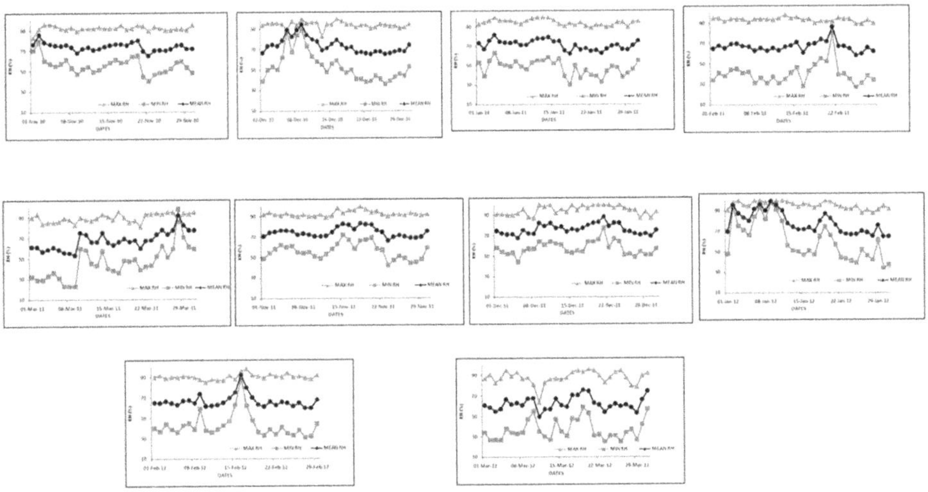

rate of transpiration, stomatal diffusion resistance as well as leaf temperature of the crop leaf (Chakraborty et al., 1991).Winter crop is generally subjected to lower humid environment which creates high transpirational loss of water from leaf. When potato was planted on 15th November, the maximum and minimum R. H.s were 92 and 58% respectively with a mean value of 75%. The D2 planting was done in lesser humid environment having the mean R. H. of 64.5% (Figure 3). The mean humidity was 71.5% when potato was planted on 29th November in 2010.

The mean maximum, minimum and average R. H.s varied widely during planting to emergence under different DOP (Table 11). The mean maximum R. H. in 2010-11, ranged from 90.9 to 96% under different DOP. In 2011-12, the mean maximum R. H. varied from 91.2 to 97.9%. This indicated that the morning humidity under all DOP remained very high which was reflected in the suspension of fog in air during this period. The afternoon humidity was found to be more consistent in 2011-12 than 2010-11. In 2011-12, the afternoon humidity ranged from 50 to 60%; whereas in 2010-11, it ranged from 50.3 to 74.6%. The low humidity in the afternoon created a possibility of increased transpiration rate as well as CO_2 absorption during photosynthesis. The photosynthetic efficiency in the winter crop is higher in the morning as well as afternoon hours (Pallas & Samish, 1974; Nayyar et al., 1990; Chakraborty, 1994).

The mean morning humidity in both the years of study remained very high ranging from 94.2 to 96.0%; whereas the afternoon humidity was less in 2010-11 (47.7 to 58.9%) than in 2011-12 (56.7 to 66.3%) under different DOP (Table 12). In 2010-11,

Table 12. The average maximum, minimum and the mean R. H.s (%) during emergence to tuber initiation under different DOP

DOP	2010-11			2011-12		
	Max R. H.	Min R. H.	Mean R. H.	Max R. H.	Min R. H.	Mean R. H.
D1	94.5	58.4	76.5	94.4	56.7	75.6
D2	94.5	53.9	74.2	95.4	58.0	76.7
D3	94.2	51.6	72.9	96.0	60.3	78.1
D4	94.3	47.7	71.0	96.8	66.3	81.5
D5	95.2	48.0	71.6	96.0	66.0	81.0

the D1 and D2 planted crop s experienced more humid environment than the late planted crops; whereas in 2011-12, the D1 planted crop experienced the minimum average humidity. It indicated that the late planted crop (D5) in 2010-11 and the early planted crop (D1) performed better in terms of photosynthetic C-fixation.

The R. H., during tuber initiation to dehaulming under different DOP, is given in Table 13. The tuber initiation started on 21st December 2010 in D1, while in D5 the date of tuber initiation was 18th January in 2011. In the second year experiment, the tuber initiation spanned from 22nd December, 2011 to 20th January, 2012. In the first year, maximum R. H. during tuber initiation stage ranged from 91 to 100% whereas the minimum R. H. 36 to 63%; in the second year, the maximum R. H. was in between 87 to 100% and the minimum R. H. 47 to 94%, showing wide variation in two experimental years (Figure 3). The mean maximum R. H. during this phenophase was always above 90% under different DOP; whereas the mean minimum R. H. ranged 40.3 to 46.3% in the first year, 44.0 to 55.7% in the second year. Higher minimum R. H. in the second year than the first year indicated the more humid environment which was reflected in the rainfall receipt during the second year experiment. This humid environment in the second year caused bacterial wilt in the late planted potato. The yield reduction under late planting obtained in the second year was due to this reason. In the first year experiment, the mean maximum and minimum R. H.s under D5 planting were 93.4 and 40.3%- the least among the different DOP. This was helpful for better productivity of potato in the first year under D5 planting.

The mean maximum and minimum R. H.s shown above revealed that the minimum R. H.s differed widely in two year experiments. The second year experienced more minimum R. H. during the early plantings as compared to the first year experiment. As a whole, the second year exhibited more humid environment than the first year during early plantings (Table 14).

Table 13. The average maximum, minimum and the mean R. H.s (%) during tuber initiation to dehaulming under different DOP

DOP	2010-11			2011-12		
	Max R. H.	Min R. H.	Mean R. H.	Max R. H.	Min R. H.	Mean R. H.
D_1	94.4	45.8	70.1	94.1	55.7	74.9
D_2	94.4	46.3	70.4	93.8	55.6	74.7
D_3	94.6	45.7	70.2	93.4	54.3	73.8
D_4	94.1	43.4	68.7	92.1	46.6	69.4
D_5	93.4	40.3	66.8	91.1	44.0	67.6

Table 14. The average maximum, minimum and the mean R. H.s (%) during dehaulming to harvesting under different DOP

DOP	2010-11			2011-12		
	Max R. H.	Min R. H.	Mean R. H.	Max R. H.	Min R. H.	Mean R. H.
D_1	94.1	37.9	66.0	91.4	55.9	73.6
D_2	94.2	37.1	65.6	91.4	53.4	72.4
D_3	94.0	39.3	66.6	90.6	35.1	62.8
D_4	93.8	44.9	69.4	89.2	32.8	61.0
D_5	91.3	40.1	65.7	89.2	32.8	61.0

Figure 4. Mean R.H. (%) during different phenophases of potato under different DOPs

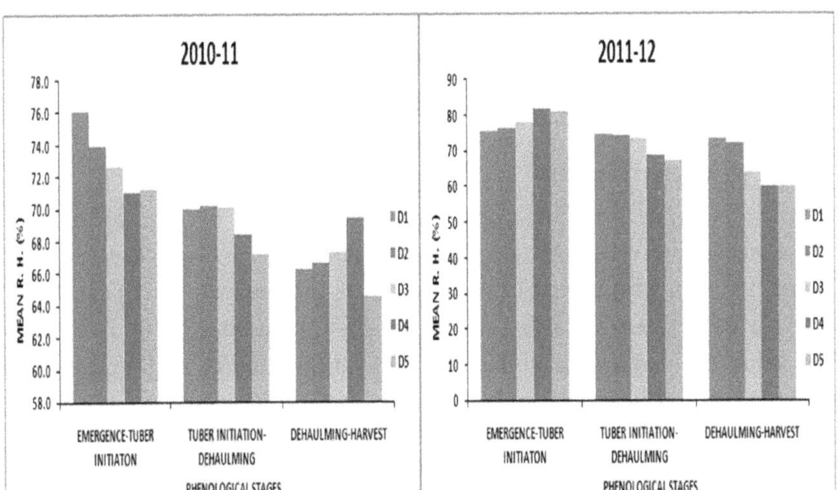

**For a more accurate representation of this figure, please see the electronic version.*

The Figure 4 shows a comparison of mean R. H. under different DOP during emergence to harvest. The mean R. H. in the first year declined gradually with the delay in planting during emergence to tuber initiation phase, whereas in the second year the minimumR. H. remained almost unaltered for different DOP. During tuber initiation to dehaulming, the mean R. H. declined gradually in D4 and D5 plantings in the first year and in the second year it declined from D3 to D5 planting. During dehaulming to harvest the mean R. H. increased from D1 to D4 with a drastic reduction in D5in the first year; whereas in the second year, the mean R. H. reduced continually from D1 to D3 and thereafter remained unaltered.

In the subsequent section, it would be observed that the D5 planted crop recorded the best yield in the first year; whereas in the second year, the D1 planted crop recorded best yield. The reason of this discrepancy was hidden in the mean R. H. during emergence to tuber initiation phase. The mean R. H. in the first year under D5 planting during this phenophase was71.2% whereas in the second year, under D1 planting it is 75.7%.Therefore, it may be concluded that the potato will show high productivity if the mean R. H. remained within a limit of 71 to 75% during emergence to tuber initiation phase.

RAINFALL RECEIVED DURING THE GROWTH PERIOD

The rain water is an important natural input which cuts the irrigation requirement. In general, the winter months of W.B. do not experience rainfall. However, the rainfall pattern has been changing due to climatic variability. The sowing or planting of winter crops has received a jolt because of non-preparation of land during the early phase of winter season (Parya et al., 2010). It is therefore important to delineate the total rainfall and number of rainy days observed during the different phenophases in winter crops.

Planting to Germination

The data revealed that the D3 and D4 plantings in 2010-11, received the rainfall of 2.2 and19.5 mm with one and three days of rainfall respectively (Table 15). The second year experiment recorded no rain in this phenophase. The D4 planting in the first year, received a quite high amount of rainfall, which might have caused an impediment in sprouting.

Table 15. Total rainfall (mm) and number of rainy days during planting to germination under different DOP

DOP	Total rainfall (mm)		No. of rainy days	
	2010-11	2011-12	2010-11	2011-12
D1	0.0	0.0	0.0	0.0
D2	0.0	0.0	0.0	0.0
D3	2.2	0.0	1.0	0.0
D4	19.5	0.0	3.0	0.0
D5	0.0	0.0	0.0	0.0

Germination to Tuber Initiation

During emergence to tuber initiation, first three date of plantings in 2010-11 received 19.5, 19.5 and 17.3 mm of rainfall respectively (Table 16). However, in the second year, the last three dates of planting received 4.1, 57 and 57 mm of rainfall during this phenophase (Figure 5).

Figure 5. Total rainfall (mm) received during different phenophases of potato under different DOPs

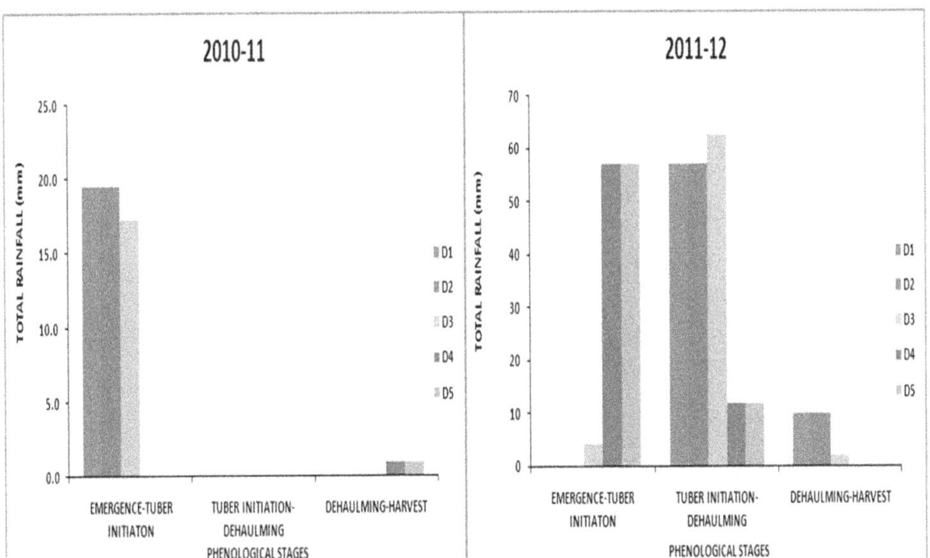

For a more accurate representation of this figure, please see the electronic version.

Table 16. Total rainfall (mm) and number of rainy days during emergence to tuber initiation under different DOP

DOP	Total rainfall (mm)		No. of rainy days	
	2010-11	2011-12	2010-11	2011-12
D1	19.5	0.0	3.0	0.0
D2	19.5	0.0	3.0	0.0
D3	17.3	4.1	2.0	2.0
D4	0.0	55.6	0.0	7.0
D5	0.0	57.0	0.0	8.0

The data presented above revealed a very contrasting nature of rainfall in two experimental years. In the first year, D1, D2 and D3 received almost similar amount of rainfall having 2 to 3 total rainy days within this phenophase. The D4 and D5 plantings received no rainfall. In the second year, the D3 received only 4.1mm rainfall distributed in 2 days. The D4 and D5 planting received 55.6 and 57.0 mm rainfall having 7 to 8 total number of rainy days. The potato loves water but it doesn't prefer humid and cloudy weather. Humid and cloudy weather increases the minimum temperature which eventually invites the late blight disease. The data presented in Table 17 clearly showed that the first three dates of planting had to face cloudy weather in the first year and the last three DOP in the second year had a similar experience.

The yield of potato crop under different DOP did not show similarity because of this reason.

Table 17. Total rainfall (mm) and number of rainy days during tuber initiation to dehaulming under different DOP

DOP	Total rainfall (mm)		No. of rainy days	
	2010-11	2011-12	2010-11	2011-12
D1	0.0	57.0	0.0	8.0
D2	0.0	57.0	0.0	8.0
D3	0.0	62.7	0.0	9.0
D4	0.0	11.2	0.0	4.0
D5	0.0	9.8	0.0	3.0

TUBER INITIATION TO DEHAULMING

The rainfall pattern observed during this phenophase widely differed in two experimental years. The first year experienced no rainfall whereas the second year experienced rainfall under all DOP (Fig.5).

The data shown in Table 17 recorded a very contradictory nature of rainfall received in the two years of experiment. The first three DOP in the second year received very high amount of rainfall, with 8 to 9 rainy days. However, the last two DOP received a meagre amount. The first year did not receive any rainfall at all. Because of this pattern, the earthing up operation in the first three DOP was delayed because of high soil moisture condition. High rainfall also reduced the soil temperature which might have an adverse effect on the process of tuber bulking. This rainfall also reduced irrigation water requirement in second year, ultimately cutting the cost of cultivation.

DEHAULMING TO HARVEST

Dehaulming is the removal of vegetative part of the potato plant thus reducing the leaf respiratory loss. Pest infestation, particularly the infestation by aphids is also checked through this process. In addition, potato skin is hardened due to curing. In the two years of investigation, the rainfall received during this period was asymmetric. In the first year, only D4 and D5 plantings received 1 mm rainfall each; whereas in the second year, D1, D2, D4 and D5 plantings received 9.8, 9.8, 2.0 and 2.0 mm rainfall respectively (Table 18).

Presence of moisture in soil during this stage reduced the soil temperature which retarded the process of hydrolysis of starch in tuber thus reducing the respiratory loss and increasing yield, although it invited a problematic harvest because of soil wetness.

Table 18. Total rainfall (mm) and number of rainy days during dehaulming to harvest under different DOP

DOP	Total rainfall (mm)		No. of rainy days	
	2010-11	2011-12	2010-11	2011-12
D1	0.0	9.8	0.0	3.0
D2	0.0	9.8	0.0	3.0
D3	0.0	0.0	0.0	0.0
D4	1.0	2.0	2.0	1.0
D5	1.0	2.0	2.0	2.0

IMPACT OF TEMPERATURE ON LEAF AREA INDEX OF POTATO

Cumulative Maximum Temperature

The LAI decreased with the increment of cumulative maximum temperature. The relationship was non significant in the first year but highly significant in the second year (Figure 6). In the second year, the LAI value sharply declined when the cumulative maximum temperature exceeded 400.0 day degree Celsius. 36.1% variation in LAI could be explained through the variation in cumulative maximum temperature. When the two year data were pooled, the LAI decreased significantly with the increase in cumulative maximum temperature. 20.4% variation in LAI could be explained through the variation in cumulative maximum temperature.

Cumulative Minimum Temperature

The impact of cumulative minimum temperature on LAI was almost similar to the cumulative maximum temperature. The LAI significantly decreased when the cumulative minimum temperature exceeded 150 day degree Celsius in the second year experiment. There was no significant relationship in between LAI and cumulative minimum temperature and LAI in the first year as well as in the pooled result (Figure 7). The coefficient of determination (R2) value was higher in case of the relationship between LAI and cumulative minimum temperature than the relationship between LAI and cumulative maximum temperature. 36.4% variation in LAI

Figure 6. Impact of cumulative maximum temperature (day 0C) on LAI under different DOP

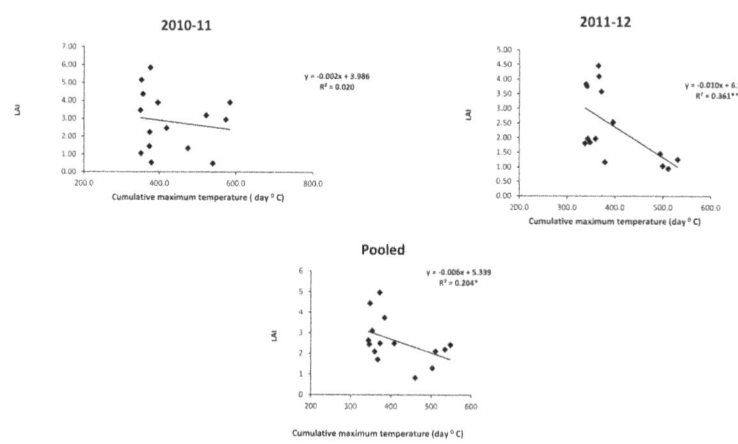

Figure 7. Impact of cumulative minimum temperature (day 0C) on LAI under different DOP

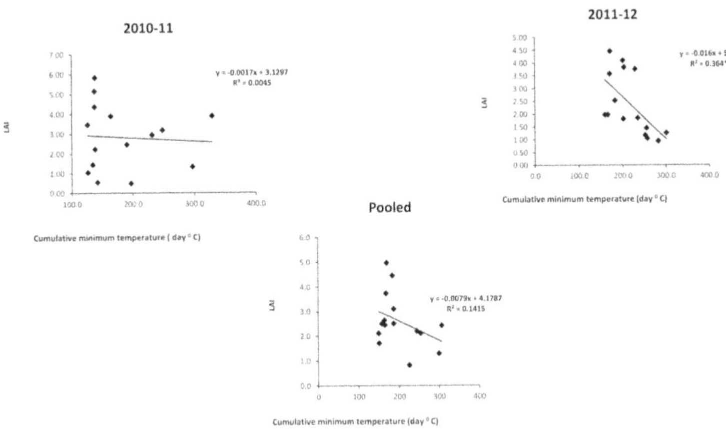

could be explained through the variation in cumulative minimum temperature in the second year experiment.

It was observed from the result that the LAI declined sharply with the increment of cumulative maximum and minimum temperatures beyond 400 and 150 day degree Celsius respectively. Parya (2009) reported that the LAI in wheat was significantly but negatively affected by the cumulative maximum and minimum temperatures. Ying et al., (1998) also recorded the decline in LAI of rice in the tropical environment of Philippines than the sub tropical environment of China.

IMPACT OF TEMPERATURE ON CGR

Cumulative Maximum Temperature

The CGR declined significantly with the increment of cumulative maximum temperature in both the year (Figure 8). 27.7% and 25.5% variation in CGR could be explained through the variation in cumulative maximum temperature. When two year data were pooled the R2 value increased, which indicated a strong and significant negative association in between the CGR and cumulative maximum temperature. 32.3% variation in CGR of potato could be explained through the variation in cumulative maximum temperature.

Figure 8. Impact of cumulative maximum temperature (day 0C) on CGR (g/m2/ day) under different DOP

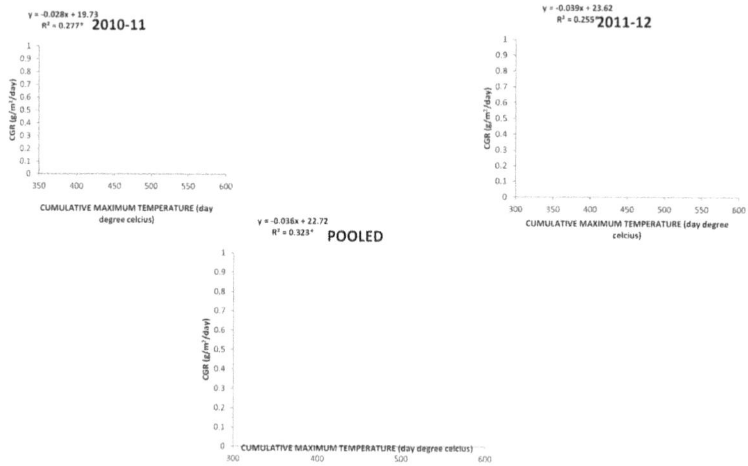

Cumulative Minimum Temperature

Increased cumulative minimum temperature significantly reduced the CGR in both the year. 20.7% and 40.6% variation in CGR could be explained through the variation in cumulative minimum temperature for the first and second year respectively. The pooled result showed that the 31.0% variation in CGR of potato could be explained through the variation in cumulative minimum temperature (Figure 9).

Figure 9. Impact of cumulative minimum temperature (day 0C) on CGR (g/m2/day) under different DOP

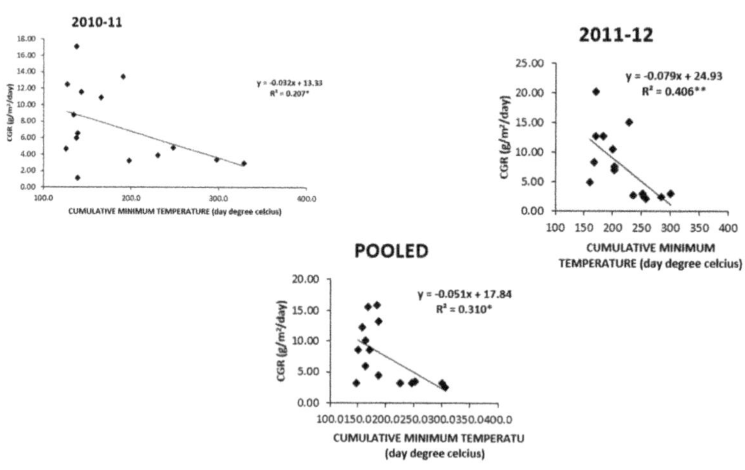

The results showed that the CGR declined significantly when the cumulative maximum and minimum temperatures exceeded 350 and 150 day degree Celsius respectively. Increased temperature declined the CGR in different crops. The CGR is a function of canopy gross photosynthesis and crop respiration (Evans, 1993). Respiration is strongly affected by temperature (Akita, 1993) whereas the canopy photosynthesis is regulated by solar radiation, temperature, LAI, canopy architecture and single leaf photosynthetic rate (Loomis & Connor, 1992). Ying et al., (1998) reported that the CGR in rice declined remarkably under tropical environment when compared with the sub tropical environment. In the present experiment, increased cumulative maximum and minimum temperatures recorded an adverse impact on the CGR of potato.

Crop Yield

The yield (t ha-1) of potato differed significantly due to variation in DOP. Year to year variation was also observed (Table 19).

In the first year, the maximum tuber yield was observed under D5 planting and the minimum yield was recorded under D2 planting. The order of tuber yield under different DOP was D5>D4>D1>D3>D2. No significant differences were observed in between D1 and D3 plantings. In the second year, the maximum yield was observed under D1 planting and the minimum yield was recorded under D4 planting. The order of tuber yield under different DOP was D1>D3>D2>D5>D4. No significant differences were observed in between D2 and D3 plantings. It was conspicuous that under D5 planting the tuber yield was almost similar in both the years. The yield increases under D1, D2 and D3 plantings in the second year, were 6.82, 7.46, 5.75 t/ ha respectively as compared to the first year. The two years pooled analysis showed the maximum yield under D1 planting. During the second year, the yield of D4 planting was the lowest because of heavy rainfall during earthing

Table 19. Effect of DOPs and N-doses on the yield (t/ha) of potato

Treatments	2010-11	2011-12	Pooled
D_1	16.8	23.62	20.21
D_2	14.53	21.99	18.26
D_3	16.6	22.35	19.48
D_4	17.87	11.65	14.76
D_5	18.6	18.22	18.41
Sem (±)	0.81	0.99	0.61
LSD at 5%	18.93	21.12	20.03

up phase. The discrepancy in tuber yield under different DOP was due to variation in maximum and minimum temperatures, rainfall and R.H. during the growing season.

INFERENCE

Cumulative maximum temperature should exceed 335 day 0C from 45 to 60 DAP for 1cm per day height increment. D1 and D5 plantings recorded similar increment in plant height in second and first year experiments. This showed that the maximum temperature should lie within 210 C during tuberization period. Cumulative minimum temperature during the same period should remain within 150 to 200 day 0 C. This implied that minimum temperature shouldn't exceed 10 to 130 C during the period of tuberization. LAI decreased significantly with the increase in cumulative maximum and minimum temperatures (2011-12). Cumulative maximum and minimum temperatures had significant effect on CGR. CGR declined linearly with increase in temperature. The strength of association is higher in the second year than the first year.

The mean R. H. should be within 71 to 75% during emergence to tuber initiation for better productivity. Rainfall during dehaulming to harvest hindered harvest yet facilitated higher yield by reducing respiratory losses and hydrolysis of starch.

REFERENCES

Akita, S. (1993). Respiration: variation and potential for manipulation. In International Crop Science I. Crop Science Society of America.

Borah, M. N., & Milthorpe, F. L. (1959). *The effect of light intensity, length of day and temperature on growth and tuber formation in the potato* (Ph.D. thesis). University of Nottingham, UK.

Chakraborty, P. K. (1994). Effect of date of sowing and irrigation and the diurnal variation in physiological process in the leaf of Indian mustard (Brassica juncea). *Journal of Oilseeds Research*, *11*(2), 210–216.

Chakraborty, P. K., Majumdar, A., & Chatterjee, B. N. (1991). Physiological processes in Indian mustard (Brassica juncea) and yellow sarson (B. napus var. glauca) and their agronomic appraisal in mild and short winter prevailing in Gangetic plains of Eastern India. *Indian Journal of Agricultural Science*, *61*(11), 815–821.

Emilsson, B., & Lindblom, H. (1963). Physiological mechanisms concerned in sprout growth. In J. D. Ivins & F. L. Milthorpe (Eds.), *Growth of Potato. Proceedings of Tenth Easter School in Agricultural Sciences, University of Nottingham* (pp. 45–62). London Butterworths Press.

Evans, L. T. (1993). *Crop Evolution, Adaptation and Yield*. Cambridge, UK: Cambridge University Press.

Ezekiel, R., & Bhargava, S. C. (2000). Potato top growth as influenced by temperature under short days. *Indian Journal of Plant Physiology*, *5*(1), 85–88.

Fischnich, O. (1959). Quoted by Emilsson and Lindblom, Physiological mechanisms concerned in sprout growth. In J. D. Ivins & F. L. Milthorpe (Eds.), *Growth of Potato* (pp. 45–62). London Butterworths Press.

Goodwin, P. B. (1967). The control of branch growth in potato tubers.II. The pattern of sprout growth. *Journal of Experimental Botany*, *18*(1), 87–89. doi:10.1093/jxb/18.1.87

Headford, D. W. R. (1961). *Sprout growth of the potato* (Ph.D. thesis). University of Nottingham, UK.

Kooman, P. L., & Rabbinge, R. (1996). An analysis of the Relation between Dry Matter Allocation on the Tuber and Earliness of a Potato Crop. *Annals of Botany*, *77*(3), 235–242. doi:10.1006/anbo.1996.0027

Loomis, R. S., & Connor, D. J. (1992). *Crop ecology: productivity and management in agricultural systems*. Cambridge, UK: Cambridge University Press. doi:10.1017/CBO9781139170161

Milthorpe, F. L. (1963). Some aspects on plant growth an introductory survey. In J. D. Ivins & F. L. Milthorpe (Eds.), *Growth of Potato. Proceedings of Tenth Easter School in Agricultural Sciences, University of Nottingham* (pp. 3–16). London Butterworths Press.

Moorby, J., & Milthorpe, F. L. (1975). Potato. In L. T. Evans (Ed.), *Crop Physiology some case histories* (pp. 225–257). London: Cambridge University Press.

Nayyar, H., Malik, C. P., Singh, P., Parmar, U., Grewal, M., & Kaur, S. (1990). Diurnal variation in photosynthetic parameters in peanut. *Photosynthetica*, *24*, 276–274.

Pallas, J. E. Jr, & Samish, Y. B. (1974). Photosynthetic response of peanut. *Crop Science*, *14*(3), 478–482. doi:10.2135/cropsci1974.0011183X001400030042x

Parya, M. (2009). *Studies on the variation in physical environment and its effect on the growth and yield of wheat cultivars (Triticum aestivum L.) under different dates of sowing* (Ph. D. Thesis). B.C.K.V., Mohanpur, India.

Parya, M., Dutta, S. K., Jena, S., Nath, R., & Chakraborty, P. K. (2010). Effect of thermal stress on wheat productivity in West Bengal. *Journal of Agrometerology*, *12*(2), 217–220.

Ying, J., Peng, S., He, Q., Yang, H., Yang, C., Visperas, R. M., & Cassman, K. G. (1998). Comparison of high-yield rice in tropical and subtropical environments I. Determinants of grain and dry matter yields. *Field Crops Research*, *57*(1), 71–84. doi:10.1016/S0378-4290(98)00077-X

Yuan, F. M. & Bland, W. (2004). Light and temperature modulated expolinear growth model of potato (Solanum tuberosum L.). *Agricultural and Forest Meterology,* (121), 141-151.

Chapter 4
Impact of Climate Change on Potato Production in India

M. K. Jatav
Central Institute for Arid Horticulture, India

P. M. Govindakrishnan
Central Potato Research Institute, India

V. K. Dua
Central Potato Research Institute, India

R. P. Sharma
National Bureau of Soil Survey and Land Use Planning, India

ABSTRACT

Potato is a temperate crop and higher day temperatures cause some areas to less suitable for potato production due to lower tuber yields and its quality. Tuber growth and yield can be severely reduced by temperature fluctuations outside 5-30 °C. The rate of warming in last 50 years is double than that for the last century. Increase in temperature and atmospheric CO2 are interlinked occurring simultaneously under future climate change and global warming scenarios. If CO2 is elevated to 550 ppm the temperature rise is likely to be 3 °C with decline in potato production by 13.72% in the year 2050. The changing climate will affect the potato production adversely due to drought, salinity, frost, flooding, erratic unseasonal rains etc. It may reduce seed tuber production, impact storage facility and potato processing industries. Therefore, the quantification of regional vulnerability and impact assessment is very important for the development of early warning on disease forecasting systems, breeding of short duration and heat, drought, salinity tolerant and disease resistant cultivars.

DOI: 10.4018/978-1-5225-1715-3.ch004

INTRODUCTION

The book chapter "Sustainable Potato Production and the Impact of Climate Change" deals with the possible impact of global warming and elevated CO_2 on Potato production. The results presented in this chapter are summarized findings of the research conducted in India; its agricultural universities and Indian council of Agricultural Research (ICAR). The level of atmospheric temperature and carbon dioxide raised under controlled conditions to some possible changes in near future to assess the impact on climate change on potato production. Findings of various researchers of India are compiled in the form of book chapter for easy understanding and in line of future work. Mitigation of impact of climate change on potato is discussed in global context.

BACKGROUND

The bottom-line conclusion of the Third Assessment Report of the Intergovernmental Panel on Climate Change (IPCC, 2001) is that the average global surface temperature will increase by between 1.4°C and 3°C above 1990 levels by 2100 for low emission scenarios and between 2.5°C and 5.8°C for higher emission scenarios of greenhouse gases and aerosols in the atmosphere. The effect of increased temperatures on potato production in specific areas will vary depending partly on the current temperature of that area. Temperatures above 30 °C can have several negative impacts on potato production like: slowing tuber growth and initiation, less partitioning of starch to the tubers, physiological damage to tubers (e.g. brown spots), shortened/non-existent tuber dormancy, making tubers sprout too early. These effects can reduce crop yield and the number and weight of tubers. As a result, areas where current temperatures are near the limits of potatoes' temperature range will likely suffer large reductions in potato crop yields in the future.

Potato farming is the most important economic activity in some parts of India. Uttar Pradesh, Punjab and West Bengal are the major potato producing states. There is direct effect of global warming and serious risk to future crop production and food security in the country. At high altitudes, global warming will probably lead to changes in the time of planting, the planting of late-maturing cultivars, and a shift of the location of potato production. In many of these regions in India, changes in potato yield are likely to be relatively small in initial stage but expected to trigger in coming era of global warming. Shifting planting time or location is less feasible at lower altitudes, and in these regions global warming could have a strong negative effect on potato production. It is likely that the currently observed trend of global warming, which has been 0.6 °C + 0.2 since 1900, will continue and that the average global temperature will increase by between 1.4 and 5.8 ºC over the period 1990 to 2100. It is shown that heat-tolerant potato cultivars could be used to mitigate effects of global warming in (sub) tropical regions. Climate change is now an acknowledged fact and reality. The evidence gathered world over using state-of-the-art technology by various national and international agencies is irrefutable. Human activities like rapid industrialization, intensive agriculture, and indiscriminate use of fertilizers, deforestation and increasing use of fossil fuels during past 150 years are the major contributing factors for climate change. The continued effect of these activities resulted in increasing emission of CO_2 and other greenhouse gases (GHG) leading to global warming as a '*greenhouse effect*' due to entrapment of back radiation from earth by these gases. The increase in temperature due to global warming is 0.76 ºC since 1850. The rate of warming in last 50 years is double than that for the last century. The rate of warming is increasing. The 20th century's last two decades were the hottest in 400 years and possibly the warmest for several millennia, according to a number of climate studies. And the United Nations' Intergovernmental Panel on Climate Change (IPCC) reports that 11 of the past 12 years are among the dozen warmest since 1850. The CO_2 concentration is projected to double from the current level of 360 ppm in the atmosphere. Global warming is occurring along with shifting pattern of rainfall and increasing incident of extreme weather events like floods, droughts and frosting. Concentration of greenhouse gases on time scale is presented in Figure 1.

Global annual average surface temperature in 2015 is looking set to reach 1°C above the pre-industrial average (as represented by the 1850-1900 reference period) for the first time, according to the HadCRUT4 dataset produced by the Met Office and the Climatic Research Unit at the University of East Anglia (Figure 2). This is based on the current January to September 2015 temperature anomaly, and is also expected to hold when the final full-year anomaly is calculated. The warmth of 2015 represents an important marker because it means we are reaching halfway to 2°C for the first time. In 2010, parties to the United Nations Framework Convention on Climate Change (UNFCCC) agreed warming should be limited to below 2°C to avoid dangerous climate change.

The growth of potato in India has been phenomenal since 1950 with increase in area production and productivity by 6, 15 and 3 times, respectively. Globally India stands at 4th and 3rd position now with respect to acreage and production, respectively. The crop is mainly confined to Indo-Gangetic plains in mild and cool winters in India. The autumn/winter planted crop in northern plains of India comprising the states of Uttar Pradesh, West Bengal, Bihar, Punjab and Haryana contributes 84% of total potato production in India. Here, the crop is grown totally under irrigated conditions. It is also grown in small scattered areas as rainfed crop in hills during summers and as rainy (*kharif*) and winter seasons crop in plateau region. However,

Figure 1. Observed changes in atmospheric greenhouse gas concentrations. Atmospheric concentrations of carbon dioxide (CO2, green), methane (CH4, orange), and nitrous oxide (N2O, red). Data from ice cores (symbols) and direct atmospheric measurements (lines) are overlaid
Source: (IPCC 2014).

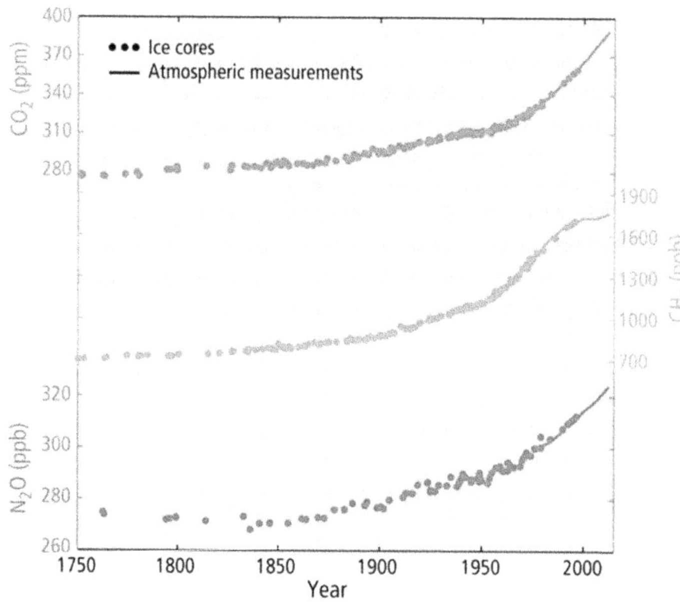

Figure 2. Observed global mean temperature difference from the 1850-1900 mean (°C) from HadCRUT4
Source: Morice et al 2012.

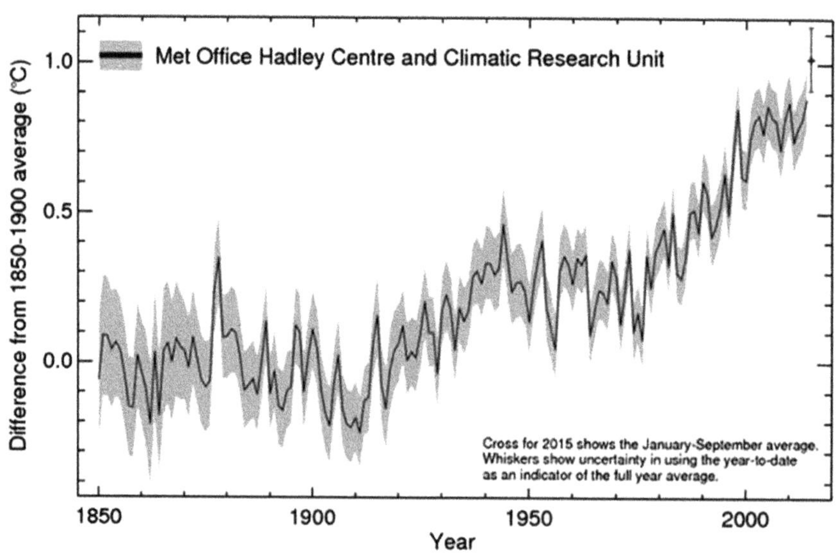

the climate change and global warming will have a profound effect on potato growth story in India impacting every aspect of not only production and profitability, but seed multiplication, storage, marketing and processing of this perishable vegetatively propagated crop. Under the impact of future scenarios of climate change, the growth projections of potato in India might be arrested or even reversed, unless effective adaptation measures are evolved for timely application.

IMPACT ANALYSIS

Increase in temperature and atmospheric CO_2 are interlinked occurring simultaneously under future climate change and global warming scenarios. Effect of their interaction on potato would be more relevant and of greater economic significance compared to their usually counteracting direct effects on crop growth, yield and quality. Potato is mostly grown in north India during winters usually receiving few scattered rains. Under future scenarios the global warming is projected to be more pronounced over land areas with maximum temperature increase over northern India. The winter and post monsoon seasons are likely to be more affected by warming. Therefore, potato in addition to direct effects on growth and yield may be subjected to indirect effects of warming. These are increasing drought due to reduction in precipitation accentuating salinity and unpredictable extreme events of erratic unseasonal rains, flooding and frosting *etc.*

Effect of Elevated CO_2 and Temperature

Productivity

Effect of Elevated CO2 and Rise in Temperature on Potato Production in India

All India estimates of potato production were made by INFOCROP-POTATO simulation model (Singh, 2005, 2008) without adaptations and assuming that area under the crop remains constant at current levels (1.2 m ha) in future climate scenarios (Table 1). Results showed that the potato production will increase by 11.12% at elevated CO_2 of 550 ppm and 1 °C rise °in temperature. However, the future climate scenarios for India indicates that at elevated CO_2 of 550 ppm the temperature rise is likely to be 3 °C (IPCC, 2007), with decline in production by 13.72% in the year 2050. The 1 °C rise in temperature is likely to be associated with only 400 ppm of CO_2 to be assumed in the year 2020 (IPCC, 2007), with a decline in potato production by 3.16% (Table 1).

Table 1. Change (%) in potato production in India from current levels as affected by elevated CO2 and rise in temperature without adaptations

Atmospheric CO_2 conc. (ppm)	Rise in temperature (°C)					
	Nil (current)	1 (2020)	2	3 (2050)	4	5 (2090)
369 (current)	0.0	-6.27	-17.09	-28.10	-42.55	-60.55
400 (2020)	3.40	-3.16	-14.57	-25.54	-58.63	-58.63
550 (2050)	18.65	11.12	-1.25	-13.72	-30.25	-49.94

(Values in parentheses are likely years for associated CO_2 levels and temperature rise) Source: (Singh, 2009)

Direct Effect of Elevated CO2

The effect of elevated CO_2 concentration in controlled experiments conducted in OTC (Open top chambers), FACE (Free air carbon enrichment) and growth chambers overwhelmingly suggests positive effect on growth and yield with only few negative influences.

The CO_2 concentration and assimilation are positively correlated. Doubling the CO_2 concentration from ambient level of 360 ppm to 720 ppm increased the total biomass by 27 to 66% (Collins, 1976; Heagle *et al.*, 2003; Olivo *et al.*, 2002; Donnelly *et al.*, 2001; Miglietta *et al.*, 1998; Van de Geijn & Dijkstra, 1995). The tuber yield increased from 32 to 85% (Collins, 1976; Wheeler *et al.*, 1991; Heagle *et al.*, 2003; Craigon *et al.*, 2002; Olivo *et al.*, 2002; Donnelly *et al.*, 2001; Miglietta *et al.*, 1998; Finnan *et al.*, 2002). The increase in tuber yield is estimated to be approximately 10% for every 100 ppm increase in CO_2 concentration (Miglietta *et al.*, 1998). These positive effects are attributed to increased photosynthesis from 10 to 40% (Collins, 1976; Katny *et al.*, 2005; Olivo *et al.*, 2002; Vandermeiren *et al.*, 2002; Schapendonk *et al.*, 2000). The increase in photosynthesis was most marked in young leaves (Katny *et al.*, 2005; Vandermeiren *et al.*, 2002). This is attributed to phenomenon of photosynthetic acclimation later in the growing season particularly in old leaves (Vandermeiren *et al.*, 2002; Schapendonk *et al.*, 2000; Lawson *et al.*, 2001). Varietal differences in response to elevated CO_2 concentration exists (Olivo *et al.*, 2002). Number of tubers remained unaffected under elevated CO_2, but mean tuber weight increased mainly through increase in number of cells in tubers without influencing the cell volume (Collins, 1976; Chen and Setter, 2003; Donnelly *et al.*, 2001). However, an increase in tuber number has been also observed (Miglietta *et al.*, 1998; Craigon *et al.*, 2002).

Elevated CO_2 concentration reduces evapotranspiration (ET) resulting in water saving to the extent of 12 to 14%(Magliulo *et al.*, 2003; Olivo *et al.*, 2002). Elevated CO_2 concentration advances the tuber initiation and flowering (Miglietta *et al.*, 1998) but hastens senescence of leaves (Miglietta *et al.*, 1998; Vaccari *et al.*, 2001). The few negative effects of elevated CO_2 concentration include reduction in chlorophyll content in leaves particularly during later growing season after tuber initiation (Bindi *et al.*, 2002; Lawson *et al.*, 2001).

Direct Effect of Temperature

Growth and development is affected at high temperatures encountered in the tropics. No potato crop growth is possible below 2 °C and above 30 °C (Van Keulen & Stol, 1995). The minimum (0-7 °C), optimum (16-25 °C) and maximum (40 °C) temperatures for net photosynthesis are reported (Kooman & Haverkort, 1995). Potato requires cool night temperature to induce tuberization (Burt, 1964; Ku *et al.*, 1977; Cutter, 1992). Although photosynthesis in potato is suppressed by high temperature (Ku *et al.*, 1977), it is not as sensitive to temperature as tuberization and partitioning of photosynthates to tuber (Reynolds *et al.*, 1990; Midmore & Prange, 1992). The radiation use efficiency (RUE) is suppressed under high temperatures (Allen & Scott, 1992). High temperature reduces tuber number and size (Ewing, 1997).

High temperature brings about marked morphological changes like etiolated growth with smaller size of compound leaves and leaflets reducing the LAI (Ewing, 1997; Fleisher *et al.*, 2006) in addition to reduction in tuber number and size (Peet & Wolfe, 2000; Khan *et al.*, 2003). However, long day conditions and high temperature prevailing in spring season in Punjab state in plains of India favoured growth of foliage at the cost of tubers and improved processing quality of tubers (Marwaha & Sandhu, 2002).

Potato requires cool night temperature to induce tuberization, which is inhibited by even moderately high temperatures (Ku *et al.*, 1977; Ewing, 1997). Tuber initiation was most affected by high temperature (Ghosh *et al.*, 2000). High temperature reduces the gross photosynthetic rate (Fleisher *et al.*, 2006). Although photosynthesis in potato is suppressed by high temperature (Ku *et al.*, 1977), it is not as sensitive to temperature as tuberization and partitioning of photosynthates to tuber (Reynolds *et al.*, 1990; Midmore and Prange, 1992). Therefore, even moderately high temperature drastically reduces tuber yield without much affecting the photosynthesis and total biomass production (Peet & Wolfe, 2000).

Interaction Effect of Temperature and CO_2

Potato tuber yield of plants exposed to high temperatures (35 °C) were extremely low regardless of CO_2 treatment, while in a non-temperature stress treatment (25 °C) doubling CO_2 increased tuber yield significantly by 71.5% (Peet & Wolfe, 2000). In another study potato was grown for 35 days under CO_2 concentrations (500, 1,000, 1,500 and 2,000 micromoles mol-1) at both 16 °C and 20 °C air temperature. The mean starch concentration increased with increasing CO_2 concentration at both 16 °C and 20 °C and was consistently higher at 16 °C than at 20 °C (Cao & Tibbitts, 1997). The SLW (g.m-2) was positively related to the foliar starch concentration on the basis of leaf area or dry weight (Cao & Tibbitts, 1997).

The CO_2 enrichment does not appear to compensate for the detrimental effects of higher temperature on tuber yield, while the quality of potato is likely to be impacted severely in terms of marketable grade of tubers and internal disorders.

Quality

Elevated CO_2 increased the amount of dry matter and starch with decrease in glycoalkaloid and nitrates improving the quality of tubers (Vorne *et al.*, 2002; Schapendonk *et al.*, 2000; Donnelly *et al.*, 2001). Nearly all the nutrient elements tend to decrease in tubersunder elevated CO_2 (Cao and Tibbitts, 1997; Fangmeier *et al.*, 2002) and the citric acid content decreases causing a higher risk of discoloration after cooking (Vorne *et al.*, 2002).

High temperature is associated with tuber disorder of internal necrosis (Sterrett *et al.*, 1991). In a pot experiment in naturally lit glass house and phytotron, the high temperature (30 °C) decreased the total dry matter and tuber yield and degraded quality by reducing specific gravity of tubers (Ghosh *et al.*, 2000). Nitrate reductase (NR) activity was also decreased by high temperature. The inhibition of tuber yield was due to limited translocation of carbohydrates from leaves to tubers following the reduction of NR activity and carbohydrate expense for dark respiration (Ghosh *et al.*, 2000). It can also affect tuber quality by causing 'heat sprouting' which is premature growth of stolons from immature tubers (Wolfe *et al.*, 1983; Struik *et al.*, 1989) and internal necrosis (Sterrett *et al.*, 1991). Potato processing requires large size tubers with high dry matter. Warming may reduce proportion of marketable and processing grade tubers for table and processing purposes, though dry matter may increase.

Indirect Effects of Climate Change and Global Warming

Draught

Optimal water supply is essential for potato, because of its shallow root system. The potato plant generally roots rather shallowly 40-50 cm (Beukema & Van der Zaag, 1990). Potato is extremely sensitive to drought particularly at tuber initiation with substantial loss in tuber yield. Dry matter partitioning to root, shoot, leaf and stem as a function of development stage (DS) and the root:shoot ratio is affected by drought stress. Drought, while reducing dry matter production increases the root:shoot ratio indicating a shift in the balance of growth in favour of roots. Roots of plants grown in drought conditions also tend to be thinner. Both responses enable drought plants to exploit the available soil moisture more effectively (Vos, 1995). Tuber initiation and maturity under drought stress conditions is hastened (Beukema & Van der Zaag, 1990).

1. **Salinity:** Potato is highly sensitive to salinity and irrigation with saline waters with even moderate residual sodium carbonate (RSC) values (Singh & Trehan, 1993).
2. **Frost:** Potato is extremely sensitive to frost. Complete loss of foliage is reported below 2 °C of ambient temperatures for 2-3 consecutive nights. More than 4-5 hrs. duration of temperature below 1 °C may result in foliage loss of 50% in even one night exposure. However, yield losses depend on crop growth stage at occurrence of frost. Frosting late in the season from 80-90 days after planting (DAP) results in yield loss of 10-15%, while that at 50-60 DAP may cause yield loss of 30-50%.
3. **Flooding:** Flooding for even short period of 2-3 days during active vegetative phase affects growth and yield. Flooding before emergence severely affects emergence due to rotting of seed tubers and soil crust formation, while that near harvesting results in rotting and rupture of tuber lenticels affecting physical appearance and marketable quality.
4. **Erratic Unseasonal Rains:** Rains of even 10-15 mm during planting or immediately after planting affects emergence due to soil crust formation and delays planting operation with consequent loss in yield. Rains during active vegetative phase may promote incidence of late blight disease.
5. **Seed Tuber Production:** In vegetatively propagated potato crop, the disease free quality seed tubers as planting material has special and added significance. Tuber as seed material is the carrier of a host of fungal, bacterial and viral diseases, responsible for rapid and drastic reduction in yield in successive generations of clonal multiplication. Most importantly the seed tubers alone accounts for half of the cost of inputs in potato cultivation and the profitability of the cropping enterprise to a large extent depends on quality of seed used. Viral diseases transmitted by aphid and other vectors are mainly responsible for rapid degeneration of planting materials in potato crop. The technology of '*seed plot technique*' was developed on the sole premise of growing seed tubers in relatively aphid free periods in plains during winters and termination of vines by dehaulming before aphid population crosses a threshold to minimize infection of viral diseases. The appearance of potato peach aphid

(*Myzus persicae*) is reported to advance by two weeks for every 1°C rise in mean temperature and population build up is positively correlated with maximum temperature and minimum relative humidity (Biswas *et al.* 2004; Dias *et al.*, 1980). Thus, under the impact of climate change and global warming the earlier appearance and increase in aphid population is likely to limit the aphid free period to the detriment of seed tuber quality and quantity, which will ultimately affect potato production in India. In many regions warming may abolish seed tuber production altogether, while in others it will involve extra cost on chemicals and pesticides treatment with increased cost of seed resulting in decline in profitability.

6. **Storage:** The harvesting of potato in plains of India coincides with onset of hot summer season. Cold storage of tubers is recommended by the end of February to prevent heavy weight loss and rotting. It is cold stored up to the end of October till withdrawal for human consumption and planting is accomplished. In certain regions relatively cooler climate permits on farm storage of tubers for short period for 80-90 days in improvised country stores, heaps and pits with acceptable losses. However, warming during the period from March to June may preclude this practice with increased cost of storage in the cold stores. On the other hand earlier cold storage than the recommended by the end of February and prolonged storage beyond October till weather is favourable for planting, might become necessary under global warming scenarios. This will increase the operational cost and energy use in cold stores with implied increase in cost of cold storage and enhanced market price of potato for table and seed purpose. More number of strategically located cold stores near production and consumption centers would be needed to maintain the supply chain during hot summer months from March to October under global warming.
7. **Potato Processing Industry:** The potato tubers stored in cold stores at 4 °C is not suitable for processing purposes due to increase in reducing sugar content imparting undesirable attributes in various processed products. Potato tubers kept in low temperature stores at 10 °C or in on farm country stores are utilized for processing. However, tubers kept in these low temperature and country stores suffer from sprouting and weight loss once the dormancy is broken rendering it relatively unfit for processing. Global warming will reduce the 'time window' of availability of potato suitable for processing and will result in enhanced cost of chemical treatment of tubers to prevent sprouting. This has implication for viability of potato processing industry for supply of raw material for extended periods to be economical.

Regional Vulnerability to Climate Change in India

The entire Indo-Gangetic plains, where irrigated potato is mainly grown is vulnerable. However, the state of West Bengal with highest productivity and second largest potato producing state in India appears highly vulnerable. Winters are mild in West Bengal and 'window' of suitable growing period is small, any rise in temperature will severely impact productivity with associated problems of storage and post harvest handling of produce in warmer conditions. Other vulnerable states are Bihar and Uttar Pradesh, which contributes maximum in total potato production. The states of Punjab, Haryana, and adjoining areas in northern Rajasthan and western Uttar Pradesh, where winters are relatively severe experiencing occasional frost might benefit from global warming to certain extent. The rainfed crop in plateau regions and other areas in south India would be most vulnerable due to warming and associated drought conditions.

Observations on Aberrant Weather and Extreme Events

- Rains in winter season received at planting affects emergence and delays planting with reduction in tuber yield.
- Heavy showers during the crops season resulting in flooding affects tuber yield.
- Heavy rains at the time of harvesting induce rottage in field and in temporary heaps of harvested potato in the field.
- Overcast sky and rains early in the crop season invariably increases the attack of late blight disease with severe reduction in yield.
- Early frosting received in last fortnight of December and first week of January damages the crop in North-Western plains and western UP.
- Relatively warmer winters in the year 2008 reduced tuber yield in West Bengal, UP and Bihar.

Adaptation Measures for Climate Change and Global Warming

- Use of crop residue mulches for some time after planting.
- Using drip irrigation in place of furrow and basin methods.
- Alter cultural management in potato based cropping systems.
- Conservation tillage and on farm crop residue management.
- Improvement and augmentation of cold storage facilities and air conditioned transportation from producing to consumption centers.
- Subsidizing additional cost of pests and water management.

- Insurance against weather for the cash crop of potato with high cost of cultivation.
- Strengthen education, research and development in warm climate production technology for ware and seed potato crop.

Mitigation Measures to Reduce Emission of CO_2/ghg and Carbon Sequestration Potential of the Crop

Potato being a short duration annual crop with readily decomposable crop residues has very limited carbon sequestration potential. Other mitigation measures are as follows.

- Resource conservation techniques
- Organic farming

FUTURE STRATEGIES FOR RESEARCH

- Quantification of regional vulnerability and impact assessment.
- Development of early warning disease forecasting systems.
- Breeding short duration and heat tolerant cultivars. Mining biodiversity to heat tolerance on priority.
- Breeding drought, salinity tolerant and disease resistant cultivars.
- Advance planning for possible relocation and identification of new areas for potato cultivation.
- Improved agronomic management for water and fertilizer use efficiency.
- Development of agro-techniques for warm weather cultivation and potato based cropping systems.

CONCLUSION

Potato a native of temperate region grown under long day conditions in mild and cool summer season in Europe and America was introduced and adapted to tropical short day conditions in India during the last century. The growth of potato in India has been phenomenal since 1950 with increase in area production and productivity by 6, 15 and 3 times, respectively. The crop is mainly confined to Indo-Gangetic plains in mild and cool winters in India. The autumn/winter planted crop in northern plains of India comprising the states of Uttar Pradesh, West Bengal, Bihar, Punjab and Haryana contributes 84% of total potato production in India, where the crop is grown totally under irrigated conditions. Growth and development is affected at high temperatures encountered in the tropics. No potato crop growth is possible below 2 °C and above 30 °C. The minimum (0-7 °C), optimum (16-25 °C) and maximum (40 °C) temperatures for net photosynthesis are reported. Potato requires cool night temperature to induce tuberization. Although photosynthesis in potato is suppressed by high temperature, it is not as sensitive to temperature as tuberization and partitioning of photosynthates to tubers.

The climate change and global warming will have a profound effect on potato growth story in India impacting every aspect of not only production and profitability, but seed multiplication, storage, marketing and processing of this perishable vegetatively propagated crop. Under the impact of future scenarios of climate change the growth projections of potato in India might be arrested or even reversed, unless effective adaptation measures are evolved for timely application and implementation. Increase in temperature and atmospheric CO_2 are interlinked occurring simultaneously under future climate change and global warming scenarios. Effect of their interaction on potato would be more relevant and of greater economic significance compared to their usually counteracting direct effects on crop growth, yield and quality. It is estimated that due to global warming potato production in India may decline by 3.16 and 13.72% from current levels by the year 2020 and 2050, respectively. The potato production will be directly affected by climate change, while there would be several indirect effects on various facets of supply, storage, utilization and acreage of the crop in future climate scenarios.

REFERENCES

Allen, E. J., & Scott, R. K. (1992). Principles of agronomy and their application in the potato Industry. In P. M. Harris (Ed.), *The potato crop. The scientific basis for improvement* (pp. 816–881). London: Chapman and Hall. doi:10.1007/978-94-011-2340-2_17

Beukema, H. P., & Van der Zaag, D. E. (1990). *Introduction to potato production.* Wageningen: Pudoc.

Bindi, M., Hacour, A., Vandermeiren, K., Craigon, J., Ojanpera, K., Selldén, G., & Fibbi, L. et al. (2002). Chlorophyll concentration of potatoes grown under elevated carbon dioxide and/or ozone concentrations. *European Journal of Agronomy, 17*(4), 319–335. doi:10.1016/S1161-0301(02)00069-2

Biswas, M. K., De, B. K., Nath, P. S., & Mohasin, M. (2004). Influence of different weather factors on the population build up of vectors of potato virus. *Annals of Plant Protection Sciences*, *12*(2), 352–355.

Burt, R. L. (1964). Influence of short periods of low temperature on tuber initiation in the potato. *European Potato Journal*, *7*(4), 197–208. doi:10.1007/BF02368251

Cao, W., & Tibbitts, T. W. (1997). Starch concentration and impact on specific leaf weight and element concentrations in potato leaves under varied carbon dioxide and temperature. *Journal of Plant Nutrition*, *20*(7-8), 871–881. doi:10.1080/01904169709365302 PMID:11541213

Chen, C. T., & Setter, T. L. (2003). Response of potato tuber cell division and growth to shade and elevated CO_2. *Annals of Botany*, *91*(3), 373–381. doi:10.1093/aob/mcg031 PMID:12547690

Collins, W. B. (1976). Effect of carbon dioxide enrichment on growth of the potato plant. *HortScience*, *11*, 467–469.

Craigon, J., Fangmeier, A., Jones, M., Donnelly, A., Bindi, M., De Temmerman, L., & Ojanpera, K. et al. (2002). Growth and marketable-yield responses of potato to increased CO2 and ozone. *European Journal of Agronomy*, *17*(4), 273–289. doi:10.1016/S1161-0301(02)00066-7

Cutter, E. G. (1992). Structure and development of the potato plant. In P. M. Harris (Ed.), *The potato crop. The scientific basis for improvement* (pp. 65–161). London: Chapman and Hall. doi:10.1007/978-94-011-2340-2_3

de Souza, D. J., Yuki, V. A., Costa, A. S., & Teixeeira, P. R. M. (1980). Study of the spread of virus diseases in a warm climate as compared to a cold climate, with a view to obtaining seed potatoes with a low rate of virus diseases. *Summa Phytopathologica*, *14*, 58–59.

Donnelly, A., Craigon, J., Black, C. R., Colls, J. J., & Landon, G. (2001). Elevated CO_2 increases biomass and tuber yield in potato even at high ozone concentrations. *The New Phytologist*, *149*(2), 265–274. doi:10.1046/j.1469-8137.2001.00015.x

Donnelly, A., Lawson, T., Craigon, J., Black, C. R., Colls, J. J., & Landon, G. (2001). Effects of elevated CO_2 and O_3 on tuber quality in potato (*Solanum tuberosum* L.). *Agriculture, Ecosystems & Environment*, *87*(3), 273–285. doi:10.1016/S0167-8809(01)00144-X

Ewing, E. E. (1997). Potato. In H. C. Wien (Ed.), *The Physiology of Vegetable crops* (pp. 295–344). Wallingford, UK: CAB International.

Fangmeier, A., De Temmerman, L., Black, C., Persson, K., & Vorne, V. (2002). Effects of elevated CO_2 and/or ozone on nutrient concentrations and nutrient uptake of potatoes. *European Journal of Agronomy, 17*(4), 353–368. doi:10.1016/S1161-0301(02)00071-0

Finnan, J. M., Donnelly, A., Burke, J. I., & Jones, M. B. (2002). The effects of elevated concentrations of carbon dioxide and ozone on potato (*Solanum tuberosum* L.) yield. *Agriculture, Ecosystems & Environment, 88*(1), 11–22. doi:10.1016/S0167-8809(01)00158-X

Fleisher, D. H., Timlin, D. J. & Reddy, V. R. (2006). Temperature influence on potato leaf and branch distribution and on canopy photosynthetic rate. *Agronomy Journal, 98,* 1442-1452.

Ghosh, S. C., Asanuma, K., Kusutani, A., & Toyota, M. (2000). Effect of temperature at different growth stages on nonstructural carbohydrate, nitrate reductase activity and yield of potato (*Solanum tuberosum*).[Japan]. *Environment Control in Biology, 38*, 197–206. doi:10.2525/ecb1963.38.197

Heagle, A. S., Miller, J. E., & Pursley, W. A. (2003). Growth and yield responses of potato to mixtures of carbon dioxide and ozone. *Journal of Environmental Quality, 32*(5), 1603–1610. doi:10.2134/jeq2003.1603 PMID:14535300

IPCC. (2001). *Third assessment report of the Intergovernmental Panel on climate change*. WMO, UNEP.

IPCC. (2007). *Fourth assessment report of the Intergovernmental Panel on climate change*. WMO, UNEP.

IPCC. (2014). Climate Change 2014: Synthesis Report. In R. K. Pachauri & L. A. Meyer (Eds.), *Contribution of Working Groups I*. Geneva, Switzerland: II and III to the Fifth Assessment Report of the Intergovernmental Panel on Climate Change.

Katny, M. A. C., Hoffmann, T. G., Schrier, A. A., Fangmeier, A., & Jager, H. J. (2005). Increase of photosynthesis and starch in potato under elevated CO2 is dependent on leaf age. *Journal of Plant Physiology, 162*(4), 429–438. doi:10.1016/j.jplph.2004.07.005 PMID:15900885

Khan, I. A., Deadman, M. L., Al Nabhani, H. S., & Al Habsi, K. A. (2003). Interactions between temperature and yield components in exotic potato cultivars grown in Oman. *Acta Horticulturae, 619*(619), 353–359. doi:10.17660/ActaHortic.2003.619.41

Kooman, P. L., & Haverkort, A. J. (1995). Modelling development and growth of the potato crop influenced by temperature and daylength: Lintul-Potato. In A. J. Haverkort & D. K. L. MacKerron (Eds.), *Potato ecology and modelling of crops under conditions limiting growth* (pp. 41–60). Dordrecht: Kluwer Academic Publishers. doi:10.1007/978-94-011-0051-9_3

Ku, G., Edwards, E., & Tanner, C. B. (1977). Effects of light, carbon dioxide and temperature on photosynthesis, oxygen inhibition of photosynthesis and transpiration in *Solanum tuberosum*. *Plant Physiology*, *59*(5), 868–872. doi:10.1104/pp.59.5.868 PMID:16659958

Lawson, T., Craigon, J., Tulloch, A. M., Black, C. R., Colls, J. J., & Landon, G. (2001). Photosynthetic responses to elevated CO2 and O3 in field-grown potato (*Solanum tuberosum*). *Journal of Plant Physiology*, *158*(3), 309–323. doi:10.1078/0176-1617-00105

Magliulo, V., Bindi, M., & Rana, G. (2003). Water use of irrigated potato (*Solanum tuberosum* L.) grown under free air carbon dioxide enrichment in central Italy. *Agriculture, Ecosystems & Environment*, *97*(1-3), 65–80. doi:10.1016/S0167-8809(03)00135-X

Marwaha, R. S., & Sandhu, S. K. (2002). Yield, growth components and processing quality of potatoes as influenced by crop maturity under short and long days. *Advances in Horticultural Sciences*, *16*, 47–52.

Midmore, D. J., & Prange, R. K. (1992). Growth responses of two *Solanum* species to contrasting temperatures and irradiance levels: Relations to photosynthesis, dark respiration and chlorophyll fluorescence. *Annals of Botany*, *69*, 13–20.

Miglietta, F., Magliulo, V., Bindi, M., Cerio, L., Vaccari, F. P., Loduca, V., & Peressotti, A. (1998). Free air CO2 enrichment of potato (*Solanum tuberosum* L.), development, growth and yield. *Global Change Biology*, *4*(2), 163–172. doi:10.1046/j.1365-2486.1998.00120.x

Olivo, N., Martinez, C. A., & Oliva, M. A. (2002). The photosynthetic response to elevated CO2 in high altitude potato species (*Solanum curtilobum*). *Photosynthetica*, *40*(2), 309–313. doi:10.1023/A:1021370429699

Peet, M. M., & Wolfe, D. W. (2000). Crop ecosystem responses to climate change: Vegetable crops. In K. R. Reddy & H. F. Hodges (Eds.), *Climate Change & Global Crop Production* (pp. 213–243). CAB International. doi:10.1079/9780851994390.0213

Reynolds, M. P., Ewing, E. E., & Owens, T. G. (1990). Photosynthesis at high temperature in tuber bearing *Solanum species. Plant Physiology*, *93*(2), 791–797. doi:10.1104/pp.93.2.791 PMID:16667538

Schapendonk, A. H. C. M., Oijen van, N., Dijkstra, P., Pot, C. S., Jordi, W. J. R. M., & Stoopen, G. M. (2000). Effects of elevated CO_2 concentration on photosynthetic acclimation and productivity of two potato cultivars grown in open-top chambers. *Australian Journal of Plant Physiology*, *27*, 1119–1130.

Singh, J. P., Govindakrishnan, P. M., Lal, S. S., & Aggarwal, P. K. (2005). Increasing the efficiency of agronomy experiments in potato using INFOCROP-POTATO model. *Potato Research*, *48*(3-4), 131–152. doi:10.1007/BF02742372

Singh, J. P., Govindakrishnan, P. M., Lal, S. S., & Aggarwal, P. K. (2008). Infocrop-Potato a Model for Simulating Growth and Yield of Potato in the Sub-Tropics. Central Potato Research Institute, Shimla.

Singh, J. P., Lal, S. S., & Pandey, S. K. (2009). Effect of climate change on potato production in India. Central Potato Research Institute, Shimla, Newsletter, 40, 17-18.

Singh, J. P., & Trehan, S. P. (1993). *A case study of soil and irrigation water related constraints in potato crop production: Cause and correction.* National Seminar on 'Developments in Soil Science, 58th Annual convention, Dehradun.

Sterrett, S. B., Lee, G. S., Henninger, M. R., & Lentner, M. (1991). Predictive model for onset and development of internal heat necrosis of 'Atlantic' potato. *Journal of the American Society for Horticultural Science*, *116*, 701–705.

Struik, P. C., Geertsema, J., & Custers, C. H. M. G. (1989). Effects of shoot, root and stolon temperatureon the development of potato plant. III. Development of tubers. *Potato Research*, *32*, 151–158. doi:10.1007/BF02358227

Vaccari, F. P., Miglietta, F., Giuntoli, A., Magliulo, V., Cerio, L., & Bindi, M. (2001). Free air CO2 enrichment of potato (*Solanum tuberosum* L.). Photosynthetic capacity of leaves. *Italian Journal of Agronomy*, *5*, 3–10.

Van de Geijn, S. C., & Dijkstra, P. (1995). Physiological effects of changes in atmospheric carbon dioxide concentration and temperature on growth and water relations of crop plants. In A. J. Haverkort & D. K. L. MacKerron (Eds.), *Potato ecology and modelling of crops under conditions limiting growth* (pp. 89–100). Dordrecht: Kluwer Academic Publishers. doi:10.1007/978-94-011-0051-9_6

Van Keulen, H., & Stol, W. (1995). Agro-ecological zonation for potato production. In A. J. Haverkort & D. K. L. MacKerron (Eds.), *Potato Ecology and Modelling of Crops under Conditions Limiting Growth* (pp. 357–372). Dordrecht: Kluwer Academic Publishers. doi:10.1007/978-94-011-0051-9_23

Vandermeiren, K., Black, C., Lawson, T., Casanova, M. A., & Ojanpera, K. (2002). Photosynthetic and stomatal responses of potatoes grown under elevated CO2 and/or O3 - results from the European CHIP-programme. *European Journal of Agronomy, 17*(4), 337–352. doi:10.1016/S1161-0301(02)00070-9

Vorne, V., Ojanpera, K., De Temmerman, L., Bindi, M., Högy, P., Jones, M. B., & Persson, K. et al. (2002). Effects of elevated carbon dioxide and ozone on potato tuber quality in the European multiple-site experiment CHIP-project. *European Journal of Agronomy, 17*(4), 369–381. doi:10.1016/S1161-0301(02)00072-2

Vos, J. (1995). Nitrogen and the growth of potato crops. In A. J. Haverkort & D. K. L. MacKerron (Eds.), *Potato ecology and modelling of crops under conditions limiting growth* (pp. 115–128). Dordrecht: Kluwer Academic Publishers. doi:10.1007/978-94-011-0051-9_8

Wheeler, R. M., Tibbitts, T. W., & Fitzpatrick, A. H. (1991). Carbon dioxide effects on potato growth under different photoperiods and irradiance. *Crop Science, 31*(5), 1209–1213. doi:10.2135/cropsci1991.0011183X003100050026x PMID:11537629

Wolfe, D. W., Fereres, E., & Voss, R. E. (1983). Growth and yield response of two potato cultivars to various levels of applied water. *Irrigation Science, 3*(4), 211–222. doi:10.1007/BF00272837

Chapter 5
Improved Agronomic Practices and Input Use Efficiency for Potato Production under Changing Climate:
Improved Practices for Potato Production

Dhiman Mukherjee
Bidhan Chandra Krishi Viswavidayalaya, India

ABSTRACT

In the emerging global economic order in which agricultural crop production is witnessing a rapid transition to agricultural commodity production, potato is appearing as an important crop, poised to sustain and diversify food production in this new millennium. Temperature and unpredictable drought are two most important factor affecting world food securities and the catalyst of the great famines of the past. Decreased precipitation could cause reduction of irrigation water availability and increase in evapo-transpiration, leading to severe crop water-stress conditions. Increasing crop productivity in unfavourable environments will require advanced technologies to complement traditional methods which are often unable to prevent yield losses due to environmental stresses. Various crop management practices such as improved nutrient application rate, mulching, raised beds and other improved technology help to raise the productivity. Conservation farming practices play important role to restore soil and enhancing soil health and play important role to combat climate change issue.

DOI: 10.4018/978-1-5225-1715-3.ch005

POTATO

Scientfic name: *Solanum tuberosum* L.
Family: Solanceae
Origin: South America.
Chromosome No.: 2n = 4x = 48.

A significant change in climate on a global scale will impact potato cultivation and agriculture as a whole, consequently affect the world's food supply. Internationally, agriculture is widely regarded as one of the sectors at most risk from a changing climate, due to the impact of increased temperatures, reduced rainfall and increased frequency of extreme events, not only in the tropics, but also in temperate environments such as the UK and Himalayan belt of India (Mukherjee, 2015). Climate transform will also impact on land suitability, the viability of rainfed potato production, and demand for supplemental irrigation (IPCC, 2007). Weather vary impacts on agriculture are being witnessed all over the world, but countries like India are more vulnerable in view of the high population depending on agriculture and excessive pressure on natural resources. The warming trend in India over the past 100 years (1901 to 2007) was observed to be 0.51°C with accelerated warming of 0.21°C per every10 years since 1970 (Kumar 2009). More erratic rainfall pattern and impulsive high temperature spells consequently reduce crop efficiency, with shifting in weed distribution pattern and its physiology system (Mukhrjee, 2007). Developing countries in the tropics will be particularly vulnerable. Latitudinal and altitudinal shifts in ecological and agro-economic zones, land degradation, extreme geophysical events, reduced water availability, and rise in temperature and deteriorating soil condition make it difficult to cultivate the potato in particular zones in the world (Burke et al., 2009). Unless measures are undertaken to alleviate the effects of climate transform, food security in developing countries will be under threat and will jeopardize the future of the potato growers in these countries (Mukhrjee, 2014 d). Potato production in developed countries, especially in Europe and the Commonwealth of Independent States, has declined on average by one percent per annum over the past 20 years. However, output in developing countries has expanded at an average rate of five percent per year (Falloon & Betts, 2010). Asian countries, particularly China and India, fuelled this growth. In recent past, the developing countries' share of global potato output stood at 52 percent, surpassing that of the developed world. This is a remarkable achievement, considering that just 20 years ago the developing countries' share in global production was little more than 20 percent (Collier et al., 2008). Even so, world potato production and consumption are currently expanding more slowly than the global population. Fresh potato consumption, once the mainstay of world potato utilization, is decreasing in many countries, especially in developed regions. This is mainly because of low production due to harsh weather condition which alarm selling price. This becomes important threat to future food security issue (Mukherjee, 2002). Potato ranks behind the cereals rice and wheat as the 3rd most important food crop worldwide. It is grown in 149 countries from latitudes of 65 ° N to 50 ° S and from altitudes ranging from sea level to 4,000 m. It is comprehensively cultivated in China, Russian Federation, Ukarine, Poland, Ireland, Great Britain, Germany, Netherlands, France, Spain, South America, India and USA (more than 149 countries). The crop is believed to have originated from Andes of Peru and Bolivia in South America, more specifically in the basin of the Lake Titicaca of Peru-Bolivia border. Potato was introduced in India in mid-17th century probably by the Portuguese traders or British missionaries. Now it is one of the principal cash crops of India. The crop covers an area of about 1.86 million ha (M ha) with an annual production of 41.46 million tonne (Mt) with an average productivity of 23.12 t/ha in India (2014-15). West Bengal, has the world's highest per day potato productivity (300 kg/day) and per capita annual availability is also highest in this state. West Bengal ranks second position in terms of area and productivity (32.96 t/ha) after U.P in the country. Potato can be regarded as a wholesome food. It contains water (75-80%), carbohydrate (22.6%), starch (14%), sugar (2%), protein (1.6%), fat (0.1%), fibre (0.4%), minerals (0.6%), vitamins (Vit C rich 17 mg) and energy: 97 k cal. Potato also contain good amount of essential amino acids like, lysine, leucine, tryptophane and isoleucine. Here, this chapter discuss with following broad heading.

PROSPECT OF POTATO CULTIVATION

Global warming is predicted to have effects on global potato production. Like many crops, potatoes are likely to be affected by changes in atmospheric carbon dioxide, temperature and precipitation, as well as interactions between these factors (Lobell & Field, 2007; Mukherjee, 2010). Shifting of climate affect the distributions and populations of many potato diseases and pests. Potato yields are predicted to benefit from increased carbon dioxide concentrations in the atmosphere upto some extent in extreme temperate zone. The major benefit of increased atmospheric carbon dioxide for potatoes (and other plants) is an increase in their photosynthetic rates which can increase their growth rates, this is quite beneficial to combat climate change issue also (Mukherjee, 2011). Crop yields are also predicted to benefit because potatoes partition more starch to the edible tubers under elevated carbon dioxide levels (Lal, 2001). Higher levels of atmospheric carbon dioxide also results in potatoes having to open their stomata less to take up an equal amount of carbon dioxide for photosynthesis, which means less water loss through transpiration from stomata. As a result, the water use efficiency (the amount of carbon assimilated per unit water lost) of potato plants is predicted to increase, this mechanism is quite beneficial in rainfed zone.

Mostly this plant prefers medium high land with sandy loam or loamy soil having proper drainage and irrigation facilities. In India, maximum area under potato is in alluvial soils, followed by hill, black and red soils. Potato prefers soils in acidic neutral range (pH 5.3 – 7.3). Saline and alkaline soils are not suitable. Potato is basically a temperate crop. It is largely grown in regions where mean temperature does not exceed 16-18^{0}C. Sprouting occurs at temp. 18-22°C. In India, it is generally grown when maximum day temperature is below 30^{0}C and night temperature not more than 20^{0}C. Highest tuberization occurs at day temperature 20°C and night temperature 14° C. Long photo period favors haulm growth but delays tuberization and maturity. Whereas short photoperiod reduces haulm growth but tuber initiation is early and maturity period is reduced. In india potato cultivation has huge prospect and this discuss in following heading.

In India Following Five Potato Growing Zone Identified Based on Agro-Climatic Condition

Plains cover 9.5%, hill cover 6.2% and plateau cover around 4.3% area.

1. **Western Himalayan Zone:** This zone covers the hills of Himachal Pradesh, Jammu and Kashmir and UP.
    a. Very high hill (2500-3000 m asl, summer crop: May- June to September-October) .
    b. High hill (1800-2500 m asl, summer crop: March-April to September.-October.
    c. Mid hills (1000-1800 m asl, spring crop-in irrigated area February - March to July-August)
2. **Plain Zone:** The crop is grown under short day conditions and suffers from mid day water stress. This zone is further divided into three sub-zones.
    a. North Western Plains: It covers Punjab, Haryana, Rajasthan and parts of neighbouring states (October to Jan/March).
    b. North central Plains: It consists of western Uttar Pradesh and Madhya Pradesh (mid October to Feb/March).
    c. North Eastern Plains: This zone consists of plains of Uttar Pradesh, Bihar, West Bengal, Assam and Orissa. (October end/2nd week of November to January/February end).
3. **North Eastern Hills:** This zone comprises hills of Meghalaya, Manipur, Tripura, Nagaland, Arunachal Pradesh and Mizoram. Darjeeling (WB). In this zone, potato is grown as summer (March-July) and autumn (August-December).

4. **Low Hill & Plateau Region:** It covers vast areas of central and peninsular India. This zone consists of parts of Gujarat, Maharashtra, Madhya Pradesh, Karnataka and Orissa. Potato is grown as Rainfed (July-Sept) and Irrigated (November - February) crops.
5. **South Indian Hills:** This zone comprises the Southern hill zone in Tamil Nadu. Potato is grown as summer, autumn and winter (valleys-irrigated) crops.

Environmental Constraints Limiting Potato Productivity

Environmental stress is the primary cause of crop losses worldwide, reducing average yields by more than 50%. Production environment is a mixture of conditions that varies with season and region. Climatic changes will influence the severity of environmental stress imposed on crops. Moreover, increasing temperatures, reduced irrigation water availability, flooding, and acidity will be major limiting factors in sustaining and increasing productivity. Extreme climatic conditions will also negatively impact soil fertility and increase soil erosion (Mukherjee, 2004 a). Thus, additional fertilizer application or improved nutrient-use efficiency of crops will be needed to maintain productivity or harness the potential for enhanced crop growth due to increased atmospheric CO_2. The response of plants to environmental stresses depends on the plant developmental stage and the length and severity of the stress. Plants may respond similarly to avoid one or more stresses through morphological or biochemical mechanisms. Environmental interactions may make the stress response of plants more complex or influence the degree of impact of climate change (Opena & Porter, 1999). Measures to adapt to these climate change-induced stresses are critical for sustainable potato production. Its concluded from obtained results of Fatma et al.(2013) that vegetative growth parameters could be predict by specific climatic factors i.e. maximum relative humidity is a predictor for plant length, minimum air temperature and maximum soil temperature are predictor for number of leaves. Moreover, total leaves area could be predicted by average relative humidity, maximum soil temperature and minimum air temperature. Concerning canopy dry weight, it could be predicted by minimum air temperature and average soil temperature. Finally potato crop yield could be predicted using Hartz and Moore (1978) prediction model which is mainly depend on revised growing degree days (RGDD), total solar radiation and daily air temperature range.

Trends in Key Weather Parameters

Rainfall is the key variable influencing crop productivity in agricultural crops in general and rainfed crops in particular. Intermittent and prolonged droughts are a major cause of yield reduction in most crops and potato in particular. Long-term data for India indicates that rain fed areas witness 3-4 drought years in every 10-year period. Of these, 2-3 are in moderate and one may be of severe intensity. However, no definite trend is seen on the frequency of droughts as a result of climate change so far. For any R&D and policy initiatives, it is important to know the spatial distribution of drought events in the country. Analysis of number of rainy days based on the IMD grid data from 1957 to 2007 showed declining trends in Chhattisgarh, Madhya Pradesh, and Jammu Kashmir. In Chhattisgarh and Eastern Madhya Pradesh, both rainfall and number of rainy days are declining which is a cause of concern as this is a rain-fed rice production system supporting large tribal population who have poor coping capabilities. Guhathakurta and Rajeevan (2008) reported that all India summer monsoons (June to September) rainfall does not show any significant trend during the last century. However, three subdivisions *viz.,* Jharkhand, Chhattisgarh, Kerala show significant decreasing trend and eight subdivisions *viz.,* Gangetic West Bengal, West Uttar Pradesh, Jammu & Kashmir, Konkan & Goa, Madhya Maharashtra, Rayalaseema, Coastal Andhra Pradesh and North Interior Karnataka show significant increasing trends. However, the amount and distribution of rainfall is becoming more erratic which is causing greater incidences of droughts and its significantly influence potato production in potato belt of India.

Parameter Which Influence Productivity Under Shifting Climate

Few of the important parameter such as high temperature and drought etc., influence greatly potato production and productivity as a whole:

- **High Temperature:** Potato is a cool season crop however, at low temperature vegetative growth is restricted and at temperatures nearing the freezing point there is permanent and often irrecoverable frost injury to the plant. For these reasons, potato is grown as a summer crop in the hills and as a winter crop in the tropical and subtropical regions. Potato is susceptible to frost. At higher temperature, the respiration rate increases and the carbohydrates produced by photosynthesis are consumed rather than stored in the tuber (Lal, 2004). In the vegetative growth period the temperature may be higher but during tuberization phase the temperature should be lower. Long day coupled with high temperature conditions promote haulm growth without formation of tuber and short day with low temperature condition induce tuberization. The optimum temperature for good plant growth is 15°C to 20°C. Temperature limits the range and production of many crops. For every one degree increase in temperature, yields of wheat, soybean, mustard, groundnut and potato are expected to decline by 3-7% (Agarwal, 2009). Analysis of climate trends in potato growing locations suggests that temperatures are rising and the severity and frequency of above-optimal temperature episodes will increase in the coming decades. High temperature stress disrupts the biochemical reactions fundamental for normal cell function in plants. It primarily affects the photosynthetic functions of the plants. High temperatures can cause significant losses in potato productivity due to reduced number of growing days, and smaller and lower quality tuber. Observation revealed in case of tomato, that pre-anthesis temperature stress is associated with developmental changes in the anthers, particularly irregularities in the epidermis and endothesium, lack of opening of the stromium, and poor pollen formation.

- **Drought:** For many agribusinesses, irrigated potato production is driving force behind farm investment, but competition between water sectors, coupled with increasing environmental regulation and the longer-term threat of climate change are limiting water supplies for irrigation. Unpredictable drought is the single most important factor affecting world potato cultivation and the catalyst of the great famines of the past. In high to mid altitude zone of Darjeeling Himalaya and its adjoining hill, we observed that, number of rainy days reduced quit often, and this drastically reduced potato yield. The world's water supply is fixed, thus increasing population pressure and competition for water resources will make the effect of successive droughts more severe. Inefficient water usage all over the world and inefficient distribution systems in developing countries further decreases water availability. Water availability is expected to be highly sensitive to climate change and severe water stress conditions will affect potato crop cultivation. Most potato genotypes do not tuberize when the prevailing drought situation is quite long (Holden et al., 2003).
- **Prospect of Varieties:** While most varieties of potatoes belong to a particular species, *Solanum tuberosum*, about 12-14 other *Solanum* species have been cultivated, and 189 uncultivated species have been recorded. Climate modification may threaten the survival of those wild relatives: it is predicting that as many as 12 percent will become wiped out as their growing conditions worsen. If climate changes radically, the area where wild potatoes grow naturally could be reduced by as much as 60 to70 percent. Since potatoes mostly propagate vegetatively, most commercial varieties of potato have a reduced ability to flower and breeders do not select for traits that make the flower attractive to pollinators.
  - Early maturing group (requires 70-80 days after planting to give economic yield): Kufri Chan dramukhi, Kufri Ashoka, Kufri Lauvkar, Kufri Kuber.
  - Medium maturing group (requires 90-100 days after planting to give economic yield): Kufri Jyoti, Kufri Bahar, Kufri Lalima, Kufri Sutlez,

Kufri Pukhraj, Kufri Chipsona 1, Kufri Chipsona 2, Kufri Anand, Kufri Himalini, Kufri Sheetman, Kufri Moti, Kufri Kundan, Kufri Red, Kufri Alankar, Kufri Sherpa, Kufri Swarna, Kurphi Megha, Kufri Thenmalai, 92-PT-27, HPS-1/13.

- Late maturing group (requires 110-130 days after planting to give economic yield): Kufri Sindhuri, Kufri Badshah, Kufri Kanchan, Kufri Giriraj, Kufri Swarna, Kufri Jeevan, Kufri Dewa, Kufri Khasigaro, Kufri Naveen, Kufri Chamatkar, Kufri Kisan, Kufri Kumar, Kufri Safed, Kufri Neela, Kufri Neelamani,
- Varieties moderately resistant to late blight disease: Kufri Alankar, Kufri Jyoti, Kufri Swarna, Kufri Megha, Kufri Naveen, Kufri Neelamani, Kufri Muthu, Kufri Jawahar, Kufri Sutlej, Kufri Pukhraj, Kufri Giriraj.
- Varieties moderately resistant / resistant to early blight disease: Kufri chamatkar, Kufri Sinduri, Kufri Jyoti, Kufri Badshah, Kufri Lalima, Kufri Jeevan, Kufri Khasigaro, Kufri Sherpa.
- Varieties resistant to wart disease: Kufri Jyoti, Kufri Sherpa, Kufri Naveen.
- Varieties resistant to PVX virus: Kufri Badshah, Kufri Sherpa.
- Varieties moderately resistant to PVY virus: Kufri Sherpa
- Varieties resistant to cyst nematode: Kufri Swarna
- Varieties suitable for peocessing: Kufri Chipsona-1, Kufri Chipsona-2, Kufri Chipsona-3, Atlantic, Diamant, Fritolaty Hybrid 1333
- Red varieties: Kufri Sindhuri, Kufri Lalima, Pimpernel, Kufri Kanchan (Suitable for Hill and Terai Regions of W.B.).
- TPS: 92-PT-27, HPS-1/13

CULTIVATION TECHNIQUES UNDER SHIFTING TEMPERATURE

The commercial viability of potato production is influenced by the spatial and temporal variability in soils and agro-climate, and availability of water resources where supplemental irrigation is required. Knowledge of potential changes in land suitability is a key determinant influencing the sustainability of potato enterprises, as any future changes will influence cultivar choice, agronomic husbandry practices and the economics of production, for both rainfed and irrigated cropping. The effect of climate change on global potato production was assessed using a simulation model developed by CIP (Peru). As per report until 2069, and depending on the climate scenario, potential potato yield is expected to decrease by 18% to 32% (without adaptation of planting time and cultivars) and by 9% to 18% (with adaptation). In high latitudes changes in potato yield are likely to be relatively small, however at low latitudes, shifting planting time or location is less feasible, and in these regions global warming could have a strong negative effect on potato production. (Manneveux et al., 2012). Day by day environmental condition become harsh for potato cultivation, however, medium high land with sandy loam or loamy soil having proper drainage and irrigation facilities would be suitable for its cultivation. Soil pH should be between 5.5-7.5 and EC: < 0.5 mmhos/cm. One deep ploughing followed by 2-3 cross and shallow ploughings are necessary for making the soil well pulverized, loose, friable, well levelled and good tilth. This become very effective, where temperature become higher in recent past, as deep ploughing help to maintain soil temperature. Apply 10-15 tonne well rotten O.M. during first ploughing, which help to maintain soil temperature and fertility also. For amelioration of acid soil, apply 1.5 tonne Dolomite or 1.0 tonne slacked lime/ha (depending upon the soil test). Apply phorate 10 G @ 15 kg/ha to control soil borne insects like cut worm, Mole cricket, and weevil etc.

Method of Planting

Seed tubers may be planted either on ridges, on flat beds, flat bed followed by ridges. Planting too early can lead to seed piece disease and rot, slow emergence, and decreased plant vigor, which can slow tuber growth rates. Planting too late delays canopy development and reduce the time available for tuber bulking. The optimal planting date varies by region, but in all areas growers should wait to plant potatoes until daytime soil temperature warms to 50°F or higher. Potato is traditionally propagated from tubers which have a dormancy of nearly 8-10 weeks. Multiple sprouting stage of the tuber is the best for planting because yield depends on the number of main stems. Collect Certified seed from Govt. approved organizations. Taking out of the whole seeds tubers from cold storage 10-15 days before planting and storage in shady place under diffused light condition in thin layer on the floor helps in sprouting the tubers. It is better to planting of potato done under light forest shed where sunny days are long (Mukherjee, 2014). Use at least 30 g sized (3-5 cm diameter) disease free healthy whole or cut tubers having 2-3 sprouts.

Proper seed treatment is must for healthy growth of crop. Following measures should be follow under good agronomic management condition:

- Potato is mainly infested with seed borne diseases. It is essential to treat the seed before planting. Treat the seed tuber with any of the following chemicals/ bioagents.
- Mancozeb @ 2.5 g / litre of water for 10 minutes.
- *Trichoderma viride* @ 4-5 g/ litre of water for 10 minutes.
- *Pseudomonas fluorescens* @ 2.0 g/ litre of water for 10 minutes.

- For treating one quintal of seed tubers, 50 litre solutions will be required and same solution can be used for 3-4 times after which fresh solution is to be prepared.
- Treated tubers should be dried under shade before planting.
- Knife or cutting instrument should be washed with phenyl/1% $KMnO_4$ soln. if tubers having brown rot infestation are cut. (infested tubers should be buried in the soil)

Average seed rate is about 25-30 quintals per hectare. Seed tubers may be planted either on ridges or on flat beds. Planting should be done at a depth of 3-5 cm keeping the sprouts upward. Planting in rows in flat beds is suitable in light soils. Tubers are planted and covered with soil to make ridge. During planting care should be taken so that the seed tubers do not come in direct contact of the fertilizers that have been applied as basal. True potato seed (TPS) technology saves considerably the large quantity of seed tubers (25-30 quintals/ha) and their storage, space and transport. However, if place become tendency to more warmer as per meteorology data, a bit high rate of seed should be apply. Approximately100-150 g TPS is sufficient to produce seedlings for one-hectare area. TPS can be stored safely at ordinary room temperature and under low humidity for 2-3 years. Planting time divided into following types based on sowing condition.

Planting in the Plains

Early Crop

- Punjab and western Uttar Pradesh: 10-20 September
- Central Uttar Pradesh: First week of October
- Bihar: Second and third week of October
- West Bengal: 2nd fortnight of October

Main Crop

- Gangetic plains: November (north-western plains and West Bengal).
- Plateau region of peninsular India: October – December

Late Crop

- Gangetic plains: Last week November to middle of December.

Spring Crop

- North-western plains: December- January

Planting in Hills

North-Eastern Hills

- Valleys and high altitudes (3000-3500m): May- June.
- High hills (2500- 3000 m): April.
- Mid hills (1000 – 1800 m): January – February (for spring crop) and August – September (for autumn crop; in Ayodhya hills of West Bengal September plantion is suitable for autumn crop.

North-Eastern Hills

- High hills: March.
- Lower hills and valleys: January – February (for spring crop) and August – September (for autumn crop).

Nilgiri Hills

- April, August and January

Spacing

Closer than optimal plant spacing has a similar effect on tuber growth as does aged seed in that it increases tuber density relative to canopy size, thereby limiting the photosynthetic capacity to bulk each tuber. Although total yields may not be reduced, bulking rates of individual tubers decrease, which results in smaller tubers and lower marketable yields. Wider than optimal spacing can lengthen the time it takes to reach full canopy, which reduces carbohydrate supply to the tubers.

- 60 x 20 cm: Medium sized tubers (40 -50 cm); Plant population: 83,000/ha
- 60 x 25 cm: Larger sized tubers (50-55 cm); Plant population: 67,000/ha
- 60 x 15 cm: Small sized tubers (35-40 cm); Plant population: 1, 11, 000/ha

Nutrient Management

Developing healthy plants necessary for maximum tuber growth requires that all essential nutrients be supplied at optimal rates. Both deficit and excess fertilizer situations can reduce tuber bulking rates. Nutrient deficiencies limit canopy growth and shorten canopy duration resulting in reduced carbohydrate production and tuber growth rates. Excessive fertilizer applications can cause nutrient imbalances that delay or slow tuber growth rates (Mukherjee, 2014 c). In light of the environmental challenges ahead, resilience of the most abundant potato crop production systems must be improved to guarantee yield stability with more efficient use of nitrogen inputs, soil and water resources. Along with genetic and agronomic innovations, diversification of northern agro-ecosystems using inter-seeded legumes provides further opportunities to improve land management practices that sustain crop yields and their resilience to biotic and abiotic stresses. Benefits of legume cover crops have been known for decades and red clover (*Trifolium pratense*) is one of the most common and beneficial when frost-seeded under winter wheat in advance of maize in a rotation. Potato responds well to nutrients. In few temperate belt of India (Darjeeling Himalaya) and in other part of world due to bit higher temperature during last decade reveals more microbial diversity, which is beneficial for soil sustainability and potato productivity (Mukherjee, 2015 a). Nitrogen is the primary limiting nutrient in potato production directly affecting the tuber yield in all soil groups. It increases roots, foliage and tuber growth. Phosphorus increases tuber yield by increasing the yield and number of medium size tubers whereas potassium increases the number of large size tubers (Godwin & Singh, 1998). Normally for healthy crop production one should apply15-20 t/ha FYM/ha during final land preparation followed by 180-200 Kg N, 120-150 Kg P_2O_5 and 120-150 Kg K_2O per hectare. As potato is very exhaustive crop, so apply nutrient based on soil test report. Specific nutrient requirement under different sowing situation mentioned in table no.1. Approximately, 30 t/ha (tuber) removes 130-60-230-09 kg NPK&S. Basal fertilizers of 100kg N, 150 kg P_2O_5 and 150 kg K_2O/ha are to be applied at the time of last ploughing. Remaining nitrogen 100 kg should be applied at 25-30 days after planting followed by earthing up and light irrigation. In light soil, N may be applied in three splits (1/2 basal + 1/4 at 1st earthing + ¼ at 2nd earthing). This kind of fertilizer application greatly improves nutrient use efficiency of potato crop and help to enhance crop yield under shifting climate ituation also. Application of FYM 30 t/ha can meet P and K needs of the crop. The application of P and K in furrows in full dose at the time of planting gives the best results. Water-soluble phosphate fertilizers like superphosphate and DAP are most suitable for potato. Similarly potassium sulphate is a better source of K than muriate of potash. The residual phosphorus and potash are generally adequate and nitrogen requirement is reduced by half in succeeding cereal crop (Mukheree,

Table 1. Fertilizer scheduling for potato crop

Type of the potato crop	Rate of nutrients to be applied (kg/ha)			Time of application	
	N	P_2O_5	K_2O	Basal application.	Top dressing
Early crop	100-120	80-100	100-130	60 N+P+K	60 N
Main crop	120-140	100-120	120-150	100 N+P+K	70 N+50K
Late crop	160-180	130-150	100-150	90 N+P+120K	90 N+30K
Hill crop	110-140	120-140	100-160	70 N+P+120K	70 N+40K
Mid hill crop	100-120	100-120	100-160	Whole amount	

2013). Farmyard manure has been found to be useful in potato production and its application @ 30 tonnes/ha has been found to meet entire P and K needs of potato and succeeding cereal crop besides meeting micro-nutrient needs.

In the highly acid soil rock phosphate and SSP should be applied in 1:3 ratio on P basis. In case of micro-nutrient deficiency (Zn and B), apply Zinc-sulphate @ 25 kg/ha and Borax @ 10.0 kg/ha at the time of final land preparation (once in 3-4 years). Micronutrient deficiency is one of the major problem under changing climate situation, so effective micronutrient application is very important for optimum productivity per unit area (Table 2). Nutrients uptake is at its greatest during tuber bulking up (intensive volume increase process). The amount of nutrients removed by a potato crop is closely related to yield. Calcium ammonium nitrate (CAN) and ammonium sulphate are good sources of nitrogen in potato than urea. Similarly, diammonium phosphate (DAP) and super phosphate is effective source of phosphorous and the better source of potassium in the potassium sulphate than the muriate of potash for potato.

Table 2. Micronutrient fertilizer scheduling for potato crop

Micro-nutrients	Soil (kg/ha)	Spray (g/100 liter of water)	Seed treatment (g 100 liter of water for 3 hours/drying)
Zinc sulphate	25	200	50
Ferrous sulphate	50	300	75
Manganese sulphate	25	200	50
Copper sulphate	25	100	50
Ammonium molybedate	2	100	20
Sodium borate	2	100	20

Water Management

Greater uncertainty in seasonal weather patterns will mean growers need to adapt and consider short term coping strategies as well as longer-term strategic developments to reduce their vulnerability to changing water availability. How they respond will depend to a large extent on their perception of risk and the opportunities that climate change presents to their business. Review of various scientific evidence, over the last couple of decades suggests that climatic conditions are changing rapidly. This trend is likely to continue, and even accelerate in future which lead to reduce availability of water and other important biotic input for crop growth (Moss et al., 2010). These anticipated changes in climate baseline, variability, and extremes will have far-reaching consequences on potato production, posing additional challenges to meeting the food security for a growing world population (Lobell et al., 2008; Roudier et al., 2011). Farmers generally have two adaptation options to reduce their water needs or try to secure additional water supplies (Mukheree, 2014 a). Options to reduce water needs include investing in improved irrigation technology (scheduling) and equipment to increase application uniformity and efficiency, using weather forecasting to increase the effective use of rainfall, encouraging deeper rooting of crops, introducing lower water use or drought tolerant crop varieties, decreasing the overall irrigated area, or modifying soil structure to improve soil moisture retention. The quality and efficiency of water management determine the yield and quality of potato products. The optimum frequency and amount of applied water is a function of climate and weather conditions, crop species, variety, stage of growth and rooting characteristics, soil water retention capacity and texture, irrigation system and management factor. Too much or too little water causes abnormal plant growth, predisposes plants to infection by pathogens, and causes nutritional disorders (Mukherjee, 2014 b). If water is scarce and supplies are erratic or variable, then timely irrigation and conservation of soil moisture reserves are the most important agronomic interventions to maintain yields during drought stress. There are several methods of applying irrigation water and the choice depends on the crop, water supply, soil characteristics and topography. Application of irrigation water could be through overhead, surface, drip, or sub-irrigation systems. Surface irrigation methods are utilized in more than 80% of the world's irrigated lands yet its field level application efficiency is often 40-50% (Takahashi et al.,1998). Water has special significance in the potato production as the plant has sparse and shallow root system. Its water requirement varies from 350-550 mm depending upon the length of growing season, atmospheric conditions, soil type, and variety. The crop is generally grown under assured irrigation. Surface (furrow) irrigation systems is widely adopted and irrigation scheduling is based on the time interval approach. In potato give one light irrigation, if necessary, before planting to ensure uniform germination, this becomes very effective under high temperature zone. Depending on the soil moisture, apply splash irrigation at 3-4 days interval up to 25-30 days before 1st earthing up. Irrigate the crop at an interval of 7-10 days after 1st earthing up. Level of irrigation water must not exceed 3/4th part of the ridges. Complete irrigation by 1 PM. There should not be water in furrow at night. Stop the irrigation 10-15 days before harvesting. Potato does not prefer moisture stress. These are most critical stages:

- Stolonization stage (25-30 DAP)
- Early tuberization (35-45 DAP)
- Tuber development stage (60-65 DAP)

Other Approaches

1. Irrigation at 0.25/0.3 bar soil-moisture Tension
2. IW/CPE ratio-2.0
3. 25 mm CPE
4. Mulching increase irrigation efficiency

Water supply and scheduling have important impacts on tuber quality - frequent irrigation reduces the occurrence of tuber malformation (Mukherjee, 2013 a). Water deficit in the early phase of yield formation increases the occurrence of spindled tubers (more noticeable in oval than in round tuber varieties) and, when followed by irrigation, may result in tuber cracking or tubers with "black hearts". This problem is quite often in the area of drought or rainfed situation, where climate shifting resulted this kind of adverse situation (Ramakrishna et al., 2007). Modern potato varieties are sensitive to soil water deficits and need frequent, shallow irrigation. A 120 to 150 day potato crop consumes from 500 to 700 mm of water and depletion of more than 50 percent of the total available soil water during the growing period results in lower yields. To reduce potato's water needs, scientists are developing varieties that are drought-resistant with longer root systems. But significant water savings can be made in cultivation of today's commercial varieties by tailoring the timing and depth of water applications to specific stages of the plant's growth cycle to combat with climate change issue. In general, water deficits in the middle to late part of the growing period – during stolonization and tuber initiation and bulking – tend to reduce yield, while the crop is less sensitive during early vegetative growth. Water savings can also be achieved by allowing higher depletion toward the ripening period so that the crop uses all available water stored in the root zone, a practice that may also hasten maturity and increase dry matter content. Varieties with few tubers are usually less sensitive to water deficit than those with many tubers. While soil should be maintained at a relatively high moisture content to maximize yield, frequent irrigation with relatively cold water may reduce the soil temperature below the optimum value for tuber formation (15 to 18°C), thus affecting yields. Also, wet and heavy soils can create soil aeration problems. The most common irrigation methods for potato use furrow or sprinkler systems. Furrow irrigation has relatively low water use efficiency and is suitable when water supply is ample. In areas with water scarcity, sprinkler or drip irrigation is preferred, especially on soils with low water retention capacity. Sprinkler and drip irrigation systems (save 40-50% water) have been evaluated and compared with the standard furrow irrigation in Indian plains. The drip system was most economical followed by sprinkler system. For the hills, several methods of water harvesting have been developed. These are used to conserve snow melt water and run off water during rainy seasons to meet the water requirements during dry period.

Conservation Agriculture Practices for Better Potato Production Under Environmental Stress Situation

Potato cultivation usually involves intensive soil tillage throughout the cropping period, which often leads to soil degradation, erosion and leaching of nitrates. During soil preparation, the entire top soil is loosened and -particularly on sticky soils – pulverized into small aggregates to avoid the formation of clods in the potato beds. Mechanical weeding and mechanized harvesting also involve intensive soil movement. Conservation agriculture (CA) plays crucial role in potato production, its aims enhancing natural biological processes both above and below ground. It is based on three principles: minimum mechanical soil disturbance, permanent organic soil cover, and diversified crop rotations for annual crops and plant associations for perennial crops. By minimizing soil disturbance, CA creates a vertical macro-pore structure in the soil, which facilitates the infiltration of excess rainwater into the subsoil, improves the aeration of deeper soil layers, and facilitates root penetration (Mukherjee, 2010 a). In conventional, tillage-based potato cropping systems, the risk of soil erosion and nitrate leaching can be reduced using the mulch planting technique. The potato beds are prepared well in advance of planting - if potato is to be planted in spring, the beds would be prepared before winter – and seeded with a green manure cover crop. The potato is later planted into the beds which, by then, are covered by the dead mulch of the manure crop. For mechanical planting, planters are equipped with special discs that cut through the mulch and split the potato beds. The mulch protects the soil from erosion during the first weeks of the crop. As the potato plants grow, the reshaping of the beds incorporates the mulch. A second green manure crop can be seeded towards the end of the potato crop, as the potato plants are drying off. The cover crop helps to dry out the potato beds, contributing to healthier tubers with reduced risk of damage during harvest. The green manure is separated from the potato by a mechanical potato harvester and is left as a mulch cover after harvest, protecting the soil from erosion. Mulch planting is being used for potatoes in parts of Germany and Switzerland, particularly in watersheds where drinking water sources might be prone to nitrate pollution from conventional cultivation methods. Nevertheless, while mulch planting of potatoes reduces the risk of erosion and nitrate leaching, it still involves major soil movement (www.potato2008.org).

Cropping System

Short duration and wide flexibility is planting and harvesting time are potato's valuable traits that help in adjusting this crop in various intensive cropping systems prevalent in the country. There are some varieties that can give sizable yield in just 65-70 days under assured irrigation and congenial night temperatures and these fit well in diversified cropping systems. Wheat, rice, maize, sugarcane, Jute, pulses and vegetables are some of the major crops in the potato growing regions. Potato based cropping systems involving these crops have been developed and evaluated. Potato can be raised as companion crop in autumn-planted sugarcane crop where 2 rows of potato can be taken between two rows of sugarcane. A companion cropping with mustard, wheat, barley can be done without much reduction in the tuber yield of potato. In potato + mustard cropping, seed tubers of 4 rows are planted in three rows by reducing plant-to-plant spacing and fourth row is used for sowing mustard. In potato intercropped with barley, wheat or maize, the later crops are grown in every alternate furrow of potato after an early earthing up of potato (25 days sowing). The important intercropping systems are given below.

In the irrigated areas, many crop rotations involving potato are feasible. Being a short duration crop, it fits well in many intensive crop rotations. Some of the important rotations are given below:

The intensive agricultural and intercropping systems take a great place in the consideration of decision makers as well as the farmers. Through climate change impacts on food security and the need to increase the agricultural production of the area unit led to change many traditional agricultural practices such as cultivation some export fruit (Orange, grape, mango and etc.) under net house instead of open field. The need to increase the soil use efficiency by using the free areas during the first three years of young trees among the rows and in between were so urgent to serve the food security options. In this study potato plants (*Solanum tuberosum* L., Valor cultivar) were cultivated as intercrop among young orange trees (first and

Table 3. Common intercropping practices in potato

Intercropping	Row ratio	Location
Potato-rajmash	3:2	Central & eastern Uttar Pradesh and north Bihar
Potato + cotton	1:1	Dharwad (Karnataka)
Potato + maize	1:1	North Bihar and Madhya Pradesh
Onion/garlic + potato	1:2	Shimla (Himachal Pradesh)

Table 4. Cropping system under various situations

Maize based cropping system	Rice based cropping system	Suitability under various cropping system
Maize-potato-sugarcane Maize- potato-greengram Maize-potato-maize Maize-potato-onion Maize-potato-wheat Maize-potato-potato	Rice-cabbage-potato, Rice-potato-sesamum, Rice-potato-greengram, Rice-potato-wheat	Cowpea-potato-wheat Potato-jute-rice Soybean- potato-okra Sesame-potato-groundnut

second year), under different five net colors for covering greenhouses (yellow, white, red, blue and black) and open field to increase the soil use efficiency through the bare areas among the citrus rows especially at the winter season. Three in-row plant spacing were applied (12.5, 25 and 50 cm) under each net house color. Trial was carried out during two growing winter seasons of 2010/2011 and 2011/2012 at El-Bossily farm, CLAC, Agricultural Research Center, El-Behira Governorate, Egypt. This study investigated the effects of different net color on the growth and production of potato in terms of light intensity, air temperature, relative humidity and plant growth were evaluated over the two seasons. Regardless of net colour, all treatments decreased maximum temperatures and increased relative humidity compared with open field conditions. The use of white and yellow nets resulted in a significant increase of the number of leaves, fresh and dry weight and tuber yield per plant compared to other treatments. Data revealed that under white and yellow nets the most appropriate microclimate for producing potato under Egyptian conditions. The net colour and in-row plant space affected on the NPK content of potato plant. Increasing in-row plant distance from 12.5 to 50 cm led to increase the tuber yield per potato plant on contrary of tuber yield per unit area. The economic consideration suggest using 12.5 or 25 cm in row plant space in case of using the bare soil between the citrus plant (Abdrabbo et al., 2013).

Table 5. Weed flora infestation in potato field

Location	Grasses	Dicots	Sedges
Plains zone	*Poa annua/pratensis* *Phalaris minor* *Polypogon monspeliensis* *Echinochloa colonum*	*Amaranthus viridis* *Avena fatua* *Anagallis arvensis* *Amaranthus viridis* *Chenopodium murale* *Coronopus didymus* *Fumaria parviflora* *Trianthema portulocastrum* *Spergula arvensis* *Nicotina plumbaginifolia*	*Cyperus rotundus* *C. iria* *Fimbristylish miliacea*
Hill area	*Digitaria spp.* *Eleusine indica,* *Erogrostis spp.* *Pennisetum clandestinum* *Cynodon dactylon*	*Amaranthus spinosus* *Cannabis sativa* *Spergula arvensis* *Chenopodium album* *Oxalis latifolia* *Melilotus indica*	*Fimbristylish miliacea* *Cyperus rotundus*

Crop-Weed Competition and Management Techniques

Weed infestation in potato causes a drastic reduction in tuber yields. Critical stage of crop-weed competition is 4-6 weeks after planting in plains and 6-8 weeks in hills (Singh et al., 2004). Frequent irrigations, coupled with heavy fertilization offer favourable conditions for weed growth. The important weed flora of potato is given below. Weeding in potato should be done as soon as weeds emerge, but preferably when potato plants are about 8-10 cm high. When manual and mechanical hoeing are not feasible, use of herbicides are inevitable (Mukherjee, 2004). The most common pre-emergence herbicides used to control weeds in potato crop are atrazine (0.3-0.5 kg/ha), simazine (0.3-0.5 kg/ha), 2, 4-D (0.5 kg/ha), metribuzin (0.7-1.0 kg/ha), oxyfluorfen (0.15 kg/ha), dichloromate (1-2 kg/ha) and pendimethalin (1.0 kg/ha). Fluchloralin (0.7-1.0 kg/ha), EPTC (1.5-2.0 kg/ha) as pre-planting may also be used to check the weeds in potato crop. For controlling perennial weeds like Cyperus (motha) and Cynodon (doob) grass paraquat may be sprayed @ 0.4-0.6 kg/ha.

Harvesting

The time of harvesting of the potato crop is decided on the basis prevailing market price, purpose for which the crop is grown. For early marketing Kufri Chandramukhi can be harvested at 60-70 DAP to fetch higher prices and for sowing succeeding crop timely. Normally under optimum condition yield varies from 20-30 t/ha depending upon the variety.

- At full maturity (when haulms start yellowing and falling on the ground)
- Harvest potato before the temperature rise above 30°C to avoid rotting of tubers due to high temperature.
- Irrigation to withheld 10-15 days before harvesting for skin hardening
- Apply mancozeb @ 2.5 g/ liter of water on cut portion of haulms.
- Haulms are cut to enforce maturity (particularly in case of seed crop)
- Tubers are harvested 15-20 days after haulm cutting and at proper soil moisture
- Utmost care should be taken for skin hardening to withstand the impact of harvesting tool and machineries.
- Harvesting can be done by spade, country plough or by tractor drawn potato digger

Measures Before Storing

- Discard cut, rotten, insect damaged and rubbed tubers after harvesting.
- Cure the tubers in heaps under shade for 10-12 days for hardening the skin of tubers. The height of the heap should be 1-1.5 m and width 3-5 m (for better healing and skin hardening). That will not only reduce the water loss but it will also prevent the entry of some microorganisms also. Hardening also ensures the tuber to withstand handling during harvesting and transport from field to store.
- As per market demand, grade the tubers into small (<25 g), medium (25-50 g) large (50-75g) & extra large (> 75 g) on the basis of their size.
- Bagging of tuber in jute bag and send to cold store for preservation.
- Store the table varieties in cold storage at temperature of 2°C - 4°C and is 80-85% relative humidity. (Potatoes should be kept in cool chamber at 15°C for 24 hours before keeping in cold storage).
- The low temperature checks sprouting and rotting of tuber and high relative humidity reduces weight loss from the tubers.

MODERN APPROACH FOR CULTIVATION UNDER SHIFTING CLIMATE

Potential impacts of climate change on potato production will depend not only on climate *per se*, but also on the internal dynamics of agricultural systems, including their ability to adapt to the changes. Success in mitigating climate change depends on how well potato crops and systems adapt to the changes and concomitant environmental stresses of those changes on the current systems. Farmers in developing countries of the tropics need tools to adapt and mitigate the adverse effects of climate change on agricultural productivity, and particularly on potato production, quality and yield. Current, and new, technologies being developed through plant stress physiology research can potentially contribute to mitigate threats from climate change on vegetable production. However, farmers in developing countries are usually small-holders, have fewer options and must rely heavily on resources available in their farms or within their communities. Thus, technologies that are simple, affordable, and accessible must be used to increase the resilience of farms in less developed countries.

New molecular biology and plant cell culture tools have enabled scientists to understand better how potato plants reproduce, grow and yield their tubers, how they interact with pests and diseases, and how they cope with environmental stresses. Those advances have unlocked new opportunities for the potato industry by boosting potato yields, improving the tuber's nutritional value, and opening the way to a variety of non-food uses of potato starch, such as the production of plastic polymers. Producing high-quality propagation material is quite beneficial under harsh environmental conditions, which help some extent to combat climate change issue. Unlike other major field crops, potatoes are vegetatively reproduced as clones, ensuring stable, "true-to-type" propagation. However, tubers taken from diseased plants also transmit the disease to their progenies. To avoid that, potato tuber "seed" needs to be produced under strict disease control conditions, which adds to the cost of propagation material and therefore limits its availability to farmers in developing countries. Micro-propagation or propagation in vitro offers a low-cost solution to the problem of pathogens in seed potato. Micro-propagation technique is beneficial mostly where climate change has become a challenging problem. Plantlets can be multiplied an unlimited number of times, by cutting them into single-node pieces and cultivating the cuttings. The plantlets can either be induced to produce small tubers directly within containers or transplanted to the field under low temperature situation, where they grow and yield low-cost, disease free tuber "seed". This technique is very popular and routinely used commercially in a number of developing and transition countries and help to combat environmental challenge effectively. For example, in Viet Nam micro-propagation directly managed by farmers contributed to the doubling of potato yields in a few years, and overcome the environmental stress problem (Kumar, 2006). The potato has the richest genetic diversity of any cultivated plant. Potato genetic resources in the South American Andes include wild relatives, native cultivated species, local farmer developed varieties, and hybrids of cultivated and wild plants. They contain a wealth of valuable traits, such as resistance to pests and diseases, nutrition value, taste and adaptation to extreme climatic conditions. Continuous efforts are being made to collect, characterize and conserve them in gene banks, and some of their traits have been transferred to commercial potato lines through cross-breeding.

Developing Climate-Resilient Production Technology

Improved, adapted potato germplasm is the most cost-effective option for farmers to meet the challenges of a changing climate. However, most modern cultivars represent a limited sampling of available genetic variability including tolerance to environmental stresses. Breeding new varieties, particularly for intensive, high input production systems in developed countries is required to be done. Superior varieties adapted to a wider range of climatic conditions could result from the discovery of novel genetic variation for tolerance to different biotic and abiotic stresses. Genotypes with improved attributes conditioned by superior combinations of alleles at multiple loci could be identified and advanced. Improved selection techniques are needed to identify these superior genotypes and associated traits, especially from wild, related species that grow in environments which do not support the growth of their domesticated relatives that are cultivated varieties (Kumar, 2006). Plants native to climates with marked seasonality are able to acclimatize more easily to variable environmental conditions and provide opportunities to identify genes or gene combinations which confer such resilience.

Impact assessment of climate change on potato productivity in Punjab for three potato cultivars of late (Kufri Badshah), medium (Kufri Jyoti) and early (Kufri Pukhraj) maturity groups was carried out by Dua et al., (2013) for A1FI scenario of temperature and atmospheric CO_2 of the years 2020 and 2055. The simulation study was done using WOFOST (World Food studies) crop growth model for potential production at 13 locations in Punjab. The results from the simulation study were interpolated using kriging technique to generate maps of potential productivity and the changes thereon. It was estimated that rise in temperature alone will result in change in productivity of Kufri Badshah from +11.6% (Amritsar) to -10% (Fatehgarh) in 2020, whereas the change in productivity of Kufri Jyoti will be from +11.6% (Amritsar) to –11.6% (Fatehgarh) and of Kufri Pukhraj from +12% (Amritsar) to -11.5% (Mansa). During this period, CO_2 fertilization is expected to increase tuber productivity from +3.9% to +4.5%, depending upon cultivar and location. However, in 2055, a mean decrease of 17.9 (Kufri Badshah), 21.1(Kufri Jyoti) and 22% (Kufri Pukhraj) is likely in the productivity due to rise in temperature only, while the expected rise in CO_2 is likely to bring about 17.3 (Kufri Badshah) to 18.5% (Kufri Jyoti) increase in potato productivity. It is estimated that under the combined influence of change in temperature and CO_2, the productivity of potato cultivars will not be affected in 2020 over the baseline scenario, but will decline in 2055 (Kufri Badshah, -2.62%; Kufri Jyoti, -4.6% and Kufri Pukhraj, -5.3%), when the total geographical area of Punjab is considered. It is further shown that if the present distribution of potato acreage within Punjab remains unaltered in future, there will be benefits from climate change as the potential productivity of Kufri

Badshah, Kufri Jyoti and Kufri Pukhraj will increase by 3.3%, 3.1% and 3.6% in 2020, although the potential productivity will again decline to baseline values in 2055 (+0.1%, -1.5% and -1.9% respectively).

Climate Proofing Through Genomics and Biotechnology

Increasing crop productivity in unfavourable environments will require advanced technologies to complement traditional methods which are often unable to prevent yield losses due to environmental stresses. In the past decade, genomics has developed from whole genome sequencing to the discovery of novel and high throughput genetic and molecular technologies. Genes have been discovered and gene functions understood. This has opened the way to genetic manipulation of genes associated with tolerance to environmental stresses. These tools promise more rapid, and potentially spectacular, returns but require high levels of investment. Many activities using these genetic and molecular tools are in place, with some successes. National and international institutes are re-tooling for plant molecular genetic research to enhance traditional plant breeding and benefit from the potential of genetic engineering to increase and sustain potato productivity.

Two main approaches are taken to create new potato varieties through improved biotechnology: 'traditional' plant breeding techniques and genetic modification. These techniques may play an important role in creating new cultivars able to maintain yields under stressors induced by climate change. Traits that may be helpful in reducing negative impacts of climate on potato production include:

1. **Heat Stress Tolerance:** Ability to maintain tuber growth and initiation under high temperatures. Developing cultivars with greater heat stress tolerance is critical for maintaining yields in countries with potato production areas near current cultivars' maximum temperature limits (e.g. Sub-Saharan Africa, India).
2. **Drought Tolerance:** This includes better water use efficiency (amount of food produced per amount of water used) as well as potatoes that can be exposed to short drought periods and recover and produce acceptable yields. Deeper root systems could also be beneficial, as most commercial potato cultivars need frequent irrigation due to their shallow roots.
3. **Fast Growth/Early Maturation:** Potatoes that grow faster could help adjust to shorter growing seasons in some areas and also reduce the number of life cycles pests such as potato tuber moth can complete in a single growing season.
4. **Disease Resistance:** Potatoes with resistances to local pests and diseases could be helpful, especially in adapting to diseases spreading into new areas.

Listed below are some examples of potato cultivars, hybrids and related species that have traits that may help reduce the negative direct and indirect impacts of climate change on potato production:

- **Zahov:** A heat tolerant polyploid hybrid between potato and several wild potato species. Several wild potato species have been found to be more heat tolerant than *Solanum tuberosum.*
- **Kufri Surya:** A heat tolerant, early maturing potato cultivar developed in India.
- **Solanum verrucosum:** A wild potato species with genes that give it good resistance to late blight.
- **Sullu:** An Andean potato variety with enhanced drought tolerance.
- **Spunta G2:** A genetically modified variety of the Spunta potato cultivar, which is resistant to Potato Tuber Moth due to having the cry1Ia1 gene from *Bacillus thuringiensis.*

FUTURE SCOPE OF POTATO CULTIVATION

Future research works need to address the effect of ecological stress on potato production. Germplasm of the major potato crops which are tolerant of high temperatures, flooding and drought is to be *identify* and advanced breeding lines will be developed. Efforts will also underway to identify nitrogen-use efficient germplasm. In addition, development of production systems geared towards improved water-use efficiency and expected to mitigate the effects of hot and dry conditions in potato production systems are top research and development priorities.

REFERENCES

Abdrabbo, M. A., Farag, A. A., & Abul-Soud, M. (2013). The intercropping effect on potato under net house as adaption procedure of climate change impacts. *Researcher, 5*(6), 48-60. Retrieved from http://www.sciencepub.net/researcher

Agarwal, P. K. (2009). *Global Climate change and Indian agriculture: Case studies from ICAR network project.* Indian Council of Agricultural Research.

Burke, M. B., Lobell, D. B., & Guarino, L. (2009). Shifts in African crop climates by 2050, and the implications for crop improvements and genetic resources conservation. *Global Environmental Change, 19*(3), 317–325. doi:10.1016/j.gloenvcha.2009.04.003

Collier, R., Fellows, J., Adams, S., Semenov, M., & Thomas, B. (2008). Vulnerability of horticultural crop production to extreme weather events. Effects of climate change on plants: Implications for agriculture. *Aspects of Applied Biology*, *88*, 3–14.

Dua, V. K., Singh, B. P., Govindakrishnan, P. M., Kumar, S., & Lal, S. S. (2013). Impact of climate change on potato productivity in Punjab- A simulation study. *Current Science*, *105*(6), 25–58.

Falloon, P., & Betts, R. (2010). Climate impacts on European agriculture and water management in the context of adaptation and mitigation: The importance of an integrated approach. *The Science of the Total Environment*, *408*(23), 5667–5687. doi:10.1016/j.scitotenv.2009.05.002 PMID:19501386

Fatma, S., Moursy, I., Sadek, S.M., Abolmaaty, & Youssef, M. A . (2013). Identification and quantify of the main climatic factors affecting potato growth and yield. *Journal of Horticultural Science & Ornamental Plants*, *5*(3), 177-182.

Godwin, D. C., & Singh, U. (1998). Nitrogen balance and crop response to nitrogen in upland and lowland cropping systems. In G.Y. Tsuji, G. Hoogenboom, & P. K. Thornton (Eds.), Understanding options of agricultural production (pp. 55-77). Kluwer Academic Publishers. doi:10.1007/978-94-017-3624-4_4

Guhathakurta, P., & Rajeevan, M. (2008). Trends in rainfall pattern over India. *International Journal of Climatology*, *28*(11), 1453–1469. doi:10.1002/joc.1640

Hartz, T. K., & Moore, F. D. (1978). Prediction of potato yield using temperature and insolation data. *American Potato Journal*, *55*(8), 431–436. doi:10.1007/BF02852146

Holden, N. M., Brereton, A. J., Fealy, R., & Sweeney, J. (2003). Possible change in Irish climate and its impact on barley and potato yields. *Agricultural and Forest Meteorology*, *116*(3-4), 181–196. doi:10.1016/S0168-1923(03)00002-9

IPCC. (2007). Climate Change 2007, The Physical Science Basis. In S. Solomon, D. Quin, M. Manning, Z. Chen, M. Marquis, K. B. Averut, & H. L. Miller et al. (Eds.), *Contribution of Working Group – I to the Fourth Assessment Report of the IPCC*. Cambridge University Press.

Kumar, K. (2009). *Impact of climate change on India's monsoon climate and development of high resolution climate change scenarios for India*. Presented at MoEF, New Delhi, India. Retrieved Jan. 2015, from http: moef.nic.

Kumar, S. (2006). Climate change and crop breeding objectives in the twenty first century. *Current Science*, *90*, 1053–1054.

Lal, R. (2001). Future climate change: Implications for India Summer monsoon and its variability. *Current Science*, *81*, 1205–1207.

Lal, R. (2004). Soil carbon sequestration in India. *Climatic Change*, *65*(3), 277–296. doi:10.1023/B:CLIM.0000038202.46720.37

Lobell, D. B., Burke, M. B., Tebaldi, C., Mastrandrea, M. D., Falcon, W. P., & Naylor, R. L. (2008). Prioritizing climate change adaptation needs for food security in 2030. *Science*, *319*(5863), 607–610. doi:10.1126/science.1152339 PMID:18239122

Lobell, D. B., & Field, C. B. (2007). Global scale climate-crop yield relationships and the impacts of recent past. *Environmental Research Letters*, *2*(1), 1–7. doi:10.1088/1748-9326/2/1/014002

Monneveux, P., Quiroz, R., Posadas, A., & Kleinwechter, U. (2012). *Facing climate change effects on potato cultivation-an integrative approach.* In *International conference on managing soils for food security and climate change, adaptation and mitigation*, Vienna, Austria.

Moss, R. H., Edmonds, J. A., Hibbard, K. A., Manning, M. R., Rose, S. K., Vuuren, D. P., & Wilbanks, T. J. et al. (2010). The next generation of scenarios for climate change research and assessment. *Nature*, *463*(7282), 747–756. doi:10.1038/nature08823 PMID:20148028

Mukherjee, D. (2002). Food security coping with nutritional challenge ahead. *Agriculture Today*, *8*, 30–32.

Mukherjee, D. (2004a). Herbicides: Safe and appropriate option for weed control. *Agriculture Today*, *11*, 26–27.

Mukherjee, D. (2004b). Influence of stresses on crop efficiency. *Kisan World*, *31*(3), 40.

Mukherjee, D. (2007). Weed management strategy in rice -A Review. *Agricultural Review*, *27*(4), 247–257.

Mukherjee, D. (2010a). Studies on climate change effect on hill agro-ecosystem. In *Abstract National Conference of Plant Physiology on "Physiological and Molecular Approaches for Crop Improvement under Changing Environment"* (pp. 1-2). Department of Plant Physiology, Institute of Agricultural Sciences, Banaras Hindu University, Varanasi.

Mukherjee, D. (2010b). *Lead Lecture on "Resource conservation through indigenous farming system in hills of west Bengal"*. Workshop on ITK in Modern Agriculture (Souvenir), Kalyani, India.

Mukherjee, D. (2011). Global warming mitigation and agricultural practices.*Proceedings of 57th Orientation Course* (pp 8-9). Banaras Hindu University.

Mukherjee, D. (2013a). Nutrient use efficiency for maximization of crop productivity. In Advances in Plant Physiology. Scientific Publishers.

Mukherjee, D. (2013b). Sustainable uses of water resources and quality with respect to agriculture. In S. R. Maloo & L. L. Somani (Eds.), *Advances in Experimental Agriculture* (Vol. 3, pp. 5–48). Udaipur, India: Agrotech Publishing Academy.

Mukherjee, D. (2014a). Climate change and its impact on Indian agriculture. In S. Nehra (Ed.), *Plant disease management and microbes* (pp. 193–206). Jaipur, India: Aavishkar Publishers.

Mukherjee, D. (2014b). Water harvesting and resource conservation technique of Darjeeling Himalaya. *Himalayan Research Journal*, *2*(1), 72–79.

Mukherjee, D. (2014c). Irrigation water quality in agriculture. In H. Rodriguez, N. C. Sarkar, R. Ramanjaneyulu, & R. Maity (Eds.), Advances in Agro-Technology: A text book (vol. 2, pp. 146-166). Puspa Publishing House.

Mukherjee, D. (2014d). Nutrient and its management: Prospect and challenges under the changing environment scenario. In A. Hemantaranjan (Ed.), Advances in Plant Physiology (vol. 15, pp. 413-442). Scientific Publishers.

Mukherjee, D. (2014e). Natural resource management under changing forest ecosystem. *MFP Newsletter, 24*(2), 15-22.

Mukherjee, D. (2015a). Current environment issue related to hill agriculture. *Himalayan Times*, *4*(10), 11–13.

Mukherjee, D. (2015b). Microbial diversity for soil sustainability and crop productivity. In B. K. Sharam & A. Singh (Eds.), *Microbial empowerment in Agriculture: A key to sustainability and crop productivity* (pp. 237–268). New Delhi: Biotech publishers.

Mukherjee, D., & Singh, R. P. (2002). Herbicide bating low-dose options. *Agriculture Today*, *12*, 44–45.

Opena, G. B., & Porter, G. A. (1999). Soil management and supplemental irrigation effects on potato II. Root growth. *Agronomy Journal*, *91*(3), 426–431. doi:10.2134/agronj1999.00021962009100030011x

Ramakrishna, Y. S., Rao, G. G. S. N., Rao, V. U. M., Rao, A. V. M. S., & Rao, K. V. (2007). Water management- water use in rainfed regions of India. In M. V. K. Sivakumar & R. P. Motha (Eds.), *Managing Weather and Climate Risks in Agriculture* (pp. 245–262). Springer. doi:10.1007/978-3-540-72746-0_15

Roudier, P., Sultan, B., Quirion, P., & Berg, A. (2011). The impact of future climate change on West African crop yields: What does the recent literature say? *Global Environmental Change*, *1*(3), 1073–1083. doi:10.1016/j.gloenvcha.2011.04.007

Singh, R. P., Singh, R. K., & Mukherjee, D. (2004). Effect of weed interference on efficiency of crop. In Abstract Seminar on biotic and abiotic stress on agro-ecosystems W.B. Institute of Agricultural Sciences (P.S.B).

Takahashi, K., Matsuoka, Y., & Harasawa, H. (1998). Impacts of climate change on water resources, crop production and natural ecosystem in the Asia and Pacific region. *Journal of Global Environmental Engineering*, *11*, 91–103.

Chapter 6
Towards the Development of Salt-Tolerant Potato

John Okoth Omondi
Ben Gurion University of the Negev, Israel

ABSTRACT

Soil salinity is a major constrain to crop production and climate change accelerates it. It reduces plant water potential, causes ion imbalance, reduce plant growth and productivity, and eventually leads to death of the plant. This is the case in potato. However, potato has coping strategies such as accumulation of proline, an osmo-regulator and osmoprotector. In addition, leaching of salts below the root zone is preferred, exogenous application of ascorbic acid and growth hormones are practiced to combat salinity. Breeding and genetic engineering also play key roles in salinity management of potato. Varieties such as: Amisk, BelRus, Bintje, Onaway, Sierra, and Tobique were tolerant in North America, variety Cara in Egypt, Sumi in Korea and varieties Vivaldi and Almera in Mediterranean region. Transgenic lines of Kennebec variety, lines S2 and M48 also proved tolerance due to transcription factor MYB4 encoded by rice Osmyb4 gene.

DOI: 10.4018/978-1-5225-1715-3.ch006

CLIMATE CHANGE AND SALINITY

Climate change is observed and predicted as the single-most event that is changing and will continue to change the cause of the world's future. It is a major challenge to agriculture. As the climate 'forcing mechanisms' shape climate through processes such as: variation in solar radiation, continental drifts and changes in greenhouse gas concentration, they encourage increase in global temperatures, melting of ice-caps, rising ocean levels, unpredictable and variable amount of rainfall, more cyclones and heat waves, and increased desertification (Abumhadi et al., 2012). There are projections of increase in precipitation in winters and overall decrease in the tropics and sub-tropics (IPCC, 2007). A decrease in precipitation means a decrease in water availability and increase in evaporative demand due to rising temperatures (Vicente-Serrano et al., 2014; Wang et al., 2012). Increase in temperatures enhance evapotranspiration especially in arid and semi-arid regions. In the event that evapotranspiration exceeds precipitation, salts accumulate on the soil surface (Sivakumar, 2007). This has an implication on irrigation requirement (Doll & Siebert, 2002). The decrease in precipitation causes demand for more irrigation (McDonald & Girvetz, 2013; Riediger et al., 2014) even though, in arid and semi-arid regions water available for irrigation is saline (Levy et al., 2013). This consequently exacerbates soil salinity. Additionally, a rise in sea-level due to global warming threatens low-lying coastal agricultural lands – this rise leads to flooding by oceans' saline water and salinization of groundwater (Gornall et al., 2010). Grenfell et al., (2016) predicts that 125 to 175 years from 2010 more wetlands will experience frequent flooding with saltwater. Flooding of the coastal area with saline water has a far-reaching impact. Indeed, this salt incursion reduces dissolved organic carbon in the coastal wetlands from 40 mg l^{-1} to 18 mg l^{-1} (Ardón et al., 2016). Furthermore, increased salinity in the coastal wetlands will cause accumulation of less stable carbon (Williams & Rosenheim, 2015), reduction in marsh biodiversity, and lead to development of more salt tolerant plant communities (Grenfell et al., 2016).

SOIL SALINITY

Salinity is a phenomenon that is causing major havoc in agricultural production (Figure 1) (D'Odorico et al., 2013). The earliest salinity effect in agriculture was recorded around 2400 to 1700 BC in the ancient Mesopotamia, currently Southern Iraq (Jacobsen & Adams, 1958). 6% of arable land around the world is affected by salinity or sodicity, this is 800 million hectares (FAO, 2009). Salt-affected soils have high levels of dissolved salts and or high concentrations of adsorbed sodium ions (Yadav et al., 2011). These soils are categorized into three classes: First are

Figure 1. Global distribution of saline soils using data from the Harmonized World Soil Database. Salinity on the map is represented by electrical conductivity (dS m^{-1}) and the coloring schemes illustrate different ranges in soil salinity corresponding to the relative degree to which soil salinity constrains plant productivity
Source: D'Odorico et al., (2013)

the saline soils having an electrical conductivity (EC) of over 4 dS m^{-1} and sodium adsorption ratio (SAR) of less than 13 or exchangeable sodium percentage (ESP) that is below 15. The second category are sodic soils which are characterized by an EC of less than 4 dS m^{-1}, SAR or ESP above 13 and 15 respectively. Finally, the saline-sodic soils, whose EC, SAR, or ESP are above 4 dS m^{-1}, 13 and 15 accordingly (United States Salinity Laboratory, 1954).

Soil salinity formation is classified into three categories (Rengasamy, 2006). The first is: groundwater associated salinity (GAS), this is prone in areas with shallow groundwater. In fact, groundwater salinity decreases with increasing groundwater depth, and as such in North-western China as an example, 2.5 m is critical (Abliz et al., 2016), although, it varies from region to region. Salts dissolved in groundwater are brought to the rhizosphere through evaporation. This is worsened by rising global temperature as it increase evaporative demand, consequently encouraging exit of salts from groundwater and their accumulation on the root-zone (Li et al., 2015). Besides, some plants have the ability to exclude salts during nutrient absorption, thus enhancing their accumulation on the root-zone (Moya et al., 2003). In other regions such as the Nile Delta, accumulation of salts on the soil surface is a factor of groundwater and rise of 3 cm $year^{-1}$ in sea-level (Geriesh et al., 2015). This movement of salts to the soil surface is richly influenced by soil hydraulic properties (Bejat et al., 2000) and climatic conditions (Rengasamy, 2006). The second is: non-groundwater-associated salinity (NAS). Some regions have deep water table and poor drainage causing accumulation of salts on the soil solum. In addition, salts are introduced by rain, weathering, and aeolian deposits. Salt accumulation through these processes are enhanced by poor hydraulic properties of the soil. The third category is: Irrigation associated salinity (IAS). Rengasamy (2006) describes this as the salts introduced by irrigation water stored within the root zone due to insufficient leaching. The author further explains that poor quality irrigation water, low hydraulic conductivity of the soil, and high evaporative demand accelerate these conditions.

Climate change causes intense, low amount and variable pattern of precipitation (Arnell, 1999; Trenberth, 2011) creating a demand for irrigation. However, as the world population soars, there is increased demand and pressure on fresh water reducing its availability (Fischer & Heilig, 1997; Hanjra & Qureshi, 2010; WWAP, 2015). Consequently, this leads to irrigation with poor quality water, highly saline, and reduction in leaching fraction, and thus accelerating the soil salinity problem (Kitamura et al., 2006).

Salinity's negative impact is grave and cannot be underestimated. It first reduces water potential (Aziz and Khan, 2001; Romero-Aranda et al., 2001), causes ion imbalance, reduction in plant growth and productivity, and eventually death of the whole-plant (Bernstein, 1975; Parida and Das, 2005). The effect on plant growth and physiological processes is exhibited by: reduction in leaf expansion, decrease in soluble protein contents (Muthukumarasamy et al., 2000; Wang and NII, 2000), photosynthesis (Allakhverdiev et al., 2002; Khavari-Nejad & Chaparzadeh, 1998), and lipid metabolism (Hassanein, 1999), and stunted growth (Takemura et al., 2000) as the stress intensifies.

SALINITY EFFECT ON POTATO

In potato, salinity delays shoot emergence and development (Levy, 1992; Levy et al., 1993). It further decreases growth with an early indication of leaf necrosis, lowers water and osmotic potential of leaves and tubers (Gao et al., 2015; Heuer and Nadler, 1998; Levy et al., 1988) and increases sodium concentration in the roots (Prasad & Potluri, 1996) and total soluble solids (Levy et al., 1988). Finally, it decreases tuber yield (Bustan et al., 2004; Levy, 1992). Levy, (1992) observed 21 – 59% decrease in tuber yield on irrigation with saline water and different varietal response to salinity (Table 1). Decreases in tuber yield of 33.7% and 79.3% on treating plants with 1.6 dS m^{-1} and 4.8 dS m^{-1} were also reported in Korea among four varieties (Kim et al., 2013). Salinity also delays tuberization, this was proven by in vitro studies. Microtuberization was delayed by 5 to 10 days in 20 and 40 mmol NaCl and inhibited completely in 80 mmol NaCl (Zhang et al., 2005). Earlier in 2001, Silva et al. (2001) observed similar results under 100 mmol l^{-1} NaCl.

Table 1. The effect of salinity on plant height, haulm, and tuber growth of four potato cultivars

Water Source	Soil Depth (cm)	EC (dS m^{-1}) at 136 DAP	Cultivar	Plant Height (cm±S.E.)	Haulm Weight (g m^{-1}±S.E.)	Tuber Weight (g m^{-1}±S.E.)
NC	0-20	1.7±0.2	Atica	42±1.2	3962±324	1975±188
	20-60	1.8±0.4	Desiree	50±1.4	4638±140	1475±0
			Cara	45±0.5	4412±429	782±32
			Alpha	42±1.0	4475±243	1175±150
NCS	0-20	4.2±0.7	Atica	37±0.9	2500±173	1863±238
	20-60	3.8±0.8	Desiree	33±1.9	4038±305	1000±75
			Cara	37±1.1	3175±304	438±138
			Alpha	33±1.8	3350±233	638±63
S	0-20	7.3±1.0	Atica	31±1.7	1825±138	925±125
	20-60	5.1±0.6	Desiree	25±0.8	2538±184	-
			Cara	32±1.4	2225±229	213±88
			Alpha	29±1.6	2500±248	294±144

NC is common irrigation water, NCS is a mixture of common irrigation water (NC) and saline water from a local well, S is saline water from the local well, DAP is days after planting. Electrical conductivity (EC) is of the extract of saturated soil samples from root zone of potato irrigated with NC, NCS, and S.

Source: Levy, (1992)

Increasing salt concentrations alters potato physiology through decrease in number of chloroplasts and cell intercellular spaces, thickening of cell walls, rapture and complete damage to mesophyll cells and chloroplasts (Gao et al., 2015). Teixeira and Fidalgo, (2009) reported increased ammonium assimilation in potato roots and decrease in the leaves due to increased accumulation of glutamate synthetize in a salinity event (Figure 2). Although, previously, Teixeira and Pereira, (2007) observed decrease of glutamate synthetize activity in leaves and roots. In addition to the aforementioned damages, salinity predisposes potato to diseases such as: verticillium wilt, early blight (Kaufman et al., 1990; Nachmias et al., 1993), and browning of the tubers (Dzengeleski et al., 2003; Kirk et al., 2006).

COPING STRATEGIES OF POTATO TO SALINITY

Despite all the losses caused by salinity, potato has coping strategies in salt-affected soils. Proline accumulation is one of the strategies. Proline is an osmoregulator and osmoprotector which accumulates during salt stress in potato (Figure 3) (Martinez

Figure 2. Total GS transferase activities in potato leaves (L) and roots (R) grown under 0, 100 mM and 200 mM NaCl, expressed as nkat γ-glutamyl hydroxamate.mg[-1] soluble protein. Columns represent mean +S.D. of triplicates ($n \geq 3$). Differences from control values are all significantly different at $P < 0.05$
Source: Teixeira and Fidalgo, 2009

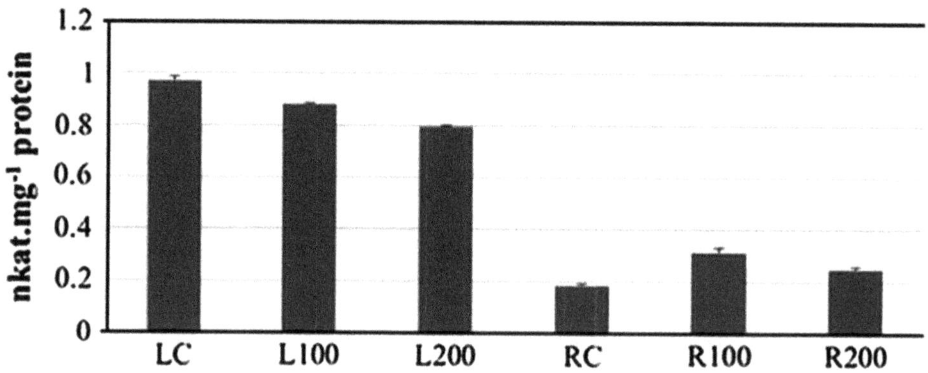

et al., 1996; Prasad and Potluri, 1996; Rahnama and Ebrahimzadeh, 2004). The second coping mechanism is: up-regulation of proteins. Aghaei et al. (2008) observed up-regulation of osmotine-like proteins, heat-shock proteins, TSI-1 protein and calreticulin proteins and down-regulation of photosynthesis related proteins in two cultivars of potato. Evers et al. (2012) would later observe up-regulation of cell rescue and transcription factor related genes and down-regulation of photosynthesis-related genes. Moreover, potato increases activity of antioxidant enzymes such as: ascorbate peroxidase, catalase, and glutathione reductase against reactive oxygen species (ROS) in salinity situations (Aghaei et al., 2009; Hamdi et al., 2009; Rahnama and Ebrahimzadeh, 2004, 2005).

AMELIORATION OF SALINITY STRESS IN POTATO

Inasmuch as potato can defend itself to an extent towards salinity, tuber yield is yet to raise under saline conditions, and hence other methodologies are used to combat salinity. Jesus et al., (2015) details several opportunities for salt phytoremediation (Figure 4). One of which is leaching the salts below the root zone (Corwin et al., 2007; Qadir & Oster, 2004). That means irrigating with excessive water (Hanson et al., 2006) although, the water resource is dwindling globally. Therefore, other strategies of dealing with salinity in potato are supposed to be used in conjunction to irrigation. A study conducted by Backhausen et al. (2005) found that high air humidity reduces salinity effect – high air humidity reduces Na^+ accumulation and

Figure 3. NaCl effects on growth and proline contents in shoot (A, C) and calli (B, D). A.B: Growth: C, D: proline content; AG: Agria; DIA: Diamant; K: Kennebec; AJ: Ajax. Agria and Kennebec: Relatively salt tolerant; Daimant and Ajax: Relatively salt sensitive. Vertical bars represent standard errors

Source: Rahnama and Ebrahimzadeh, (2004)

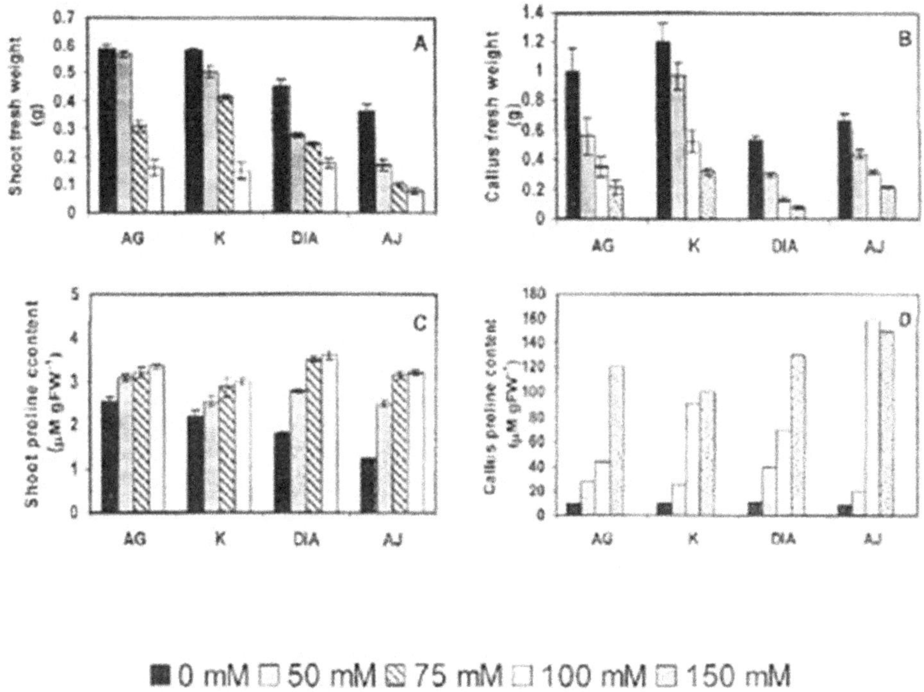

increases non-photochemical quenching (NPQ) and thus reducing the overall rate of photosynthetic electron flow. Application of 60 Mt ha^{-1} of poultry manure was observed to reduce salinity stress in potato and increase tuber yield (Oustani et al., 2014). In addition, Richardson et al. (2001) noted that supplemental Ca^{++} could alleviate salinity stress effects by inhibiting degradation of nuclear of the root cell. Application of low to medium concentrations of Ca^{++} are encouraged, however, higher concentrations of 2 gm Ca^{++} per plant in saline conditions reduce potato yields (Abdel-Naby et al., 2001). Similarly, potassium deficiency was shown to encourage salt damage in micropropagated potato plantlets (Alhagdow et al., 1999) and therefore its supply reduces salinity effect (Elkhatib et al., 2004).

Exogenous application of ascorbic acid was noted to increase tolerance of potato towards salinity and protected chloroplasts from salt damage (Sajid & Aftab, 2009). Salicylic acid application succeeded in reducing salinity effect in potato (Sajid & Aftab, 2012) as it lowered oxidative damage caused by salinity. Further-

Figure 4. Techniques for enhanced salt phytoremediation grouped by type. AMF - arbuscular mycorrhizal fungi; PGPB - plant growth-promoting bacteria
Source: Jesus et al., (2015)

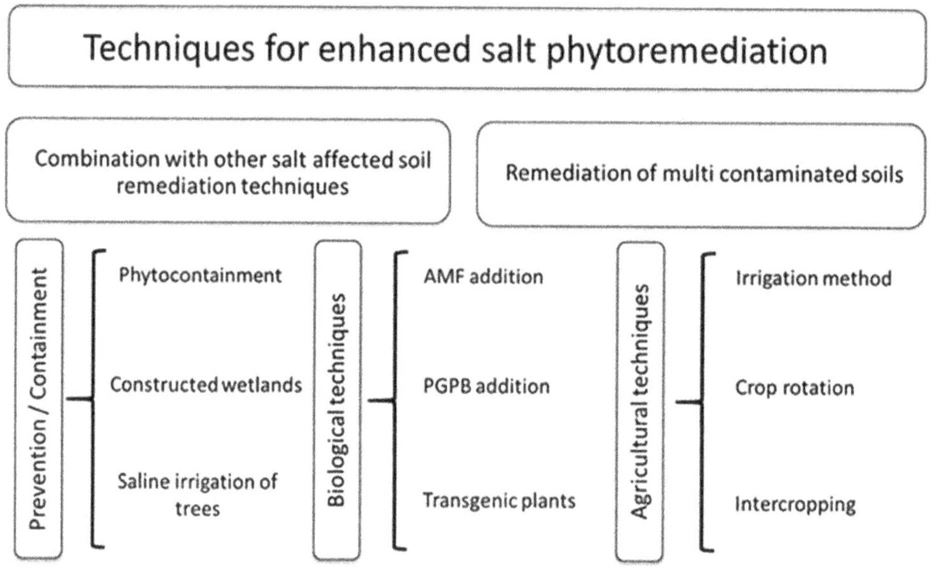

more, exogenous application of abscisic acid and or its method of application modified the response of potato to salinity (Etehadnia et al., 2008).

Besides exogenous application of growth hormones, breeding and genetic engineering is being used to alleviate salinity effects in potato. In tetrasomic tetraploid potatoes, DREBIA gene expression was observed to significantly reduce salinity (Celebi-Toprak et al., 2005). Another gene from rice, Osmyb4 encoding transcription factor MYB4 found in two transgenic lines of potato S2 and M48 (Variety Kennebec) caused reduction of salinity effects (Aydin et al., 2014). Further gene studies indicate that bacterial mannitol 1-phosphate dehydrogenase (mtlD) gene expression in potato increases mannitol which lead to enhanced salt tolerance (Rahnama et al., 2011). Glutamate synthetase (GS) plays a major role in regulating proline accumulation in salt stress potato, and in turn GS1 gene has been shown to be responsible for GS activity in salt affected potato (Teixeira et al., 2006). Overexpression of GalUR gene which is associated with ascorbate pathway was observed to enhance ascorbic acid content in transgenic potato variety, Taedong Valley, by minimizing oxidative stress caused by salt (Upadhyaya et al., 2011). In 2012, potato variety 'Marfona' was created via gamma irradiation (Yaycili & Alikamanoglu, 2012). This variety which was 27.5% genetically different from the control plants was proven to tolerate salinity.

Selection of potato cultivars tolerant to salinity is an age-old practice. Khrais et al. (1998) assessed 130 European and North America potato varieties for salinity. They eventually asserted that varieties: Amisk, BelRus, Bintje, Onaway, Sierra, and Tobique were the most tolerant to salinity. Other similar studies were conducted across the globe. In Egypt, variety Cara was tolerant among the four tested varieties by Elkhatib et al. (2004); in Korea, variety Sumi was tolerant among five tested varieties (Kim et al., 2013) and in the Mediterranean region, ten varieties were tested and only Vivaldi and Almera were found to be tolerant to salinity (Levy & Tai, 2013). Cultivars Desiree and Russett Burbank were also found to express tolerance towards salinity in the Netherlands (Jaarsma et al., 2013).

FUTURE RESEARCH DIRECTIONS

Potato is the third most important food crop in the world after rice and wheat. It is relatively sensitive to abiotic stress of excessive salt content in the soil. However, comparatively little work has been done with respect to abiotic stress of soil salinity on potato (Londhe, 2016). Further, it is expected to increase in salt affected soils across the globe due to climate change which influence on salinity is immense. As these changes occur, potato cultivation may be severely influenced and need coping strategies to be enhanced. Potato is means of food security and livelihood of many poor and marginal farmers. Increase in further agriculture area is not possible and cultivation of potato on abiotic stress affected soils due to salts is essential. The further research in this direction should be concentrated on development of salt tolerant potato varieties and better management practices for its cultivation on salt affected soils.

Precision nutrient application packages need to be developed and tested for various agro-ecological regions of salt affected soils. Further it is necessary to synchronizing potato growth stages with their nutrient requirement for optimal growth and yield and thereafter point application of nutrients either through drip irrigation or foliar could reduce accumulation salts in the soils. Research is needed on studying the pathway through which excess salts affects potato quality is crucial; the salt absorbed by potato could be channeled and not accumulate in the tubers such that during cooking it enhances taste and flavor.

CONCLUSION

Climate change exacerbate salinity leading to decrease in potato tuber yield. As this happens, the world population is also increasing thereby worsening the already impaired food security situation. This situation is of high concern owing to the fact that potato is a major food crop around the globe and a staple food in many countries. Therefore, addressing the climate change factors that intensify salinity is the first step towards reducing or alleviating salinity, followed by control or remediation measures in areas already experiencing the problem, and finally using the potato itself – its tolerance to salinity, breeding, selecting, and engineering it towards that.

REFERENCES

Abdel-Naby, A., El-Beltagy, M. S., Abou-Hadid, A., Helmy, Y. I., & El-Abd, S. O. (2001). Effect of Salinity Stress Modified by Calcium Application on Potato Plants. *Egyptian Journal of Horticulture*, *28*(4), 519–530.

Abliz, A., Tiyip, T., Ghulam, A., Halik, U., Ding, J. L., Sawut, M., & Abliz, A. et al. (2016). Effects. of shallow groundwater table and salinity on soil salt dynamics in the Keriya Oasis, Northwestern China E*nviron.mental. Earth Science*, *75*(3), 1–15. doi:10.1007/s12665-015-4794-8

Abumhadi, N., Todorovska, E., Assenov, B., Tsonev, S., Vulcheva, D., & Vulchev, D. (2012). Agricultural Research in 21 st century : Challenges facing the food security under the impacts of climate change. *Bulgarian Journal of Agricultural Science*, *18*(6), 801–818.

Aghaei, K., Ehsanpour, A. A., & Komatsu, S. (2008). Proteome Analysis of Potato under Salt Stress. *Journal of Proteome Research*, *7*(11), 4858–4868. doi:10.1021/pr800460y PMID:18855355

Aghaei, K., Ehsanpour, A. A., & Komatsu, S. (2009). Potato responds to salt stress by increased activity of antioxidant enzymes. *Journal of Integrative Plant Biology*, *51*(12), 1095–1103. doi:10.1111/j.1744-7909.2009.00886.x PMID:20021557

Alhagdow, M. M., Barthakur, N. N., & Donnelly, D. J. (1999). Salinity stress and sodium-potassium micropropagated potatoes interactions. *Potato Research*, *42*, 73–78. doi:10.1007/BF02358392

Allakhverdiev, S. I., Nishiyama, Y., Miyairi, S., Yamamoto, H., Inagaki, N., Kanesaki, Y., & Murata, N. (2002). Salt stress inhibits the repair of photodamaged photosystem II by suppressing the transcription and translation of psbA genes in synechocystis. *Plant Physiology*, *130*(3), 1443–1453. doi:10.1104/pp.011114 PMID:12428009

Ardón, M., Helton, A. M., & Bernhardt, E. S. (2016). Drought and saltwater incursion synergistically reduce dissolved organic carbon export from coastal freshwater wetlands. *Biogeochemistry*, *127*(2-3), 411–426. doi:10.1007/s10533-016-0189-5

Arnell, N. W. (1999). Climate change and global water resources. *Global Environmental Change*, *9*(1), 31–49. doi:10.1016/S0959-3780(99)00017-5

Aydin, G., Yucel, M., Chan, M. T., & Oktem, H. A. (2014). Evaluation of abiotic stress tolerance and physiological characteristics of potato (Solanum tuberosum L. cv. Kennebec) that heterologously expresses the rice Osmyb4 gene. *Plant Biotechnology Reports*, *8*(3), 295–304. doi:10.1007/s11816-014-0322-7

Aziz, I., & Khan, M. A. (2001). Experimental assessment of salinity tolerance of Ceriops tagal seedlings and saplings from the Indus delta, Pakistan. *Aquatic Botany*, *70*(3), 259–268. doi:10.1016/S0304-3770(01)00160-7

Backhausen, J. E., Klein, M., Klocke, M., Jung, S., & Scheibe, R. (2005). Salt tolerance of potato (Solanum tuberosum L. var. Desiree) plants depends on light intensity and air humidity. *Plant Science*, *169*(1), 229–237. doi:10.1016/j.plantsci.2005.03.021

Bejat, L., Perfect, E., Quisenberry, V. L., Coyne, M. S., & Haszler, G. R. (2000). Solute Transport as Related to Soil Structure in Unsaturated Intact Soil Blocks. *Soil Science Society of America Journal*, *64*(3), 818. doi:10.2136/sssaj2000.643818x

Bernstein, L. (1975). Effects of Salinity and Sodicity on Plant Growth. *Annual Review of Phytopathology*, *13*(1), 295–312. doi:10.1146/annurev.py.13.090175.001455

Bustan, A., Sagi, M., De Malach, Y., & Pasternak, D. (2004). Effects of saline irrigation water and heat waves on potato production in an arid environment. *Field Crops Research*, *90*(2-3), 275–285. doi:10.1016/j.fcr.2004.03.007

Celebi-Toprak, F., Behnam, B., Serrano, G., Kasuga, M., Yamaguchi-Shinozaki, K., Naka, H., & Watanabe, K. N. et al. (2005). Tolerance to salt stress of the transgenic tetrasomic tetraploid potato, Solanum tuberosum cv. Desiree appears to be induced by the DREB1A gene and rd29A promoter of Arabidopsis thaliana. *Breeding Science*, *55*(3), 311–319. doi:10.1270/jsbbs.55.311

Corwin, D. L., Rhoades, J. D., & Simunek, J. (2007). Leaching requirement for soil salinity control: Steady-state versus transient models. *Agricultural Water Management*, *90*(3), 165–180. doi:10.1016/j.agwat.2007.02.007

D'Odorico, P., Bhattachan, A., Davis, K. F., Ravi, S., & Runyan, C. W. (2013). Global desertification: Drivers and feedbacks. *Advances in Water Resources, Elsevier Ltd*, *51*, 326–344. doi:10.1016/j.advwatres.2012.01.013

Doll, P., & Siebert, S. (2002). Global modeling of irrigation water requirements. *Water Resources Research*, *38*(4), 1–10. doi:10.1029/2001WR000355

Dzengeleski, S., Da Rocha, A. B., Kirk, W. W., & Hammerschmidt, R. (2003). Effect of soil salinity and Fusarium sambucinum infection on development of potatoes cultivar "Atlantic". *Acta Horticulturae*, *19*(619), 251–261. doi:10.17660/ActaHortic.2003.619.28

Elkhatib, H. A., Elkhatib, E. A., Khalaf Allah, A. M., & El-Sharkawy, A. M. (2004). Yield Response of Salt-Stressed Potato to Potassium Fertilization: A Preliminary Mathematical Model. *Journal of Plant Nutrition*, *27*(1), 111–122. doi:10.1081/PLN-120027550

Etehadnia, M., Waterer, D. R., & Tanino, K. K. (2008). The method of ABA application affects salt stress responses in resistant and sensitive potato lines. *Journal of Plant Growth Regulation*, *27*(4), 331–341. doi:10.1007/s00344-008-9060-9

Evers, D., Legay, S., Lamoureux, D., Hausman, J. F., Hoffmann, L., & Renaut, J. (2012). Towards a synthetic view of potato cold and salt stress response by transcriptomic and proteomic analyses. *Plant Molecular Biology*, *78*(4-5), 4–5, 503–514. doi:10.1007/s11103-012-9879-0 PMID:22258187

FAO. (2009). Advances in the Assessment and Monitoring of Salinization and Status of Biosaline Agriculture. *World Soil Resources Reports, 104.*

Fischer, G., & Heilig, G. K. (1997). Population momentum and the demand on land and water resources. *Philosophical Transactions of the Royal Society of London. Series B, Biological Sciences*, *352*(1356), 869–889. doi:10.1098/rstb.1997.0067

Gao, H. J., Yang, H. Y., Bai, J. P., Liang, X. Y., Lou, Y., Zhang, J. L., & Chen, Y. L. et al. (2015). 'Ultrastructural and physiological responses of potato (Solanum tuberosum L) plantl.ets to gradient saline stress. *Frontiers in Plant Science*, *5*, 1–14. doi:10.3389/fpls.2014.00787

Geriesh, M. H., Balke, K. D., El-Rayes, A. E., & Mansour, B. M. (2015). Implications of climate change on the groundwater flow regime and geochemistry of the Nile Delta, Egypt. *Journal of Coastal Conservation*, *19*(4), 589–608. doi:10.1007/s11852-015-0409-5

Gornall, J., Betts, R., Burke, E., Clark, R., Camp, J., Willett, K., & Wiltshire, A. (2010). Implications of climate change for agricultural productivity in the early twenty-first century. *Philosophical Transactions of the Royal Society of London. Series B, Biological Sciences*, *365*(1554), 2973–2989. doi:10.1098/rstb.2010.0158 PMID:20713397

Grenfell, S. E., Callaway, R. M., Grenfell, M. C., Bertelli, C. M., Mendzil, A. F., & Tew, I. (2016). Will a rising sea sink some estuarine wetland ecosystems? *Science of the Total Environment. Elsevier B.V., 554-555*, 276–292.

Hamdi, M. M., Bettaieb, T., Harbaoui, Y., Mougou, A. A., & Jardin, P. (2009). Insight into the role of catalases in salt stress in potato (Solanum tuberosum L.). *Wild, 13*(3), 373–379.

Hanjra, M. A., & Qureshi, M. E. (2010). Global water crisis and future food security in an era of climate change. *Food Policy, Elsevier Ltd, 35*(5), 365–377. doi:10.1016/j.foodpol.2010.05.006

Hanson, B. R., Grattan, S. R. & Fulton, A. (2006). Agricultural Salinity and Drainage, Division of Agriculture and Natural Resources, *Publication, 3375*.

Hassanein, A. M. (1999). Alterations in protein and esterase patterns of peanut in response to salinity stress. *Biologia Plantarum, 42*(2), 241–248. doi:10.1023/A:1002112702771

Heuer, B., & Nadler, A. (1998). Physiological response of potato plants to soil salinity and water deficit. *Plant Science, 137*(1), 43–51. doi:10.1016/S0168-9452(98)00133-2

IPCC. (2007). Climate Change 2007, Mitigation of Climate Change. In B. Metz, O. R. Davidson, P. R. Bosch, R. Dave, & L. A. Meyer (Eds.), *Contribution of Working Group III to the Fourth Assessment Report of the Intergovernmental Panel on Climate Change, 2007*. Cambridge University Press. Retrieved July 31, 2016, from https://www.ipcc.ch/pdf/assessment-report/ar4/wg3/ar4_wg3_full_report.pdf

Jaarsma, R., de Vries, R. S. M., & de Boer, A. H. (2013). Effect of Salt Stress on Growth, Na+ Accumulation and Proline Metabolism in Potato (Solanum tuberosum) Cultivars. *PLoS ONE, 8*(3), e60183. doi:10.1371/journal.pone.0060183 PMID:23533673

Jacobsen, T., & Adams, R. M. (1958). Salt and silt in ancient Mesopotamian Agriculture. *Science, 128*(3334), 1251–1258. doi:10.1126/science.128.3334.1251 PMID:17793690

Jesus, J. M., Danko, A. S., Fiúza, A., & Borges, M. T. (2015). Phytoremediation of salt-affected soils: A review of processes, applicability, and the impact of climate change. *Environmental Science and Pollution Research International, 22*(9), 6511–6525. doi:10.1007/s11356-015-4205-4 PMID:25854203

Kaufman, Z., Nachmias, A., Livescu, L., Meiri, A., & Tibor, M. (1990). Verticillium wilt of potatoes under irrigation with saline water. *Hassadeh, 70*(6), 898–901.

Khavari-Nejad, R. A., & Chaparzadeh, N. (1998). The effects of NaCl and $CaCl_2$ on photosynthesis and growth of alfalfa plants. *Photosynthetica, 35*(3), 461–466. doi:10.1023/A:1006928721986

Khrais, T., Leclerc, Y., & Donnelly, J. D. (1998). Relative Salinity Tolerance of Potato Cultivars Assessed By I n Vitro Screening Multiplication of In Vitro Material. *American Journal of Potato Research, 75*, 207–210. doi:10.1007/BF02854214

Kim, S., Yang, C., Jeong, J., Choi, W., Lee, K., & Kim, S. (2013). Physiological Response of Potato Variety to Soil Salinity. *Korean Journal of Crop Science, 58*(2), 85–90. doi:10.7740/kjcs.2013.58.2.085

Kirk, W., da Rocha, A., Hollosy, S., Hammerschmidt, R., & Wharton, P. (2006). Effect of Soil Salinity on Internal Browning of potato Tuber Tissue in Two Soil Types. *American Journal of Potato Research, 83*(3), 223-232. Retrieved 30, 2016, from http://link.springer.com/article/10.1007/BF02872158

Kitamura, Y., Yano, T., Honna, T., Yamamoto, S., & Inosako, K. (2006). Causes of farmland salinization and remedial measures in the Aral Sea basin-Research on water management to prevent secondary salinization in rice-based cropping system in arid land. *Agricultural Water Management, 85*(1-2), 1–14. doi:10.1016/j.agwat.2006.03.007

Levy, D. (1992). The response of potatoes (Solanum tuberosum L.) to salinity: Plant growth and tuber yields in the arid desert of Israel. *Annals of Applied Biology, 120*(1), 547–555. doi:10.1111/j.1744-7348.1992.tb04914.x

Levy, D., Coleman, W. K., & Veilleux, R. E. (2013). Adaptation of Potato to Water Shortage: Irrigation Management and Enhancement of Tolerance to Drought and Salinity. *American Journal of Potato Research, 90*(2), 186–206. doi:10.1007/s12230-012-9291-y

Levy, D., Fogelman, E. & Itzhak, Y. (1988). The effect of water salinity on potatoes (Solanum tuberosum L.): Physiological indices and yielding capacity. *Potato Research, 31*(4), 601–610.

Levy, D., Fogelman, E., & Ytzhak, Y. (1993). Influence of water and soil salinity on emergence and early development of potato (Solanum tubersum L.) cultivars and effect of physiological age of seed tubers. *Potato Research, 36*(4), 335–340. doi:10.1007/BF02361800

Levy, D., & Tai, G. C. C. (2013). Differential response of potatoes (Solanum tuberosum L.) to salinity in an arid environment and field performance of the seed tubers grown with fresh water in the following season. *Agricultural Water Management. Elsevier B.V., 116*, 122–127.

Li, H., Yi, J., Zhang, J., Zhao, Y., Si, B., Hill, R., … Liu, X. (2015). Modeling of Soil Water and Salt Dynamics and Its Effects on Root Water Uptake in Heihe Arid Wetland Gansu, China. *Water, 7*(5), 2382–2401. Retrieved July 30, 2016, from http://www.mdpi.com/2073-4441/7/5/2382

Londhe, S. (2016). Cultivation of potato on abiotic stress-affected soils of India. *Current Science, 111*(1), 21-22. Retrieved July 30, 2016, from http://www.current-science.ac.in/Volumes/111/01/0021.pdf

Martinez, C. A., Maestri, M., & Lani, E. G. (1996). In vitro salt tolerance and proline accumulation in Andean potato (Solanum spp.) differing in frost resistance. *Plant Science, 116*(2), 177–184. doi:10.1016/0168-9452(96)04374-9

McDonald, R. I., & Girvetz, E. H. (2013). Two Challenges for U.S. Irrigation Due to Climate Change: Increasing Irrigated Area in Wet States and Increasing Irrigation Rates in Dry States. *PLoS ONE, 8*(6), 1–10. doi:10.1371/journal.pone.0065589 PMID:23755255

Moya, J. L., Gómez-Cadenas, A., Primo-Millo, E., & Talon, M. (2003). Chloride absorption in salt-sensitive Carrizo citrange and salt-tolerant Cleopatra mandarin citrus rootstocks is linked to water use. *Journal of Experimental Botany, 54*(383), 825–833. doi:10.1093/jxb/erg064 PMID:12554725

Muthukumarasamy, M., Gupta, D. S., & Panneerselvam, R. (2000). Enhancement of peroxidase, polyphenol oxidase and superoxide dismutase activities by triadimefon in NaCl stressed Raphanus sativus L. *Biologia Plantarum, 43*(2), 317–320. doi:10.1023/A:1002741302485

Nachmias, A., Kaufman, Z., Livescu, L., Tsror, L., Meiri, A., & Caligari, P. D. S. (1993). Effect of salinity and its interactions with disease incidence on potatoes grown in hot climates. *Phytoparasitica, 21*(3), 245–255. doi:10.1007/BF02980946

Oustani, M., Halilat, M. M. T., & Chenchouni, H. (2014). Effect of poultry manure on the yield and nutriments uptake of potato under saline conditions of arid regions. *Emirates Journal of Food and Agriculture, 27*(1), 106–120. doi:10.9755/ejfa.v27i1.17971

Parida, A. K., & Das, A. B. (2005). Salt tolerance and salinity effects on plants: A review. *Ecotoxicology and Environmental Safety*, *60*(3), 324–349. doi:10.1016/j.ecoenv.2004.06.010 PMID:15590011

Prasad, P. V., & Potluri, S. D. P. (1996). Influence of proline and hydroxyproline on salt-stressed axillary bud cultures of two varieties of potato (Solanum tuberosum). *In Vitro Cellular & Developmental Biology. Plant*, *32*(1), 47–50. doi:10.1007/BF02823013

Qadir, M., & Oster, J. D. (2004). Crop and irrigation management strategies for saline-sodic soils and waters aimed at environmentally sustainable agriculture. *The Science of the Total Environment*, *323*(1-3), 1–19. doi:10.1016/j.scitotenv.2003.10.012 PMID:15081713

Rahnama, H., & Ebrahimzadeh, H. (2004). The effect of NaCl on proline accumulation in potato seedlings and calli. *Acta Physiologiae Plantarum*, *26*(3), 263–270. doi:10.1007/s11738-004-0016-9

Rahnama, H., & Ebrahimzadeh, H. (2005). The effect of NaCl on antioxidant enzyme activities in potato seedlings. *Biologia Plantarum*, *49*(1), 93–97. doi:10.1007/s10535-005-3097-4

Rahnama, H., Vakilian, H., Fahimi, H., & Ghareyazie, B. (2011). Enhanced salt stress tolerance in transgenic potato plants (Solanum tuberosum L.) expressing a bacterial mtlD gene. *Acta Physiologiae Plantarum*, *33*(4), 1521–1532. doi:10.1007/s11738-010-0690-8

Rengasamy, P. (2006). World salinization with emphasis on Australia. *Journal of Experimental Botany*, *57*(5), 1017–1023. doi:10.1093/jxb/erj108 PMID:16510516

Richardson, K. V. A., Wetten, A. C., & Caligari, P. D. S. (2001). Cell and nuclear degradation in root meristems following exposure of potatoes (Solanum tuberosum L.) to salinity. *Potato Research*, *44*(4), 389–399. doi:10.1007/BF02358598

Riediger, J., Breckling, B., Nuske, R. S., & Schröder, W. (2014). Will climate change increase irrigation requirements in agriculture of Central Europe? A simulation study for Northern Germany. *Environmental Sciences Europe*, *26*(18), 1–13.

Romero-Aranda, R., Soria, T., & Cuartero, J. (2001). Tomato plant-water uptake and plant-water relationships under saline growth conditions. *Plant Science*, *160*(2), 265–272. doi:10.1016/S0168-9452(00)00388-5 PMID:11164598

Sajid, Z., & Aftab, F. (2012). Role of Salicylic Acid in Amelioration of Salt Tolerance in Potato (Solanum Tuberosum L.) Under in Vitro Conditions. *Pakistan Journal of Botany*, *44*, 37–42.

Sajid, Z. A., & Aftab, F. (2009). Amelioration of salinity tolerance in Solanum tuberosum L. by exogenous application of ascorbic acid. *In Vitro Cellular & Developmental Biology. Plant*, *45*(5), 540–549. doi:10.1007/s11627-009-9252-4

Silva, J. A. B., Otoni, W. C., Martinez, C. A., Dias, L. M., & Silva, M. A. P. (2001). Microtuberization of Andean potato species (Solanum spp.) as affected by salinity. *Scientia Horticulturae*, *89*(2), 91–101. doi:10.1016/S0304-4238(00)00226-0

Sivakumar, M. V. K. (2007). Interactions between climate and desertification. *Agricultural and Forest Meteorology*, *142*(2-4), 143–155. doi:10.1016/j.agrformet.2006.03.025

Takemura, T., Hanagata, N., Sugihara, K., Baba, S., Karube, I., & Dubinsky, Z. (2000). Physiological and biochemical responses to salt stress in the mangrove, Bruguiera gymnorrhiza. *Aquatic Botany*, *68*(1), 15–28. doi:10.1016/S0304-3770(00)00106-6

Teixeira, J., & Fidalgo, F. (2009). Salt stress affects glutamine synthetase activity and mRNA accumulation on potato plants in an organ-dependent manner. *Plant Physiology and Biochemistry. Elsevier Masson SAS*, *47*(9), 807–813.

Teixeira, J., & Pereira, S. (2007). High salinity and drought act on an organ-dependent manner on potato glutamine synthetase expression and accumulation. *Environmental and Experimental Botany*, *60*(1), 121–126. doi:10.1016/j.envexpbot.2006.09.003

Teixeira, J., Pereira, S., Queirós, F., & Fidalgo, F. (2006). Specific roles of potato glutamine synthetase isoenzymes in callus tissue grown under salinity: Molecular and biochemical responses. *Plant Cell, Tissue and Organ Culture*, *87*(1), 1–7. doi:10.1007/s11240-006-9103-5

Trenberth, K. E. (2011). Changes in precipitation with climate change. *Climate Research*, *47*(1-2), 123–138. doi:10.3354/cr00953

United States Salinity Laboratory. (1954). Diagnosis and Improvement of Saline and Alkali Soils. *Agricultural Handbook, 60*.

Upadhyaya, C. P., Venkatesh, J., Gururani, M. A., Asnin, L., Sharma, K., Ajappala, H., & Park, S. W. (2011). Transgenic potato overproducing l-ascorbic acid resisted an increase in methylglyoxal under salinity stress via maintaining higher reduced glutathione level and glyoxalase enzyme activity. *Biotechnology Letters*, *33*(11), 2297–2307. doi:10.1007/s10529-011-0684-7 PMID:21750996

Vicente-Serrano, S. M., Lopez-Moreno, J. I., Beguería, S., Lorenzo-Lacruz, J., Sanchez-Lorenzo, A., García-Ruiz, J. M., . . . Espejo, F. (2014). Evidence of increasing drought severity caused by temperature rise in southern Europe. *Environmental Research Letters*, *9*(4), 1-9. Retrieved July 30, 2016 from http://digital.csic.es/bitstream/10261/95049/1/BegueriaS_EnvResLett_2014.pdf

Wang, K., Dickinson, R. E., & Liang, S. (2012). Global atmospheric evaporative demand over land from 1973 to 2008. *Journal of Climate*, *25*(23), 8353–8361. doi:10.1175/JCLI-D-11-00492.1

Wang, Y., & Nii, N. (2000). Changes in chlorophyll, ribulose bisphosphate carboxylase-oxygenase, glycine betaine content, photosynthesis and transpiration in Amaranthus tricolor leaves during salt stress. *The Journal of Horticultural Science & Biotechnology*, *75*(6), 623–627. doi:10.1080/14620316.2000.11511297

Williams, E. K., & Rosenheim, B. E. (2015). What happens to soil organic carbon as coastal marsh ecosystems change in response to increasing salinity? An exploration using ramped pyrolysis. *Geochemistry Geophysics Geosystems*, *16*(7), 2322–2335. doi:10.1002/2015GC005839

WWAP. (2015). *The United Nations World Water Development Report 2015: Water for a Sustainable World*. Paris: UNESCO.

Yadav, S., Irfan, M., Ahmad, A., & Hayat, S. (2011). Causes of salinity and plant manifestations to salt stress: A review. *Journal of Environmental Biology*, *32*(5), 667–685. PMID:22319886

Yaycili, O., & Alikamanoglu, S. (2012). Induction of salt-tolerant potato (Solanum tuberosum L.) mutants with gamma irradiation and characterization of genetic variations via RAPD-PCR analysis. *Turkish Journal of Biology*, *36*(4), 405–412.

Zhang, Z., Mao, B., Li, H., Zhou, W., Takeuchi, Y., & Yoneyama, K. (2005). Effect of salinity on physiological characteristics, yield and quality of microtubers in vitro in potato. *Acta Physiologiae Plantarum*, *27*(4), 481–489. doi:10.1007/s11738-005-0053-z

KEY TERMS AND DEFINITIONS

Amelioration: To reduce the effects of bad condition/move bad conditions to better.

Desertification: Degradation of land especially arid and semi-arid lands due to climate changes and human influence mostly deforestation and agricultural activities.

Exacerbates: Increasing the severity of a bad situation for example climate change exacerbates salinity.

Remediation: To control or rectify problems caused, in this case salinity.

Salinity: Soils whose electrical conductivity are over 4 dS m^{-1} and sodium absorption ratio are less than 13.

Salt-Tolerance: The ability of a plant to endure high electrical conductivity and still be able to give reasonable yields.

Sodicity: Soils whose electrical conductivity are less than 4 dS m^{-1} and sodium absorption ratio are above 13.

Chapter 7
Scenario of Quality Potato Production in Rajasthan

Lokesh Kumar Jain
College of Agriculture, Agriculture University, Jodhpur, India

ABSTRACT

Potato (Solanum tuberosum L.) is one of the most important non-traditional tuber crops of Rajasthan. The potato tuber is a modified stem developed underground on a specialized structure called stolen. It contributes to food and nutritional security and provide cheap source of vegetable. It is used either alone or intermingled with other vegetables. It is also consumed as many fried salted food items. Potato is a highly nutritious, easily digestible, wholesome food. In Rajasthan, where varied climatic conditions promoting cultivation of almost every crops and vegetables, the economic conditions of growers, lack of storage facilities and lack of improved technologies for the state remain as bottleneck for its cultivation. In this chapter I tried to elaborate the constraints and possible suggestion for increasing cultivation of potato which is fairly to highly responsive to inputs supplied and gave cash returns in short periods.

DOI: 10.4018/978-1-5225-1715-3.ch007

INTRODUCTION

Potato (*Solanum tuberosum* L.) is one of the most important food crop after wheat, maize and rice, contributing to food and nutritional security in the world and produces more food per unit area per unit time than cereals (http://cpri.ernet.in). Potato is a highly nutritious, easily digestible, wholesome food containing carbohydrates, proteins, minerals, vitamins and high quality dietary fiber with high calorific value. Similarly, the quality of protein is better and balanced than that of cereals protein from nutrition point of view. Potato is a capital and energy intensive crop.

The potato tuber is a modified stem developed underground on a specialized structure called stolen. It contains all the characteristics of a normal stem like dormant bud (eye) and scaly leaf (eyebrow). Thus, potato provides more nutrition than cereals and vegetables. Potato tuber is a bulky commodity which responds strongly to its prevailing environment so it needs proper storage. Most of the people in India have either no knowledge or wrong notions about the nutritive value of potato. It does not cause obesity because it contains low fat (0.1 per cent) (Singh, 2013). Potato is fairly to highly responsive to inputs supplied and gave cash returns.

In the present chapter I tried to explore some possibilities for intensive and profitable production of potato in the Rajasthan on sustain basis.

CLIMATE OF RAJASTHAN

The success of Indian agriculture is largely dependent on behavior of SE monsoon as the country's almost 60% of total cultivated area is still rainfed. The mostly Rajasthan state and hot arid western zone of India comprises of Thar Desert always faces an uncertainty of weather condition. Improved agronomical and engineering practices lead to sustainable income from this drought prone areas. The environmental degradation adversely affects the livelihoods of rural poor as they relies on the available natural resources. Thus, agriculture or particularly rainfed agriculture in Rajasthan is very complex, diverse and risk prone and characterized by low levels of productivity with low input usage. The Rajasthan state represents almost all types of climatic conditions including extremes of weather variables, soils, water availability and biodiversity. The extreme of western parts having light sandy to sandy loam soils while other parts of the state have fairly deep vertisols. Similarly, the soils provide varying types of underground environment to crops.

AREA AND PRODUCTION OF POTATO IN THE STATE

In Rajasthan, it is grown on an approximate 10000 ha of land as vegetables and it occupies nearly 0.5 per cent of total area in the country. The productivity in the state is 116.96 q/ha (2012-13) (and was almost 40% of the nation (227.6 q/ha) productivity on such harsh and non-traditional areas (WWW.agriexchange.apeda.gov.in/production/). The crop is grown by many vegetable growers for local household consumption either in very small area or in kitchen gardens but these are not in black and white in the state production statistics. The contribution was totally rainfall dependent and production varies from 54.82 to 178.02 (000'tonnes) from 2005 to 2013-14 (Agricultural Statistics at a Glance 2014). Under such harsh climatic condition (non-traditional areas for potato cultivation) where soil and water along with climate pose a great threaten to cultivation, crop has great potential for increase. It is sure that the production may be boost up and may be touch the national status if managed properly by following improved agronomic practices under changing climate and global warming situations. Potato is grown in almost all the districts excluding Barmer, Jaiselmer and Jodhpur of Rajasthan. The major potato producing districts are Dholpur, Bharatpur, Kota, Sirohi, Jhalawar etc. in the state. The production is largely dependent of rainfall pattern (Rajasthan Agriculture Statistics at a Glance 2012-2013).

CONSTRAINTS

The Crop is recently introduced in non-traditional and non-conventional areas of the state. The main constraints are as follows:

1. The potato in Indian plains is grown during winters having short photoperiod (with about 10-11 hours of bright sunshine) and the crop duration is also limited to 90-100 days because of short and mild winter.
2. The winters in mornings usually have fog, which further reduces the sunshine hours posing severe constraints on photosynthetic activity.
3. The post-harvest period consists of long hot summer, which creates storage problems as the tuber re-grow and decay.
4. The quality of underground water available for irrigation is fairly saline and limits its cultivation.
5. The organic matter contents of soils are poor in some places.
6. The winter rains also hinder the normal potato development.
7. The lack of adequate storage facility.

8. Higher cost of cultivation because seed alone contributes 33% to total cost of production.
9. The higher demands of fertilizers and manures limit the profitable production.
10. The crop requires intensive care during crop growing season and at other essential intercultural operations like ridging.
11. The economically poor small and marginal farmers are in difficult position to adopt the crop abundantly due to scarcity in bullock power and steep hike in daily wages of laborers.

STRATEGIES TO OVERCOME THE CONSTRAINTS

1. The development of packages of practices for different agro-climatic zones of the state. The efforts was made, but there is uncertainty of production in the changing climate. The biotic and a biotic stresses, too, need further improvement.
2. In order to harvest a good crop of potato, selection of suitable varieties is one of the most important pre-requisites. The use of varieties that are better adapted to local biotic conditions (e.g. pests and diseases, climatic stress) shall be promoted. The suitable varieties are still lacking to expand the area under cultivation.
3. The potato cultivation has also been mechanized through the fabrication and development of cost-effective tools and implements. These practices lower the production cost and reduce the drudgery. Also shrinkage the duration from preparation to sowing helps in better utilization of resources.
4. The recent advancement in irrigation technology like sprinkler/ drip may help to overcome the problem of saline water.
5. The quality of soils like alkali and salinity may be improved by using green manures, use of gypsum, leaching of salts, sowing on ridges etc.
6. The organic matter content may be increased by incorporating either haulm of the crop or adding organic manures.
7. The timely availability of slow release manures and fertilizer along with their method of application will improve the nutrient use efficiency that will in turn, result in production of higher biomass, partitioned efficiency to improve the harvest index/economic yield.
8. The suppression of diseases by periodic spray of cost effective chemical or locally available plant extracts should be explored.
9. Basic requirements like suitable areas, soils, fertilizers, manures and suitable varieties etc. play a vital role in a successful potato production programme. For instance, a quality crop can be produced in areas and fields which are free from serious soil borne pathogens and pests.

SCOPE TO BOOST POTATO PRODUCTION IN STATE

1. The state follows a variety of farming systems like mono cropping, intercropping, mixed cropping and double cropping. Recent advances in development of short duration high yielding varieties make it suitable for Rajasthan conditions. It will help in increasing existing cropping intensity as the crop produce good yields under well and assured irrigation conditions.
2. Several profitable potato-based inter-cropping and crop rotations have also been identified for different regions of the country. Potato can be profitably intercropped with wheat, mustard and sugarcane. These cropping systems are prevailing in the state and have helped in the maintenance of soil fertility and have improved the fertilizer economy, crop yield and gross returns.
3. There is decrease in the infestation of soil borne insects and pathogens due to use of new varieties resistant to biotic and abiotic stress. The development of photo insensitive geneotypes will also increase the area under cultivation along with wide sowing time which is suited for different phonological phases.
4. The various types of soils existing in the state promote the production of crop in the state.
5. The availability of cold storages in the state also favours the production of potato. It has more than 100 numbers of cold storage facilities with the capacity of about 314747 MT (2008).
6. At present 90% of total cultivated area in the state is under irrigation, and the area under irrigation is increasing day by day.
7. The benefit from prevailing cropping system (cereal based) is diminishing day by day. In these circumstances the cultivation of potato may improve the livelihood of farmer by increasing the net returns.
8. Production cost may be decreased if cooperative efforts are practiced (Inclusion of NREGA activities).

IMPACTS OF CLIMATE CHANGE ON POTATO PRODUCTION

It is likely that the currently observed trend of global warming, which has been $0.6^{0}C \pm 0.2$ since 1900, will continue and that the average global temperature will increase by between 1.4 and $5.8^{0}C$ over the period 1990 to 2100 (Houghton *et al.* 2001). The impact of this type of climate change will probably lead to a decrease in crop productivity, but with important differences between regions. Climate change and global warming is now an acknowledged fact and reality. Global warming is witnessing shifting pattern of rainfall and increasing incidents of extreme weather events like floods, droughts and frosting along with increasing soil salinity and impaired quality of irrigation water. The heat sensitive potato crop is mostly confined to Indo-Gangetic plains under irrigated conditions due to climatic constraints. The climate change and global warming will have a profound effect on potato growth story in India affecting not only production and profitability, but seed multiplication, storage, marketing and processing of this perishable vegetatively propagated crop (Singh et al., 2009). Currently the winters are severe in Western India witnessing frosting in December and January. In future climate scenarios warming may ease the chilling conditions in these regions to favour potato productivity, while in other regions with cooler winter season the warming from current levels may prove detrimental. Under the impact of future scenarios of climate change and global warming the growth projections of potato in India might be arrested or even reversed, unless effective adaptation measures are evolved for timely intervention. The radiation use efficiency (RUE) is suppressed under high temperatures. High temperature reduces tuber number and size (Ewing, 1997).

Specific even within small states and varies appreciably according to local weather conditions, soil and cropping systems. Therefore, a general recommendation to advance or delay in future climate scenarios is impractical. However, adaptation studies on change in date of planting indicate possibility and extent of sustainable potato production in future climate scenarios by modification in date of planting. In Climate change will also affect the distributions and populations of many potato diseases and pests. Potato production must be adapted to climate change to avoid reductions in crop yields. Possible adaptations like change in planting time, breeding heat tolerant varieties, efficient agronomic and water management and shifting cultivation to new and suitable agro-climatic zones can significantly arrest the decline in the production.

Potato plants and potato crop yields are predicted to benefit from increased carbon dioxide concentrations in the atmosphere. The major benefit of increased atmospheric carbon dioxide for potatoes is an increase in their photosynthetic rates which can increase their growth rates. Potato crop yields are also predicted to benefit because potatoes partition more starch to the edible tubers under elevated carbon dioxide levels. Higher levels of atmospheric carbon dioxide also results in potatoes having to open their stomata less to take up an equal amount of carbon dioxide for photosynthesis, which means less water loss through transpiration from stomata. As a result, the water use efficiency (the amount of carbon assimilated per unit water lost) of potato plants is predicted to increase (Haverkort & Verhagen 2008).

AGRO-TECHNIQUES FOR POTATO PRODUCTION

The package of practices for potato production in different agro-climatic zones has helped in improving potato productivity. The crop is input intensive and requires up to date technology for achieving higher productivity. The advanced era of multi-media and communication will provide latest technologies to the farmers. Adoption of correction measures (agro-advisory) will strengthen the production process. Optimum cultural practices depend on delineated phenological phases of crop growth and development viz. pre-emergence, emergence to tuber initiation, tuber initiation to tuber bulking and tuber bulking to termination of bulking (Grewal et al., 1992).

Thus, the cultural practices adjusted in the Indian plains in such a way so that tuber initiation and development coincide with the period when night temperature is less than 20°C and day temperature is below 30°C. The phenological phase of tuber initiation to tuber bulking is mainly conditioned by supply of nutrition and moisture adequately. For this purpose, it's advisable to follow the fertilizer recommendations and irrigation scheduling according to variations in status of agro-climatic zones. The termination of tuber bulking coincides with onset of foliage senescence. It should be delayed for ensuring continuation of linear tuber bulking phase resulting in higher yield by manipulating the nutrition and moisture status. The following improved practices will help in boosting the production in Rajasthan.

Climate

Potato is basically a cool season crop and grows well where sufficient moisture and fertile soil are available. Potato is a long day plant and matures early in plains as compared to hills. In autumn, the potato is produced in sub-tropical plains under short days and irrigated conditions from October to March. In Rajasthan, it is grown during winter season under assured irrigation facilities. It requires favourable environmental conditions such as low temperature and short day conditions at the time of tuberization for rapid bulking rate. The optimum temperature for foliar growth is 18-22°C and for tuberization is 10-16oC. About 20°C temperature is good for tuber formation and it reduces as the temperature increases. Tuberization is badly affected at about 300C temperature (Govindakrishnan, 2013).

Temperature, especially soil temperature also affects the establishment phase. The optimum temperature for emergence has been reported to be between 22-24^0 C (Sale, 1979) and up to this temperature, the emergence rate increases linearly with increase in temperature. At higher temperature, the respiration rate increases and the carbohydrates produced by photosynthesis are consumed rather than stored in tuber. High temperatures at any part of growing period affect the size of leaflets, thereby reducing the tuber formation. These effects can reduce crop yield and the number and weight of tubers. It grows best under long day conditions. At low temperatures potatoes are at risk of frost damage, which can reduce growth and badly damage tubers. Sunshine along with cooler nights is essential for reducing the spread of diseases. The favorable climatic conditions are found in most parts of Rajasthan except arid areas.

Soil

Soil structure and texture has a marked effect on the quality of the tuber. Potato can be produced on almost all types of soils, ranging from sandy loam, silt loam, loam to clay soils. The potato needs friable, well aerated, fairly deep soil well supplied with organic matter. Well- drained sandy loam and medium loam soils are most suitable for successful and profitable potato production. Similarly, light soils are preferred, because they tend to promote more uniform soil temperatures and make harvesting of the crop easier besides uniform development of tubers. Such soils ensure availability of sufficient oxygen for the growth of roots, stolons and tubers, retain moisture and are helpful in drainage of excess water that allows production of beautiful tubers. Alkaline or saline soils are not suitable for potato cultivation.

Field Preparation

The field should be well leveled with adequate drainage as the crop does not thrive on wet and un-drained soil. Plough the fields with a mouldboard plough or disc harrow followed by one or two tilling with a tiller or a desi plough to get the desired tilth. It is better to plank the soil after each round of tilling. Plough the fields during the summer months to keep the land open in May and June. It will reduce incidence of soil borne diseases and control perennial weeds. If the soils are problematic, supply either green manuring during Kharif or add gypsum before planting.

Planting Time

The state varies in climate as well as in soil conditions along with availability of resources, the crop may be planted as follows:

- **Early Crop:** Third week of September to first week of October.
- **Main crop:** First week of October to third week of October.
- **Late Crop:** Third week of October to first week of November

Seed Rate, Methods of Sowing and Spacing

The seed requirements for a hectare vary according the characteristics of cultivars, time of sowing and method of sowing. Tubers having 30 to 50 g weight are the most economical and give the highest yield. Whole tubers should be planted for early crop. This will avoid rotting of tubers. Due to high temperature and moisture in soil, there is always more rotting of cut tubers in early plantings. Large seeds can be effectively used by increasing plant to plant spacing and smaller tubers by decreasing it. In general about 20-25 quintals of seed are sufficient for planting one hectare area but it may vary on the basis of seed size as follows:

Large size- 25-30 q/ha; Medium size- 15-20 q/ha; Small size- 10-15 q/ha; Out tubers- 8-12 q/ha

Potato is planted mainly by two methods:

1. **Ridge and Furrow Method:** In this method, the ridges are prepared. The length of the ridges depends on slope of the plot. Too long ridges and furrows arc not supplied with irrigation water conveniently. The potato tubers are planted into furrows.
2. **Flat Bed Method:** In this method, the whole plot is divided into beds of convenient length and width. The shallow furrows are opened and potato tubers are planted at recommended distance. The tubers are covered with the original soil of furrows. When the germination is completed and plants become 10 to 12 cm height, earthing up operation should be done. Place the seed in the furrows already made for the application of fertilizers.

To get the optimum plant population a row to row spacing of 45-60 cm and between the tubers at 15-20 cm should be maintained in the main crop. Cover the tubers with soil after planting using a ridger. For main crop, cut tubers can be planted. While cutting the tubers, care should be taken that each piece has two to three eyes and weighs at least 25 g. If any diseased tuber is observed, it should be discarded. A row to row distance of 45 to 60 cm and plant to plant 15 to 20 cm should be maintained in the main crop.

Seed Treatment

For planting the crop, the seed potatoes after removing from the cold storage are kept in a cool and shady place for one to two weeks to allow the emergence of sprouts. The sprouted tubers should be used as planting material. Both the whole and cut tubers should be treated with 0.25 per cent Aretan / Tafasan (6% mercury) solution for at least 2 minutes against black scurf disease and rotting of seed potatoes (Sharma 2013). Dipping of cut seed tubers in 0.5% Dithane M-45 for ten minutes is also effective in avoiding rotting in early planting. If the seed was raised from the autumn crop, it is to be used for spring planting its dormancy should be broken before sowing. Dormancy can be broken by treating the potatoes with 1% thiourea (1 kg thiourea in 100 litres water) plus 1 ppm gibberellic acid (1 mg in one litre water) for one hour followed by treatment with 3 per cent ethylene chlorohydrins solution and keeping the tubers in an air tight space for 72 hours.

Varieties

Potato variety for cultivation can be selected according to the soil and climatic conditions as well as market demand and susceptibility to disease like potato blights, which causes maximum damage to the potato crop. The high yielding varieties recommended for North-Western plains having different maturity periods are listed below along with important characters.

- **Early Maturating Varieties (80-90 days):** Kufri Ashoka, Kufri Chandramukhi, Kufri Jawahar, Kufri Khyati, Kufri Pukhraj, Kufri Alankar
- **Medium Maturating Varieties (90-110 days):** Kufri Anand, Kufri Arun, Kufri Chipsona-1, Kufri Chipsona-2, Kufri Chipsona-3, Kufri Pushkar, Kufri Sadabahar, Kufri Sutlej, Kufri Surya, Kufri Jyoti, Kufri Garima, Kufri Gaurav, Kufri Lalima, Kufri Sheetman.
- **Late Maturating Varieties (>110 days):** Kufri Badshah, Kufri Sindhuri, Kufri Bahar

Kufri Alankar

It is especially suited for sandy soils for plains of Western India. It matures 75 days in plains with a potential of 300 q/ha in plains. This variety is having moderate field resistance to late blight with immunity to race 1 and '0' susceptible to common scab.

Kufri Ashoka

It is a wider adaptable variety matures in 70-80 days and has a yield potential of 400q/ha. It is susceptible to late blight and not suitable for processing.

Kufri Badshah

It matures in plains 90-100 days. It is tolerant to frost, resistant to late blight, early blight and potato virus 'X' but susceptible to soft.

Kufri Bahar

It is mid maturity variety (90-100 days). Yield is 250-300 q/ha. Resistant to late blight, early blight and potato virus 'X', 'Y' and leaf roll. It is susceptible to insect pests, drought and frost.

Kufri Chamatkar

It is grown in areas where one crop of long duration is raised. It is late maturing variety, which matures 110-120 days in plains. The yield potential of this variety is 250 q/ha. It is resistant to early blight but susceptible to viruses, late blight, brown rot, charcoal rot, wilts and common scab.

Kufri Chandramukhi

It is mid-season variety, matures 80-90 days in plains. An average yield is 200 q/ha in plains. It is susceptible to common scab, late blight, brown rot, nematodes, charcoal rot and wilts.

Kufri Jawahar

It is an early maturing variety (80-90 days) yielded 400 q/ha. This variety is resistant to late blight. It is not suitable for processing.

Kufri Pukhraj

It is a wider adaptable variety. It is an early maturing variety (70- 90 days). The yield potential is 400 q/ha. It is resistant to early blight and moderately resistant to late blight.

Kufri Sheetman

It was released by Central Variety Release Committee for plains, especially frost affected areas of Punjab, Rajasthan, Haryana and Western Uttar Pradesh. It is resistant to frost and drought. It is susceptible to late blight and leaf roll.

Kufri Sutlej

It is medium maturing (90-100 days) variety yielded 400 q/ha. This variety is moderately resistant to late blight. It is recommended for cultivation in Bihar, Haryana, Madhya Pradesh, Punjab and Uttar Pradesh. This variety has good consumer quality because of easy to cook, waxy texture, and mild flavour and free from discolouration after cooking.

Manures and Fertilizers and Their Management

The status of major nutrients availability varies widely and the requirement of manures or fertilizers depend largely upon the soil type, previous crop history and duration of cultivar. Potato plant is a heavy feeder, so soils poor in organic matter content should be supplied with 250 - 500 q/ha of farmyard manure or compost during land preparation, preferably a fortnight before planting. When it is grown in medium type of soils, it needs 100 to 150 kg nitrogen, 80 to 100 kg phosphorous and 80 to 100 kg potassium per hectare. The crops need highly mobile nutrients likes N throughout the growing period, need to be split as per climatic conditions and soil types. Two - third to three fourth quantity of nitrogen along with whole quantity of phosphorus and potassium is applied at the time of planting. Remaining one fourth to one third of nitrogen is applied 30 to 35 days after planting i.e. at the time of first earthing up or when plants become 25 to 30 cm in height (Dua, 2013). It may be applied either in the form of top dressing or as a foliar feeding. Management of fertilizer phosphorus (P) is a critical component of potato production systems as potato has a relatively high P requirement and inefficiently uses soil P. Phosphorus promotes rapid canopy development, root cell division, tuber set, and starch synthesis. Adequate P is essential for optimizing tuber yield, solids content, nutritional quality, and resistance to some diseases. Most research determined that fertilizer P was most efficiently used when band-applied at planting (e.g., 5 cm to each side of the seed piece). Now –a-days, application of DAP @ 2% as foliar spray is popular and more effective in increasing the efficiency of Phosphorus.

The potato plant needs potassium from early stage of plant growth because of its positive effect on root growth, therefore K application at planting is needed. It is a common practice to surface broadcast K fertilizer before planting with incorporation into the soil during seedbed preparation (Roberts and McDole, 1985). In sandy soils, where K may be lost by leaching, it is recommended to apply K in two splits (half at planting and half at earthing up). This practice may give better results than the entire dose applied at planting (Grewal et al, 1991). Banding potassium fertilizers near the potato seed is the most efficient method of application. However, if a soil test recommends the addition of a large quantity of potassium then it should be split between a pre-plant broadcast and an at-plant banding to avoid potential problems with salt toxicity and/or K leaching.

Spraying of essential micronutrients such as boron, zinc, copper, iron, manganese, molybdenum etc. is done when crop is showing deficiency symptoms. The preferable sources for supply of major nutrients are ammonium sulphate, single supper phosphate and muriate of potash. Fertilizers are applied in the furrows so that the tubers do not come in direct contact with the fertilizers.

Intercultural Operations and Weed Management

In potato, both types of weeds are found i.e. broad-leaved weeds as well as narrow leaved weeds. Weeds compete for nutrients, moisture, light and space and cause considerable loss in potato yields. They also harbour a few pathogens and act as host to a number of insects and pests. Important weeds of potato fields in plains are *Anagallis arvensis, Chenopodium album, Trianthema monogyna, Vicia sativa, Cyperus rotundus, Spergula arvensis, Melilotus spp., and Oxalis spp.* The use of weedicides in potato crop in general is not essential because earthing up operation destroy almost all weeds, if somehow, weed plants are growing on ridges, they may be pulled out by hands .Proper development of tubers depends upon aeration, moisture availability and proper soil temperature. Therefore, proper earthing up is necessary. Earthing should be done when the plants are 10 to 15 cm in height. Generally earthing is done at the time of top dressing of nitrogenous fertilizers. The ridges should be high enough to cover up tubers. If necessary, a second earthing may be done after two -week of the first one. A mould board plough or a ridger may be used for earthing up in large area. Weeds are effectively managed by cultural or chemical methods or combination of both the methods. They are effectively controlled by hoeing and weeding when the crop is about a month old followed by earthing up. Use of mulch helps in conserving soil-moisture, reducing soil-temperature and inducing quick germination in winter. It also suppresses weed growth. Among the herbicides pre-planting application of Fluchloralin and Pendimethalin and pre-emergence application of nitrofen @ 1.0 kg a.i./ha or alachlor @.2.0 kg a.i./ha or post emergence application of propanil @ 1.0 kg a.i./ha may be used in solution form (800-1000 litre/ha) (Lal, 2013). Among post-emergence herbicides, Paraquat at about 5% emergence is quite effective. Care should be taken while spraying of post- emergence herbicides that they should not come in the contact to potato plants. Other pre emergence herbicides are Linuron, Metribuzin, Oxyfluorfen, Ametryn, Simazine, etc., are the most effective herbicides for weed control.

Water Management

Potato (*Solanum tuberosum* L.) is an herbaceous plant with sparse and shallow rooting system, it requires light and frequent irrigation throughout period of crop growth. The plants vary in their water requirements according to their size and growth stage as well as the length of their maturity for maximum growth. Possibly no other major crop varies in its sensitivity to water stress based on growth stage than potato. The water requirement of crop varies from 400-600 mm depending upon soil type/texture, atmospheric conditions, duration of variety, length of growing period, cropping pattern and management practices etc. Drought at any stage can prove detrimental, the excess water is also equally harmful as it creates aeration problem and favours certain diseases and pests. Too little water will create stress making plants susceptible to opportunistic diseases, promote common scab, and drastically reduce yields and increase culls. Excessive water will increase water rots of vines and tubers, and create conditions for late blight infestation. Potatoes need frequent irrigation, especially while tubers are growing. Reduced rainfall in many areas is predicted to increase the need for irrigation of potato crops. As well as reductions in overall rainfall, potato crops also face challenges from changing seasonal rainfall patterns. It should be kept always in mind that the sufficient soil moisture must be available for satisfactory sprouting. Pre-planting irrigation is advantageous for uniform germination. If not then light pre-irrigation or just after planting may be given. Second irrigation is given after about a week and subsequent irrigations as and when required. Light and frequent irrigations are better than heavy and less frequent irrigations. When irrigation water is in short supply, water is applied efficiently and economically at critical stages in crop development, i.e. at stolen formation, tuber initiation and tuber development stages of crop. In North-Western parts of India, where frost is a problem, crops to be irrigated even on alternate days to prevent frost damage. The last irrigation to the main crop could be timed 10-14 days before the haulms die down or are killed deliberately. It is desirable to withhold the last irrigation a few days before harvest of crop to allow firming of tuber skin. Lifting the crop when the soil is too moist impairs the transportability and keeping quality of the produce. As regards method of irrigation in potato, the furrow method is commonly followed.

The rate of water use is low till 30-35 days after planting; it means that the first irrigation is essentially done within 30-35 days after planting. However, when soil moisture seems insufficient for sprouting, intervals of first irrigation should be reduced. Further, irrigation is done as and when crop needs. Among various irrigation methods viz. surface, sub-surface, drip and sprinkler methods of irrigation, the furrow irrigation is commonly used for potato cultivation. Under the circumstances (prevailing in western India) where water is poor in quality and is a very scarce resource, the potato equivalent yield was significantly higher under drip method of irrigation along with 40% water saving followed by sprinkler irrigation that is a climate smart technique (Singh et al., 2005). The studies based on drip irrigation approach, revealed that the irrigation scheduled at 125-150% CPE water level (cumulative pan evaporation) at alternate day were needed for best yield of potatoes. The drip method economized 40-50% water and increased crop yield by 25-35% along with saving of 25% NPK fertilizer nutrients in comparison to furrow method of irrigation (Singh, 2005).

Harvesting, Yield, and Storage

Harvested potatoes are heaped under shade for a couple of days, so that their skin becomes hard and soil adhering with them is also separated out. Under good crop management, 350-450 quintals of marketable potatoes of good quality can be produced from one hectare land. The sorting operation is the most important, in that all cut tubers, bruised, injured by insects-pest and disease are removed. Sorted healthy tubers are graded into different grades based on diameter to fetch better prices in the market. Oversized tubers are great in demand for chips making. Very small sized tubers are also not remaining unsold. These tubers are purchased by poor people for making vegetable by partially crushing them before cooking. However, both the over sized and under sized are quite unsuitable for seed purposes. Potatoes can be stored in the cold storage at the temperatures of 4 to 7^0C.

FUTURE RESEARCH NEED

The research recommendations through on-farm testing must be applied for hopefully adoption by farmers. The knowledge gained in basic research can be used in practical ways in devising new or modified cultural practices. Conducting research into more biological aspects using the latest research techniques is employed. Under the influence of climate change, it is very important to change the planting pattern according to prevailing climate conditions. The diversification in cropping systems will helps to grower to grow more potato by manipulating the traditional planting techniques. The role of remote sensing and weather forecasting will definitely help to sustain the production also. The impact of crop museum (varietal or other recent technology display) at different units like KVK, Research station and state agricultural university farms and the frequent farmers visit will helps in diversion of mindset for sustained potato production. An emphasis towards ringing down cost of cultivation and increasing output prices will definitely turn the growers more and more potato. Additional room for growth and opportunity for potato farmers lie in alternative sectors – such as processing and packaging as potential drivers for potato industry growth.

CONCLUSION

It may be concluded that the diversity in the state will increase the cultivation of potato under changing climate scenario. As the potato is most popular and low cost food, most prime among the needs is making awareness and popularization of technology because the growers mindset is revolving around other food crops most popular in the region. As the farmers of the state are poor in resources along with poor economic in conditions, they have very low investing capacity. The financial support along with low cost storage availability helps the farming community. In this context, the role of state agriculture department is vital in boosting up the GDP of the nation with a view to uplift livelihood status of them. The vast scope of post-harvest value addition will open the doors for industries and helps in generation of employments throughout the year.

REFERENCES

Agricultural Statistics at a Glance. (2014). Government of India, Ministry of Agriculture, Department of Agriculture and Cooperation, Directorate of Economics and Statistics.

Dua, V. K. (2013). *Nutrient management in potato production*. Compendium on Summer School on Advances in Quality Potato Production and Post-harvest Management.

Ewing, E. E. (1997). Potato. In H. C. Wien (Ed.), *The Physiology of Vegetable crops* (pp. 295–344). Wallingford, UK: CAB International.

Govindakrishnan, P. M. (2013). Climate requirement for ware & seed potato cultivation. Compendium on Summer School on Advances in Quality Potato Production and Post-harvest Management, Shimla, India.

Grewal, J. S., Sharma, R. C., & Saini, S. S. (1992). *Agro techniques for intensive potato cultivation in India*. New Delhi: ICAR Publication.

Grewal, J. S., Trehan, S. P., & Sharma, R. C. (1991). *Phosphorus and potassium nutrition of potato*. Technical Bulletin No. 31. Central Potato Research Institute.

Haverkort, A. J., & Verhagen, A. (2008). Climate Change and Its Repercussions for the Potato Supply Chain. *Potato Research, 51*(3-4), 223–237. doi:10.1007/s11540-008-9107-0

Houghton, J. T., Ding, Y., Griggs, D. J., Noguer, M., van der Linden, P. J., Xiaosu, D., … Johnson, C. A. (Eds.). (2001). *Climate Change 2001. The Scientific Basis*. Contribution of Working Group I to the Third Assessment Report of the Intergovernmental Panel on Climate Change (IPCC). Cambridge University Press. Retrieved January 09, 2016 from http://cpri.ernet.in/

Lal, S. S. (2013). *Potato based cropping systems and weed management in potato production*. Compendium on Summer School on Advances in Quality Potato Production and Post-harvest Management, Shimla, India.

Rajasthan Agriculture Statistics at a Glance. (2012-13). Commissionerate of Agriculture, Rajasthan, Jaipur (Statistical Cell).

Roberts, S., & Mc Dole, R. E. (1985). Potassium Nutrition of Potatoes. In R. S. Munson (Ed.), *Potassium in Agriculture* (pp. 800–818). Madison, WI: ASA-CSSA-SSSA.

Sale, P. J. M. (1979). Growth of potatoes (*Solanum tuberosum* L) to the small tuber stage as related to soil temperature. *Australian Journal of Agricultural Research, 30*, 667–675.

Sharma, A. K. (2013). *Practices for Quality Seed Potato Production*. Compendium on Summer School on Advances in Quality Potato Production and Post-harvest Management, Shimla, India.

Sharma, S. (2013). *Fungal diseases of potato and their management*. Compendium on Summer School on Advances in Quality Potato Production and Post-harvest Management, Shimla, India.

Singh, B. (2013). *Potato scenario- past, present and future*. Compendium on Summer School on Advances in Quality Potato Production and Post-harvest Management, Shimla, India.

Singh, J. P., Lal, S. S., & Pandey, S. K. (2009). Effect of climate change on potato production in India. Central Potato Research Institute, Shimla, Newsletter, 40, 17-18.

Singh, N. (2013). *Micro irrigation system and water management in potato production*. Compendium on Summer School on Advances in Quality Potato Production and Post-harvest Management, Shimla, India.

Singh, N., Sood, M. C., & Lal, S. S. (2005). Evaluation of potato based cropping sequences under drip, sprinkler and furrow methods of irrigation. *Potato Journal, 32*(3-4), 175–176. Retrieved from WWW.agriexchange.apeda.gov.in/production/

Chapter 8
Phytotoxicity of Oxidised and Reduced Nitrogen Aerosols on Potato (*Solanum Tuberosum L.*) Crop

Bhagawan Bharali
Assam Agricultural University, India

Bhupendra Haloi
Assam Agricultural University, India

Zafar Ullah
Assam Agricultural University, India

Jayashree Chutia
Assam Agricultural University, India

Sonbeer Chack
Assam Agricultural University, India

ABSTRACT

In a field trial (2012), simulated aerosols: NH_4Cl (reduced) and $NaNO_2$ (oxidised) @ 10 & 20 kg $ha^{-1}y^{-1}$ (≈ 100 ppm & ≈ 200 ppm respectively), 1000 cm^3m^{-2} of each along with a control were misted to population of Kufri Jyoti at different growth stages viz., vegetative (10-60 DAS), tuber initiation (60-90 DAS) and tuber bulking >90DAS). The higher dose of aerosols lowered nitrate reductase activity, nitrogen use efficiency, cell membrane stability, tuber yield, but increased photosynthesis, peroxidise activity significantly. The mechanisms of injury in terms of higher peroxidase activity and lower membrane stability of leaf cells have been elucidated. Foliar feeding of nitrogenous pollutant in the form of aerosols to plants at juvenile stage is important in addition to basal use of recommended fertilizers.

DOI: 10.4018/978-1-5225-1715-3.ch008

INTRODUCTION

Potato (*Solanum tuberosum* L.) is one of the finest sources of starch, vitamins, minerals and dietary fiber. 100 g of it provides 70 calories, however, they contain very little fat (just 0.1 g per100 g) and no cholesterol. Potatos are very good natural sources of both soluble and insoluble fiber. The dietary fiber in them increases the bulk of the stool, thus, it helps prevent constipation, decreases absorption of dietary cholesterol and thereby, lowers plasma LDL cholesterol. Additionally, its rich fiber content also helps protect from colon polyps and cancer. The tubers are one of the richest sources of B-complex group of vitamins such as pyridoxine (vitamin B6), thiamin, niacin, pantothenic acid and folates. Fresh potato along with its skin is one of a good source of antioxidant vitamin, vitamin-C. 100 g of fresh tuber provides 11.4 mg or 20% of daily required levels of this vitamin. They also contain adequate amounts of many essential minerals like iron, manganese, magnesium, phosphorous, copper and potassium. Red potato contains good amount vitamin A, and antioxidant flavonoids like carotenes and zeaxanthins (USDA, National Nutrient Database revised May, 2016).

On the other hand, global warming and climate change are the facts relating the observed century-scale rise in the average temperature of the Earth's climate system, and its impacts on environments. The rate of global warming in last 50 years is double than that for the last century. As many as 11 of the past 12 years were warmest since 1850, when records began. The threshold value of temperature rise is 2^0C for devastating, dangerous and irreversible consequences of warming to manifest world over. Global warming is witnessing shifting pattern of rainfall and increasing incidents of extreme weather events like floods, droughts and frosting along with increasing soil salinity and impaired quality of irrigation water. The current level of CO_2 (369 ppm) in the atmosphere, the main GHG is 35.4% more than the pre-industrial level, and is rising unabatedly. The CO_2 level is predicted to be 393, 543 and 789 ppm in year 2020, 2050 and 2080 respectively. In potato crop, a C_3 group of plant, yields are presumed to benefit from increased carbon dioxide concentrations in the atmosphere by an increase in their photosynthetic rates, and more starch to the edible tubers under elevated carbon dioxide levels (Haverkort & Verhagen, 2008). Potatoes grow best under temperate conditions (Hijmans, 2003). For instance, the corresponding rise in temperature would be 1, 3 and 5 ^{0}C approximately during main potato growing winter season in India. The climate change and global warming will have a profound effect on potato growth story in India affecting not only production and profitability, but seed multiplication, storage, marketing and processing of this perishable vegetatively propagated crop (Singh *et al.*, 2009). India produces about 24 million tonnes of potato from 1.32 million hectares under the crop. The Northern plains contribute about 84% of the total produce. Under the impact of future scenarios of climate change and global warming, the growth projections of potato in India might be arrested or even reversed, unless effective adaptation measures are evolved for timely intervention.

Now question rises whether air pollution, specifically particulate matter aerosols affect global warming. Air pollution occurs when the air contains gases, dust, fumes or odor in harmful amounts—aerosols are a subset of air pollution that refers to the tiny particles suspended everywhere in our atmosphere. These particles can be both solid and liquid and are collectively referred to as 'atmospheric aerosol particles'. Most are produced by natural processes such as erupting volcanoes, and some are from human industrial and agricultural activities. In clear air, particles of sizes of approximately 0.1 to 1 micron (comparable to the dominant wavelengths in the solar spectrum) interact with the solar beam. Particles containing little carbon are effectively 'white.' They reflect solar radiation, making the air and Earth surface below them a bit cooler than they would otherwise be. In contrast, particles containing substantial amounts of black carbon (e. g., soot, which is typically produced from combustion of fossil fuels, biofuels, and biomass burning) warm their surroundings by absorbing solar radiation before it reaches the ground (Alpert & Kischka, 2008; Ramanathan & Carmichael. 2008). When water vapor clings to water soluble particles in the same size range (~0.1 to 1 micron) it creates cloud droplets in the lower troposphere. In clean air, the concentrations of these droplets range from 10 to several 100 per cubic centimeter. At lower temperatures certain aerosol particles facilitate the formation of cloud ice. In and near urban areas, where the concentration of aerosol particles is high, the concentration of droplets can be as high as several thousand per centimeter cubed. The increased number of little drops causes the reflectivity of clouds to increase, so that clouds near polluted areas are often brighter than those above cleaner regions. Water droplets and ice particles are basically white, so they reflect solar radiation; on the other hand, the condensed water also traps and emits long wave radiation, producing heat. Thus clouds can have either cooling or warming effects on a local area, depending on whether the reflecting or trapping effect is strongest. Because of many unknowns relating to aerosol particles, and in particular, to the possible effects of particles on cloud stability, the magnitude of aerosol impacts on climate remains among the most uncertain factors in climate projections (IPCC, 2007).

Chameides *et al.* (1998) described that the fine particle typically reside in atmosphere from days to weeks, they may be transported over thousands of kilometers before being removed. As a result, large regions of the globe with sufficient industrial activity and urbanization and / or can be covered by a contiguous layer of air containing enhanced concentrations of fine particles. Under the appropriate meteorological conditions, the affected areas can extend over 106 square kilometers or more. The fine particles has also the characteristic ability to affect the flux of solar radiation passing through the atmosphere directly by scattering and absorbing solar radiation, and indirectly, by acting as cloud condensation nuclei, and thereby influencing the optical properties of clouds. The two characteristics collectively lead to the phenomenon known as haze (reduction in visibility), and reduction in flux of solar radiation reaching the earth's surface over large geographic areas. The magnitude of diminishing surface solar visible radiation may be 30% or more. However, the magnitude of indirect effect is more uncertain than the direct effects on reducing the amount of solar radiation reaching the earth's surface. The effects of aerosols on atmospheric temperature and climate are studied (IPCC, 1995). Considerable research has already been carried out on the effects of air pollutants on crop yields and the mechanisms by which these effects are induced. Moreover, significant reductions of solar radiation that can occur as a result of regional haze, reduces crop yield, as it is sensitive to the amount of sunlight they receive. Assessment of direct impact of N-aerosols on potato crop is lacking in Northeast India.

As concerned to the sources of nitrogen, the responses of potato crop to rate of soil applied nitrogen have been studied extensively elsewhere (Meyer & Marcum, 1998; Bellanger *et al.*, 2000; Rodrigues *et al.*, 2005; Sparrow & Chapman, 2003). The crop may be deprived of nitrogen due to immobilization in the soil, it's gaseous (NH_3, N_2) emission to the environment or leaching of NO_3^- (Paulo *et al.*, 2010). On the other way, nitrogen oxides (NO, NO_2, called NO_x) and N_2O have increased in importance as atmospheric pollutants in rapidly growing urban and its surrounding areas (Bharali *et al.*, 2012). The NOx emitted practically as nitric oxide (NO), reacts rapidly in the atmosphere, and in a complex cycle with light, ozone, hydrocarbons, produces nitric acid. These materials may interact with plants and soil locally or be transported from the site, and interact with atmospheric particulates to form aerosols. These salts and aerosols return to fertilize terrestrial and aquatic systems in wet and dry deposition. A small fraction of this N may biologically converted to N_2O. About 5% of the total anthropogenic greenhouse effect is attributed to N_2O from which 70% of annual global anthropogenic emissions come from animals and crop production (Arvind, 2001). As a gas phase, on entry into the large airspace and surface areas of mesophyll cells, the oxides of nitrogen dissolve into the extra cellular water of the sub-stomatal cavity, form weak acids (i.e. nitrous and nitric acids), which then dissociate to form nitrate, nitrite and protons (Zeevart, 1976). The nitrogen oxides at concentration $<10\ \mu ll^{-1}$ disrupt many physiological processes including net photosynthesis and yields (Darrall, 1989). The deposition of gaseous NH_3 and particulate NH_4^+ (reduced nitrogen: collectively NH_y) may also acidify ecosystem (Mohan and Kumar, 1998; McClean *et. al.* 2011). The reduced nitrogen at high concentration (>1mM) causes toxicity in plants (Mehrer & Mohr, 1989). The possibilities of nitrogen nutrition by foliar feeding with aerosols of nitrogen pertaining to the growth periods, dose response of potato crop and its productivity are explored in the present investigation for future climate change scenarios elsewhere.

MATERIALS AND METHODS

Field Culture

Potato genotype (*Kufri Jyoti*) was collected from the AICRP on potato at Assam Agricultural University, Jorhat. The recommended doses of NPK fertilizers (@ 60:50:50 per hectare as Urea, SSP and MOP) were applied as basal before sowing of seeds of potato. The experiments were laid in Randomised Block Design and replicated thrice. Earthling-up at sixty days after sowing and other prophylactic measures to prevent diseases were undertaken timely.

Preparation of the Nitrogenous Aerosols and Their Application

In field, aerosols of nitrogen (1000 cm^3 stage wise) viz., NH_4Cl @ 10 kg $ha^{-1}y^{-1}$ (≈100 ppm) and $NaNO_2$ @ 20 kg $ha^{-1}y^{-1}$ (≈200 ppm) along distilled water (control) were misted to population of the crop (variety: *Kufri Jyoti*) at the three different growth stages viz., 10-60 days after sowing (DAS): vegetative (stage 1), 60-90 DAS: tuber initiation (stage 2), >90 DAS: tuber bulking (stage 3). Care was taken using a hard board as partition to avoid drifting of the solutions from one plot to another while spraying.

Incubation of Plants and Measurement of Photosynthesis

We used transparent, airtight acrylic assimilation chambers (60x60x60 cm^3). Temperature (28-30°C), Relative humidity (50-65%) and Light intensity (44-70 K lux) prevailing inside the assimilation chambers during measurements of net photosynthesis were recorded by Hygrometer and Light meter. The plants were incubated in presence of normal ambient carbon dioxide concentration (380 ppm) inside the chambers during mid-day under sunlight for half an hour. Air samples (10 cm^3) were collected by clinical syringe through the rubber port of the chambers, and injected into the Environmental Gas Monitor (EGM-4) for measuring the carbon dioxide concentration after incubation. The rate of net photosynthesis was expressed as ppm CO_2 absorbed per gram plant dry weight per hour as suggested by Larson and Karsaw, 1975 as follows:

(a). CO_2 concentration (C) of the chamber at the end of the incubation period

C (ppm)= (PSxCCY)/PCY

Where, PS= Peak height given by chamber sample

PCY= Peak height given by the known ambient CO_2 concentration (base line concentration)

CCY= Known CO_2 concentration (ppm) of the ambient air

Rate of CO_2 absorption, R= ppm CO_2 g^{-1} dw h^{-1} = (CCY-C)x (V/1000)x(60/T)x(1/dw)

Where, V= Chamber volume (cm^3) used in the CO_2 measurement

T= Incubation time in minutes

dw= Dry weight of the plants (g)

In Vivo Estimation of Nitrate Reductase (NR) Activity in Plants

The NR estimation was based on conversion of nitrate to nitrite and inhibition of nitrite reduction to ammonia in anaerobic condition (Srivastava and Ormrod, 1984). Green leaf samples (300 mg) of 10-15 mm square were put into 2.5 ml each of solutions containing 200 mM phosphate buffer (pH 7.5), 30 mM $KNO_{3,}$ 5% (v/v) propanol in assay tubes. The tubes were incubated in a water bath at 30°C for 30 minutes, incubated further for 2 minutes in 100°C and allowed them to cool to room temperature. To detect nitrite in the assay tubes, color development reagent i.e. 1 ml each of 1% sulfanilamide in 1N HCl and 0.02% N-(1-naphthyl)-ethylenediaminedihydrochloride were added to the solution. Mixed thoroughly and placed it in the dark at room temperature for 15 min. To determine the NR activity, the absorbency readings obtained at 540 nm in spectrophotometer were plotted on a standard curve, which was prepared from a stock of 25 n mol nitrite per ml using KNO_2 in water.

In Vivo Estimation of Peroxidase Activity in Plants

The level of lipid peroxidation was measured in terms of Malondialdehyde content (MDA), a product of lipid peroxidation following the method of Heath and Packer (1968). The leaf sample of 0.5 g was homogenized in 10 ml of 0.1 per cent trichloroacetic acid (TCA). The homogenate was centrifuged at 15,000g for 5 minutes. Two milliliter of aliquot of the supernatant and 4 ml of 0.5% thiobarbuteric acid (TBA) in 20 per cent of TCA were mixed. The mixture was heated at 95°C for 30 minutes and cooled in ice bath. It was centrifuged at 10,000g for 5 minutes, and the absorbance of supernatant was recorded at 532 nm. The value for non-specific absorption at 600 nm was subtracted from the value of 532 nm. The absorption coefficient of 155 n mol per cm was used to calculate MDA content as: MDA (n mol per g fresh weight) = (OD x 6)/0.155 x volume extract / (2 x weight of sample)

Nitrogen Use Efficiency (NUE)

NUE in potato leaf at different growth stages under treatment were calculated from their per cent Nitrogen and dry weight as NUE % in leaf = (Nitrogen % in leaf x total leaf dry weight per plant). Total Nitrogen content was determined by Kjeldhal method, which is based on catalytic conversion of organic nitrogen into ammonia and its subsequent estimation by acid base titration (Yoshida *et al.,* 1976).

Measure of Membrane Permeability

Cell Membrane stability (CMS) is a measure of changes in membrane permeability due to cellular injury caused by the external agents as suggested by Sullivan and Rose (1972), and it was calculated corresponding to the experimental treatments. Twenty pieces of young leaves from the treated plants were cut into about one centimeter square and immersed them first into 20 cm^3 distilled water taken in plastic bottles (60 ml capacity). The mouth of the bottles was closed tightly to avoid leaking of the solution and checked gently using magnetic stirrer. Thus, freely water soluble ions in intercellular spaces of leaf were removed by the three serial washes with distilled water (each 10 min, 20 cm^3). Now, to extract the cations present in the exchangeable sites of the cellular locations, the same leaf discs were eluted by two treatments (each 1 h 20 cm^3) with 25 mM Sr_2Cl (Bharali & Bates 2002). The solutions were collected into other plastic bottles. The plant samples were oven dried at 60°C and weighed. The electrical conductivity readings of these solutions against the samples collected from the experimental treatments were used to compute CMS as follows.

CMS = [1-(1-T1/T2)/(1-C1/C2)x100]/ Dry weight of leaf samples

Where,

T_1= Conductivity reading of 25mM Sr_2Cl (20 cm^3) without leaf samples

T_2= Conductivity readings of 25mM Sr_2Cl (20 cm^3) with leaf samples

C_1= Conductivity reading of double distilled water without leaf samples

C_2= Conductivity readings of double distilled water with leaf samples

Economic Yield

Tubers on maturity of the crop under respective treatments at different growth stages were harvested and tones recorded as per hectare.

Statistical Analysis

Data were analysed following Generalized Linear Model (GLIM) program of Royal Society of London (Crawley, 1995). Significant differences between two mean values due to treatments or varieties and their interaction at a crop growth stage were computed by comparing their significant levels at $P < 0.05$.

RESULTS

Net Photosynthesis (P^n)

In the field trial, there were significant effects of the aerosol treatments and their time of application on P^n rates of potato crop (Figure 1). The ammonium chloride and sodium nitrite increased P^n rates by 16.85%, 29.31% respectively as compared to control. The oxidised aerosol increased 14.99% more P^n rate than the reduced aerosol. Concerning the time of application of the aerosols, higher P^n rates were obtained at stage 1(45.15%) > stage 3 (36.94%) > stage 2.

Nitrate Reductase Activity (NR)

In field, there was significant variation in NR activity due to the stages of application of the aerosols, but not for the concentration of the aerosols (Figure 2). NR activity in vegetative stage was 42.71% higher than in the tuber initiation stage and it was 23.8% more in the tuber bulking stage following application of the aerosols.

Nitrogen Use Efficiency (NUE)

Leaf nitrogen use efficiency varied significantly due to the time of aerosol application but not for the treatments (Figure 3). In general, the NUE per cent was lowered in ammonium chloride (by 10.13%) and sodium nitrite (by 4.88%) treatments as compared to control. The NUE% was the highest at Stage3 followed

Figure 1. Net photosynthesis of potato (Variety: Kufri jyoti) following treatments at different days after sowing (DAS)

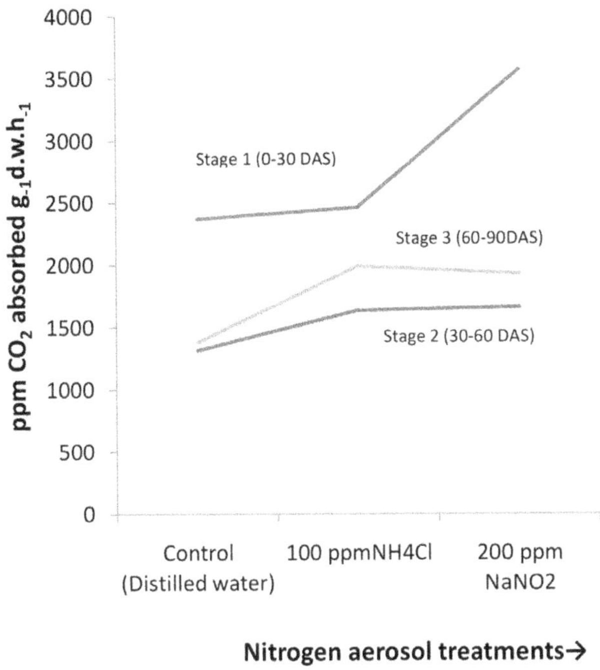

Figure 2. Nitrate reductase activity of potato leaf (Variety: Kufri jyoti) at different days after sowing (DAS)

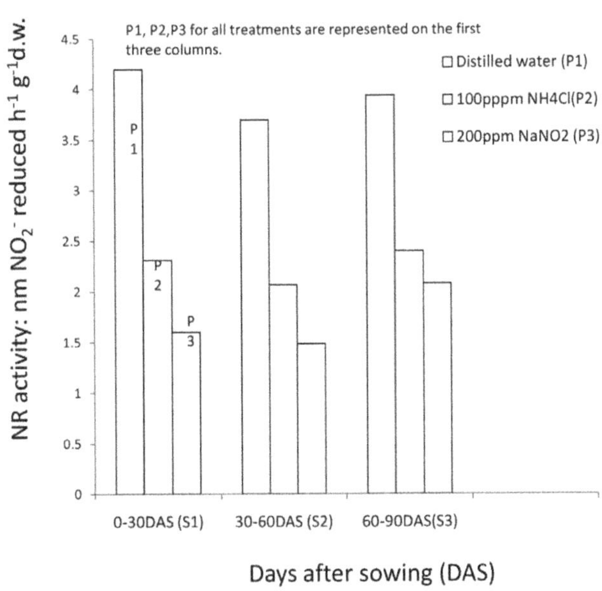

Figure 3. Tuber nitrogen use efficiency of potato (Variety: Kufri jyoti) following nitrogen aerosol treatments at different days after sowing (DAS)

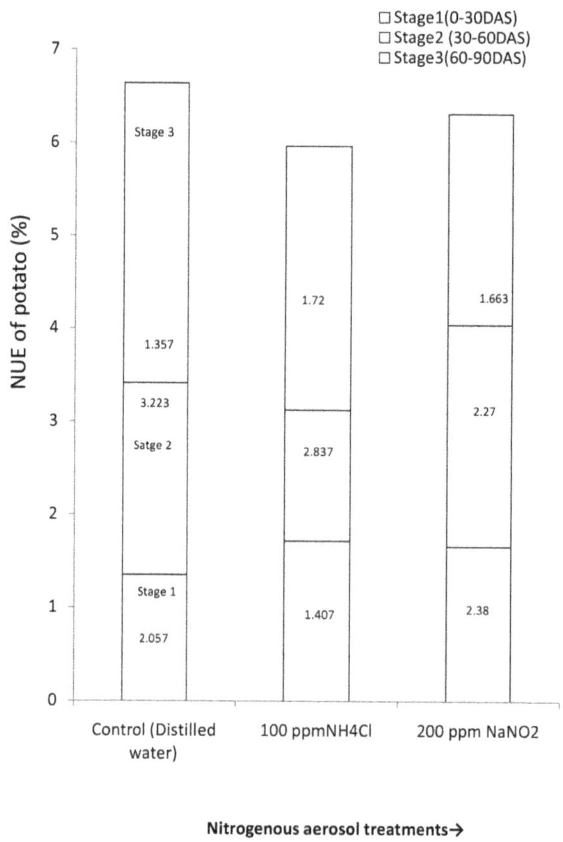

by Stage2 > Stage1 by 6.79%, 40.45% respectively. Lower was the dose of the aerosols; higher was the beneficial impact of the pollutants.

Cell Membrane Stability (CMS)

The nitrogen aerosol treatments and their time of application changed CMS significantly (Table 1). Cell membrane stability was significantly altered by the aerosol treatments and their time of application. The decrease in CMS was lesser in case of the reduced aerosol (13.77%) than the oxidized one (38.39%) as compared to control. The membrane stability was higher with the application of the aerosols at Stage1 (13.63%) > Stage3 (38.80%) > Stage2.

Table 1. Variation of Cell membrane permeability (CMS) of potato crop (Variety: Kufri Jyoti) following treatments with reduced (NH_4Cl) and oxidised nitrogen ($NaNO_2$) at different days after sowing (DAS)

Treatments (T)→ Stages (S)↓	Control (Distilled water)	100 ppm NH_4Cl	200 ppm $NaNO_2$
S1: 0-30 DAS	-1.503	-1.619	-2.656
S2: 30-60 DAS	-4.086	-2.700	-4.146
S3: 60-90 DAS	-2.840	-0.8740	-2.976
		SEDiff (±)	CD (0.05)
T		0.332	0.730
S		0.640	1.394
T x S		-	Not significant

Peroxidase Activity (PO)

Both the aerosol treatment and the application time had significant effects on the peroxidise activity of potato crop (Table 2). The PO activity was the lowest following treatments at the vegetative stage (by 6.62%), tuber initiation stage (by 43.47%) than at the tuber bulking stage as compared to control. The oxidised aerosol increased 43.47% more PO activity than the reduced nitrogen. The application of the aerosols at earlier stages caused lower PO activity at Stage 1(63.67%) < Stage 2(27.65%) < Stage 3.

Table 2. Variation of Peroxidase (PO) activity of potato crop (Variety:Kufri Jyoti) following treatments with reduced (NH_4Cl) and oxidised nitrogen ($NaNO_2$) at different days after sowing (DAS)

PO Activity (nmol Meldialdehyde g^{-1} f.w.)			
Treatments (T)→ Stages (S)↓	Control (Distilled water)	100 ppm NH_4Cl	200 ppm $NaNO_2$
S1: 0-30 DAS	35.81	35.81	37.74
S2: 30-60 DAS	83.23	89.03	128.71
S3: 60-90 DAS	44.52	50.32	122.90
		SEDiff (±)	CD (0.05)
T		25.17	54.85
S		19.96	43.92
T x S		16.86	38.14

Economic Yield

Although the aerosol treatments had no significant effects (Figure 4), but their application time impacted upon tuber yield of potato crop at stage 1(10.60%) > stage2 (25.90%) > stage 3. So, earlier the application of the aerosols, better was the tuber production. In general, there were 17.56% and 3.52% reductions in yield of tuber by the reduced and the oxidised aerosols as compared to control respectively.

DISCUSSION

The present study deals with the impacts of nitrogenous aerosols viz., ammonium chloride (reduced aerosol) and sodium nitrite (oxidised aerosol) on some physiological parameters (e.g. net photosynthesis, nitrate reductase activity, nitrogen use

Figure 4. Tuber yield of potato (Variety: Kufri jyoti) following nitrogen aerosol treatments at different days after sowing (DAS)

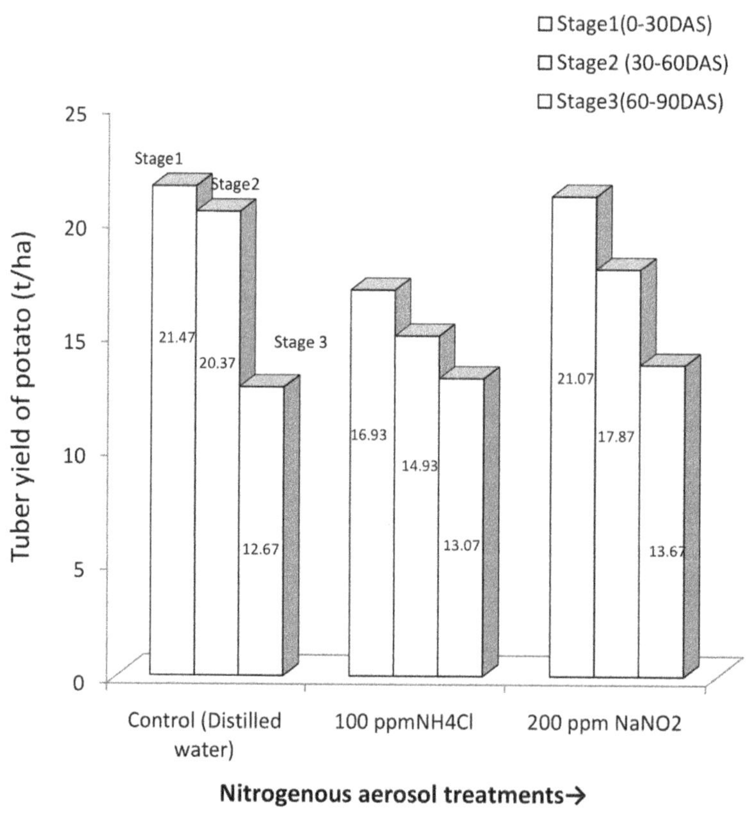

efficiency, cell membrane stability, peroxidase activity, and tuber yield of potato). The mechanism(s) of injury to the crop by the nitrogenous compounds are discussed in this paper. The question whether the nitrogenous aerosols may be phytotoxic or potential fertilizer for potato crop with respect to the time of its application is answered.

The aerosols influenced rates of net photosynthesis in the field grown potato at all the growth stages. The rate was higher commensuration with the nitrogen dose of oxidized aerosol ($NaNO_2$). The key enzyme of carbon assimilation (ribulose-1-5-bisphosphate carboxylase/oxygenase) is pH dependent. Perhaps, in potato crop the number of protons in the chloroplast was not allowed to exceed than required (six) for NO_2 reduction (Heldt *et. al.*, 1986). The aerosols might not cause ultra-structural changes (protrusion, swelling of thylakoids) and P^n depression in plants (Lopata and Ulrich, 1975; Wellburn *et al.,* 1972). Should sodium and chloride ions enter into cells, they were neglected as non-physiological at relatively alkaline pH (> 5.0) of the aerosols in our work.

In general, Nitrate reductase activity in potato leaf was lowered by ammonium chloride and sodium nitrite. Nitrate reductase catalyzes the reduction of nitrate to nitrite, and its levels of activity are determined by the supply of nitrate. Inhibition of NR activity may be crop specific also. The NR activity is substrate induced (Beevers & Hegeman, 1969). Barley mutants exposed to a concentration of $0.3 \mu l l^{-1}$ for 9d in a nutrient culture experiment, had lower NR activity (Rowland *et al.*, 1989). Squash cotyledons on exposure to high levels of NO_2, inhibition of NR is caused by accumulating larger amount of ammonium ions and certain amino acids in squash cotyledons during fumigation (Hisamatsu *et al.*, 1988).

In the present study, nitrogen use efficiency (NUE) in potato crop depleted with aerosol treatments. It decreased in leaf with higher dose of nitrogen. The later stages had exhibited higher NUE% as the nitrogen content decreased in leaf probably with mobilization of it towards the developing tuber. Plant receiving low nitrogen has higher NUE than plants with high nitrogen deposition rates. The differences between accessions in the response to N for physiological and phonological variables exists in case of Arabidopsis lyrata petrae (Vergeer *et al.*, 2008). Therefore, the potato variety having high NUE with lower dose of nitrogen (e.g. 100 ppm ammonium chloride) might have faster growth and higher turnover rates. Bharali *et al., 2015ab* also found that NUE in wheat grains decreased with the higher concentration of nitrogen aerosols in both pot culture and in field experiments. Moreover, nitrogen nutrition enhanced grain yield of wheat varieties irrespective of their growth stages. The enhancement is more prominent in $NaNO_2$ than NH_4Cl fed plants. Higher nitrogen use efficiency with lower quantity of nitrogen from the source might improve nutritive quality of the crop varieties (Prasad and Rao, 1980).

In the current study, the aerosol treatments lowered cell membrane stability in potato crop. The oxidised aerosol lessened CMS more than the reduced aerosol. The fall of CMS was higher in the later growth stages of aerosol application. The oxides of nitrogen causes lipid breakdown in membrane, induces cellular plasmolysis and lowers CMS (Pryor & Lightsey, 1981; Bharali *et al., 2015ab*. Exposure of plants to ammonium chloride reduces inter cellular and exchangeable cations viz., calcium, magnesium, and potassium in plants (Boxman *et al.,* 1991). Calcium being integral component of membrane helps maintain CMS (Legge *et al.*, 1982; Bharali & Bates 2004). The plant cell may become vulnerable to solute leakage due to the ammonia attack on the membrane.

In a recent study on wheat Bharali *et al.* (2015b) stated that Cell membrane permeability was increased (with lower CMS) by both NH_4Cl and $NaNO_2$. As NH_4Cl treated plants had higher leakage of ions from the cells, they possessed higher quantum of the cations in the intercellular and exchangeable sites. Similarly, $NaNO_2$ treated plants had higher CMS and lower membrane leakage than NH_4Cl treated plants, a lower amount of the cations were recovered from the cellular locations. The rate of Peroxidase activity of wheat crop treated with $NaNO_2$ was higher than the rate shown by the NH_4Cl treatment as compared to the control. Therefore, the membrane damages caused by NH_4Cl and $NaNO_2$ were brought by two different mechanisms. The former depleted the cations from the membrane directly and the later caused peroxidation of lipids present in the membrane. Hence, the membrane became leaky for the cations, and their quantum was higher in the intercellular and exchangeable sites irrespective of varieties, which were detected in the extraction processes with water and $SrCl_2$ solutions respectively. Although, nitrite causes swelling of thylakoids and changes membrane stability, direct interference of free radicals with critical enzymes (Murray and Wellborn, 1985; Welburn, 1990), may be responsible for reduction in growth and yield of crops. The oxides of nitrogen following the lipid breakdown in membrane cause cellular plasmolysis (Pryor and Lightsey, 1981). Apart from uncoupling electron transport chain in chloroplast (Lilley *et al.,* 1975), ammonia reduces cations viz., calcium, magnesium, and potassium (Boxman *et al.,* 1991). In plant cells, calcium is one of the integral components of plasma membrane, where it helps maintain stability (Legge *et al.,* 1982). Calcium ions binds with modulator proteins e.g. calmodulin (Dieter, 1984), and serves as chemical signaling that in some cases equips the plant to resist external stresses (Bharali & Bates, 2004). These possibilities have not been explored meticulously in the present studies.

In the study, nitrogen nutrition enhanced tuber yield of potato crop with earlier application of the aerosols at vegetative or tuber initiation stage. Higher nitrogen use efficiency with lower quantity of nitrogen from the source might improve nutritive quality of the crop varieties (Prasad & Rao, 1980). Nitrogen aerosols brought significant changes in the peroxidise activity of potato crop in respect of dose and time of their application. Higher lipid peroxidation of membrane with higher dose of aerosol applied at later growth stages caused cellular disruption. It perhaps prevented photo assimilates (accumulated by higher P^n) from effective participation through impaired symplast. As NO^- is a reactive nitrogen species, it causes cell injury (Nathan, 1995). One of the fastest reactions of NO^- within biological systems is the combination with ROS, a process that is well described as either toxic or protective. At high NO^-donor concentrations (e.g. $0{\cdot}5^{-1}$ mM sodium nitroprusside), NO is generally toxic. In these conditions, when combined with low amounts of superoxide anion (O_2-), the formation of peroxynitrite ($ONOO^-$) was reported to be deleterious to lipids, proteins and DNA (Lipton *et al.,* 1993).

FUTURE RESEARCHES

Several climate engineering (so-called 'geoengineering') strategies are important for reducing global warming propose using atmospheric aerosol particles to reflect the sun's energy away from Earth. Because, aerosol particles do not stay in the atmosphere for very long—and global warming gases stay in the atmosphere for decades to centuries—accumulated heat-trapping gases, will overpower any temporary cooling due to short-lived aerosol particles (Latham, 1990).

In cultivation of potato, the optimum date of planting is highly location specific even within small states and varies appreciably according to local weather conditions, soil and cropping systems. However, adaptation studies on change indicate possibility of extension of sustainable potato production in future climate scenarios by modification of its date of planting by 5-10 days in general to escape cold temperature effects on the crop (Singh *et al.,* 2009).

One of the cumulative strategies is the biodiversity which can mitigate climate change by conservation of ecosystems with high carbon stocks, and establishment of agro-ecosystems sequestering more carbon. Exploration of genes in the existing wild relatives of crop plants, transfer of genes by conventional breeding or non-conventional approaches like genetic engineering, somatic hybridization etc. may protect from the climate change impacts on vegetable crops like potato (Hightower *et al.,* 1991; Rambabu *et al.,* 2016).

Significant reductions of solar radiation can occur as a result of regional haze. Crop yield is sensitive to the amount of sunlight they receive. Assessment of direct impact of different aerosols and regional haze on crop yields are lacking. The expected output are 'crop yield model' pertaining to regional haze, crop sensitivity to aerosols, Phenotypes of crop germplasm, Crop cultivars suitable for cultivating under aerosol deposited and demining conditions are some of the promising future research areas in the context of climate change in the present decades.

CONCLUSION

Considering all the physiological events together, it can be inferred that the oxidised and reduced nitrogen play dual roles in potato crop in respect of dose and application time. Therefore, washing of the aerosols from polluted air and their application at lower doses during the earlier growth stages may be implicated as alternative fertilizer for potato crop production. Otherwise, they become phytotoxic inducing more peroxidation of membrane, lowering CMS and reducing tuber yield in potato crop.

ACKNOWLEDGMENT

The experimental work on potato (substitute to cereal) was a part of the Ad-hoc research project entitled 'Atmospheric deposition of Nitrogen and productivity of winter cereal crops in North-east India'. It was funded by the Ministry of Environment & Forests, Govt. of India. We acknowledge all facilities received from Assam Agricultural University, Jorhat for smooth running of the scheme.

REFERENCES

Alpert, P., & Kischka, P. (2008). Quantification of the effect of urbanization on solar dimming. *Geophysical Research Letters*, *35*(8), L08801. doi:10.1029/2007GL033012

Arvind, R. M. (2001). Exchange of gaseous nitrogen compounds between agricultural systems and the atmosphere. *Plant and Soil*, *228*, 17–27. doi:10.1023/A:1004821205442

Beevers, L., & Hegeman, R. H. (1969). Nitrate reduction in higher plants. *Annual Review of Plant Physiology*, *20*(1), 495–522. doi:10.1146/annurev.pp.20.060169.002431

Belanger, G., Walsh, T. R., Richards, J. E., Milburn, P. H., & Ziadi, N. (2000). Yield responses of two potato cultivars to supplemental irrigation and N fertilizer in New Brunwick. *American Journal of Potato Research*, *77*(1), 11–21. doi:10.1007/BF02853657

Bharali, B., & Bates, J. W. (2002). Soil cations influence Bryophyte susceptibility to bisulphyte. *Annals of Botany*, *90*(3), 337–343. doi:10.1093/aob/mcf192 PMID:12234145

Bharali, B., & Bates, J. W. (2004). Influences of extracellular calcium and iron on membrane sensitivity to bisulphiote in the mosses *Pleurozium schreberi* and *Rhytidiadelphus triquetrus*. *Journal of Bryology*, *26*(1), 53–59. doi:10.1179/jbr.2004.26.1.53

Bharali, B., Haloi, B., Chutia, J., Chack, S., & Hazarika, K. (2015a). Susceptibility of some wheat (*Triticum aestivum* L.) varieties to aerosols of oxidised and reduced Nitrogen. *Advances in Crop Science and Technology*, *3*, 4. doi:10.4172/2329-8863.1000182

Bharali, B., Haloi, B., Chutia, J., & Dey, P. C. (2012). *Atmospheric deposition of Nitrogen in North-east India and productivity of winter cereals.* In The 5th International Conference on Water, Climate and Environment, Ohrid, Macedonia. Retrieved June 5, 2016, from http//www. BALWOIS.Com

Bharali, B., Haloi, B., Dey, P. C., Chutia, J., Chack, S., & Hazarika, K. (2015a). Roles of oxidized and reduced nitrogen aerosols on productivity of winter rice (*Oryza sativa* L.). *Intl. J. of Agric. and Environment*, *1*, 1–15.

Boxman, A. W., Krabbendam, H., Bellemakers, M. J. S., & Roelofs, G. M. (1991). Effects of ammonium and aluminium on the development and nutrition of *Pinus nigra* in hydroculture. *Environmental Pollution*, *73*(2), 119–136. doi:10.1016/0269-7491(91)90018-R PMID:15092085

Chameides, W.L., Yu, H., Liu, S. C., Bergin, M., Zhou, M., Meanrns, L., ... Giorg, F. (1998). Case study of the effects on agriculture: An opportunity to enhance crop yields on China through emission controls. *Inaugural articles*: *National Academy of Science*.

Crawley, M. J. (1995). *Methods in Ecology: GLIM for Ecologists*. London: Oxford Blackwell Scientific Publications.

Darrall, N. M. (1989). The effects of air pollutants on physiological processes in plants. *Plant, Cell & Environment*, *12*(1), 1–30. doi:10.1111/j.1365-3040.1989.tb01913.x

Dieter, P. (1984). Calmodulin and calmodulin-mediated processes in plants. *Plant, Cell & Environment*, *7*(6), 371–380. doi:10.1111/j.1365-3040.1984.tb01426.x

Haverkort, A. J. & Verhagen, A. (2008). Climate Change and Its Repercussions for the Potato Supply Chain. *Potato Research, 51*, 223–237. doi:10.1007/s11540-008-9107-0

Heath, L. R., & Packer, L. (1968). Photo-oxidation in isolated chloroplast. I kinetics and stochiomentry of fatty acid peroxidation. *Archives of Biochemistry and Biophysics*, *125*, 189–198. doi:10.1016/0003-9861(68)90654-1 PMID:5655425

Heldt, H.W., Flügge, U.I. and Stitt, M. (1986). Kohlenhydratstoffwechsel der pflanzlichen photosynthese. Biologie in unserer Zeitschrift, 16: 97-105.

Hightower, R., Badon, L., Penzes, E., Lund, P., & Dunsumuir, F. (1991). Expression of antifreeze proteins in transgenic plants. *Plant Molecular Biology*, *17*(5), 1013–1021. doi:10.1007/BF00037141 PMID:1932678

Hijmans, R. J. (2003). The Effect of Climate Change on Global Potato Production. *American Journal of Potato Research, 80,* 271–280. doi:10.1007/bf02855363

Hisamatsu, S., Nihira, J., Takeuchi, Y., Satoh, S., & Kondo, N. (1988). NO_2 suppression of light-induced nitrate reductase in squash cotyledons. *Plant & Cell Physiology*, *29*, 395–401.

IPCC. (1995) Climate Change 1994. Cambridge University Press.

IPCC. (2007). The Physical Science Basis. In S. Solomon., D. Qin, M. Manning, Z. Chen, M. Marquis, K. B. Averyt, M. Tignor, & H. L. Miller (Eds.), Contribution of Working Group I to the Fourth Assessment Report. Cambridge University Press.

Larsson, D., & Kershaw, K. A. (1975). Measurement of CO_2 exchangeable in lichens, a new method. *Canadian Journal of Botany*, *53*(15), 1535–1541. doi:10.1139/b75-184

Latham, J. (1990). Control of Global warming? *Nature*, *347*(6291), 339–340. doi:10.1038/347339b0 PMID:2215647

Legge, R. L., Thompson, J. E., Baker, J. E., & Lieberman, M. (1982). The effect of calcium on the fluidity of phase properties of microsomal membranes isolated from postclimacteric golden delicious apples. *Plant & Cell Physiology*, *23*, 161–169.

Lilley, R. M., Fitzgerald, M. P., Rienits, K. G., & Walker, D. A. (1975). Criteria of intactness and the photosynthetic activity of spinach chloroplast preparations. *The New Phytologist*, *75*(1), 1–10. doi:10.1111/j.1469-8137.1975.tb01365.x

Lipton, S. A., Yun-Beom, C. H., Pan, Z. H., Le, S. Z., Vincent Chen, H. S., & Sucher, N. J. (1993). A redox-based mechanism for the neuroprotective and neurodestructive effects of nitric oxide and related nitroso-compounds. *Nature*, *364*(6438), 626–632. doi:10.1038/364626a0 PMID:8394509

Lopata, W. D., & Ullrich, H. (1975). Untersuchungen zu stoffichen und strukturellen Veränderungen an Pflanzen under NO_2-Einfluss. *Staub Reinhaltung der Luft*, *35*, 196–200.

Mc Clean, C. J., van den Berg, L. J. L., Ashmore, M. R., & Preston, C. D. (2011). Atmospheric nitrogen deposition explains patterns of plant species loss. *Global Change Biology*, *17*(9), 2882–2892. doi:10.1111/j.1365-2486.2011.02462.x

Mehree, I., & Mohr, H. (1989). Ammonium toxicity:description of the syndrome in Sinapsis alba and the search for its causation. *Physiologia Plantarum*, *77*(4), 545–554. doi:10.1111/j.1399-3054.1989.tb05390.x

Meyer, R. D., & Marcum, D. B. (1998). Potato yield, petiole nitrogen and soil nitrogen response to water and nitrogen. *Agronomy Journal*, *90*(3), 420–429. doi:10.2134/agronj1998.00021962009000030017x

Mohan, M., & Kumar, S. (1998). Review of Acid rain potential in India: Future threats and remedial measures. *Current Science*, *75*(6), 579–593.

Murray, A. J., & Wellborn, A. R. (1985). Differences in nitrogen metabolism between cultivars of tomato and pepper during exposure to glasshouse atmospheres containing oxides of nitrogen. *Environmental Pollution*, *39*(4), 303–316. doi:10.1016/0143-1471(85)90112-6

Nathan, C. (1995). Natural resistance and nitric oxide. *Cell*, *82*(6), 873–876. doi:10.1016/0092-8674(95)90019-5 PMID:7553846

Paulo, C. F. S., Heder, B., Camilo, B., & Paulo, R. C. (2010). Economic optimum nitrogen fertilizer rates and nitrogen fertilizer rate effects on tuber characteristics of potato cultivars. *Potato Research*, *53*(3), 167–179. doi:10.1007/s11540-010-9160-3

Prasad, B. J., & Rao, D. N. (1980). Alteration in metabolic pools of nitrogen dioxide exposed wheat plants. *Indian Journal of Experimental Biology*, *18*, 879–882.

Pryor, W. A., & Lightsey, J. W. (1981). Mechanisms of NO_2 reaction: Initiation of lipid peroxidates and the production of nitrous acid. *Science*, *214*, 435–437. doi:10.1126/science.214.4519.435 PMID:17730242

Ramanathan, V., & Carmichael, G. (2008). Global and regional climate changes due to black carbon. *Nature Geoscience*, *1*(4), 221–227. doi:10.1038/ngeo156

Rambabu, E., Chandini, R., & Pradeepthi, L. (2016). Climate change challenges in vegetables and their adaptations through bio-diversity. *Readers Shelf, 12*, 9–10.

Rodrigues, M. A., Coutinho, J., Mertins, F., & Arrobas, M. (2005). Quantitative sidedress nitrogen recommendations for potatoes based upon crop nutritional indices. *European Journal of Agronomy, 23*(1), 79–88. doi:10.1016/j.eja.2004.10.001

Rowland-Bamford, A. J., Lea, P. J., & Wellburn, A. R. (1989). NO_2 flux into leaves of nitrite reductase deficient barley mutants and corresponding changes in nitrate reductase activity. *Environmental and Experimental Botany, 29*(4), 439–444. doi:10.1016/0098-8472(89)90022-1

Singh, J. P., Lal, S. S., & Pandey, S. K. (2009). Effects of climate change on potato production in India. Central Potato Research Institute. *Shimla Newsletter, 40*, 17–18.

Sparrow, L. A., & Chapman, K. S. R. (2003). Effects of nitrogen fertilizer on potato (Solanum tuberosum L.) in Jasmanica 1. Yield and quality. *Australian Journal of Experimental Agriculture, 43*, 631–641. doi:10.1071/EA02099

Srivastava, H. S., & Ormrod, D. P. (1984). Effect of nitrogen dioxide and nitrate nutrition on growth and assimilation in bean leaves. *Plant Physiology, 76*(2), 418–423. doi:10.1104/pp.76.2.418 PMID:16663857

Sullivan, C. Y., & Rose, M. W. (1972). Selection for draught and resistance in grain sorghum In H. Mussel & R. Staples (Eds.), Stress Physiology in Crop Plants (pp. 263-289). Wiley.

Vergeer, P., Van Den Berg, L. L. J., Bulling, M. T., Ashmore, M. R., & Kunin, W. E. (2008). Geographical variation in the response to nitrogen deposition in Arabidopsis lyrata petraea. *The New Phytologist, 179*(1), 129–141. doi:10.1111/j.1469-8137.2008.02445.x PMID:18422899

Wellborn, A. R. (1990). Why atmospheric oxides of nitrogen usually phytotoxic and not alternative fertilizers? *The New Phytologist, 115*, 395–429. doi:10.1111/j.1469-8137.1990.tb00467.x

Wellborn, A. R., Mayernik, O., & Wellborn, F. (1972). Effect of SO_2 and NO_2 polluted air upon the ultrastructure of chloroplasts. *Environmental Pollution, 3*(1), 37–49. doi:10.1016/0013-9327(72)90016-X

Yoshida, S., & Douglas, A., Forno, J., Cock Kwanchai, A., & Gomez, A. (1976). *Laboratory Manual for Physiological Studies of Rice*. Intl. Rice Res. Inst.

Zeevaart, A. Z. (1976). Some effects of fumigating plants for short periods with NO_2. *Environmental Pollution, 11*(2), 97–108. doi:10.1016/0013-9327(76)90022-7

Chapter 9
Issues of Climate Change, Impact, and Adaptation Strategies in Nigeria

Nwakor Flora Ngozi
NRCRI Umudike, Nigeria

Okwusi Moses Chukwunwike
NRCRI Umudike, Nigeria

Amadi C. Okey
NRCRI Umudike, Nigeria

Adiele Ezekiel Chinyere
NRCRI Umudike, Nigeria

ABSTRACT

Climate change is a global problem affecting agricultural production, a good adaptation strategy for this phenomena should be sought for increase agricultural production. The study was conducted in Nigeria to assess the Impact of Climate Change on root and tuber crops production among farmers in Nigeria. Secondary data were used for the study, they were collected from NRCRI Umudike and other individual publications. The result shows that climate change had negative impact on root and tubers crops production including potato. Adaptation of Agriculture to climate change in the areas of crop and animal production, post harvest activities and capacity building, divers friction of livelihood sources through the use of different farming methods and improved agricultural practices will help to reduce the impact of climate change. Examples are establishment of forestry, generation of improved and disease resistance crop varieties addition of value into agricultural products and post harvest activities for climate change adaptation and sustainable development.

DOI: 10.4018/978-1-5225-1715-3.ch009

INTRODUCTION

Climate commonly refers to the average weather conditions of a particular place/location/country/region of the world over a long period of time, usually thirty years or more with regard to climatic elements such as temperature, rainfall, air pressure, etc.(Obioha, 2002) while climate change refers to a change in the state of the climate that can be identified (e.g., by using statistical tests)by changes in the mean and/or the variability of its properties, and that persists for an extended period, typically decades or longer (IPCC, 2014). In other words climate change means any systematic change in the long-term statistics of these climatic elements sustained over several decades or longer. Climate Change may result from factors such as changes in orbital elements (eccentricity, obliquity of the ecliptic, precession of equinoxes), natural internal processes of the climate system or anthropogenic forcing (for example, increasing concentration of carbon dioxide and other greenhouse gases) (Agbola & Ijeleye., 2007).

Gaseous emissions from human activities are substantially increasing the concentrations of atmospheric greenhouse gases like Carbon Dioxide, Methane, Chlorofluorocarbon and Nitrous Oxide and this global warming has significant impacts on agriculture through the interaction of these elements (Wikipedia, 2016). The earth's average surface temperature has increased by 1 degree Fahrenheit just over the last century as a result of greenhouse emissions (Fischer et al, 2002). However, the rising concentrations of greenhouse gases (GHGs) in the earth's atmosphere, resulting from both economic and demographic growth since the industrial revolution are overriding the equilibrium for natural variability, leading to potentially irreversible climate change making GHG's accounting for about 55% in the intensity of the greenhouse effect. Since the industrial revolution mid-1700s activities that increased the concentration of CO_2 in the atmosphere increased in scale and distribution. Natural ecosystems like forests & wetlands permanently hold up to 20 to 100 times more of CO_2 /unit area for very long periods (carbon sink) because agricultural crops are harvested and carbon released back into the atmosphere. Anaerobic conditions in paddy rice flooding, grazing animals and termites release Methane (CH_4) to the environment as a result of herbaceous digestion.Also, gaseous emissions from human activities such as burning of fossil fuels, coal mines, gas and oil drilling through production, bush clearing including deforestation and animal rearing, result into emissions of GHG.

ANNUAL GREENHOUSE GAS EMISSIONS BY SECTORS

Annual GHG emission by sectors is shown in Figure 1(Wikipedia, 2016). Power stations (21.3%) and Industrial processes (16.8%) were the highest culprits while waste disposal and treatment was the least.

Figure 1. Figure showing the relative fraction of man-made greenhouse gases coming from each of eight categories of sources, as estimated by the Emission Database for Global Atmospheric Research version 3.2, fast track 2000 project
Source: Wikipedia, 2016

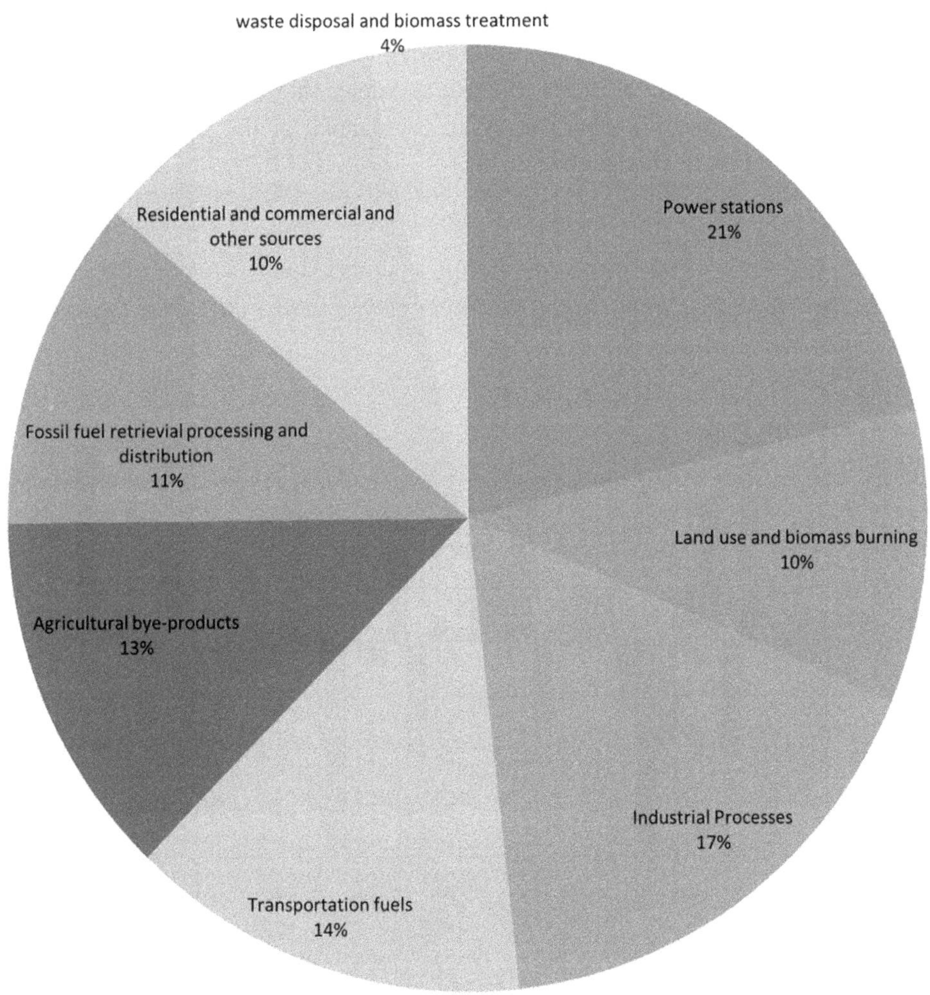

In the long run, as gleaned from various reports (Heagle et al. 1983; Ludlow & Smit, 1987; Smit et al 1988) climate change could affect crop production in three main ways

- Effect on crop yield by alteration of the crop's immediate growing environments.
- Effect on area and location of lands suitable for agricultural production.
- Effect on multiple agricultural activities and on the functioning of the agriculture food sector, including prices, trade pattern, and employment.

Climate Variability depicted in Figure 2 is any deviation in the long-term statistics of climate elements over a short period of time: diurnal, seasonal, year-to-year and decade-to-decade differences in climate but not up to 30 years. All key sectors of the economy are already being affected by climate variability. Whilst African agriculture has always had to, and continues to, adapt to changing environmental circumstances (including climate), there is a danger that this adaptation with its focus on climate variability, will not take cognizance of the trends imposed by anthropogenic climate change (SEI, 2008)

Figure 2. Variability: Fluctuations of weather elements/variables from the average e.g. temperature, rainfall etc
Source:SEI, 2008

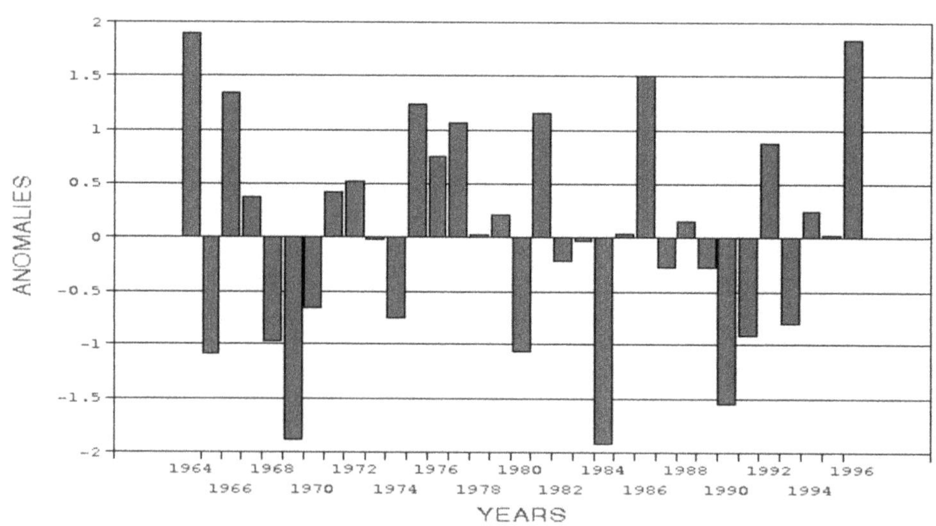

CLIMATE CHANGE AND NIGERIA AGRICULTURE

Climate change exerts multiple stresses on the biophysical as well as the social and institutional environments that underpin agricultural production (IPCC, 2007). It is perhaps the most serious environmental threat to the fight against hunger, malnutrition, disease and poverty in Africa, mainly through its impact on agricultural productivity (Enete & Amusa, 2010).It is projected that crop yield in Africa may fall by 10-20% by 2050 or even up to 50% due to climate change (Jones & Thornton, 2002), particularly because African agriculture is predominantly rain-fed and hence fundamentally dependent on the vagaries of weather. Climate change is expected to present a high tended risk, new combinations of risks and potentially grave consequences, particularly in Africa due to its direct dependence on rain-fed agriculture as noted above (Enete & Amusa, 2010),

Nigeria is one of the developing countries (Odjugo, 2010). As the planet warms, rainfall patterns shift, and extreme events such as droughts, floods, and forest fires become more frequent (Zoellick, 2009), which results in poor and unpredictable yields, thereby making farmers more vulnerable. Agriculture is the main source of food and employment in Nigeria with over seventy percent of the population earning their livelihood from the sector and therefore anything that affects agricultural sector affects every other sector.

The climate of Nigeria is threatened by global warming and the key sectors affected by this climate change include Health, Agriculture, Water Resources, Forestry, Ecosystem and Tourism (Adejuwon, 2006. Increased incidences of some diseases like malaria, schistosomiasis, trypanosomiasis, yellow fever, cholera and respiratory infections. Decreased production as a result of floods and droughts, increased soil erosion, reduced arable land, disruption of the cropping calendar and conflict between farmers and pastoralists are constraints to agricultural production (Adejuwon, 2006)., Variations in water availability, impact on power generation, depletion of underground water reserves, frequent flooding in flood prone areas, increased water pollution, reduction in the biodiversity in tropical forests which could lead to migration of species, loss of important medicinal and gene resources, reduction in the moderating impacts of forests on climate, loss of regeneration capacity of forests putting entire forest ecosystems at risk, destruction or damage to wetlands and estuaries, loss of wildlife habitats which will affect tourism (Ngaira, et.al. 2009, Watson, 2005) are climatic deteriorations that can cause devastating social and economic problems to all the sectors. Despite technological advances such as improved crop varieties, genetically modified organisms and irrigation systems, weather is still a key factor in agricultural productivity (Fischer et al., 2002, Wikipedia, 2010). There has also been an impact on agriculture in sub-Saharan Africa generally seen as prolonged drought which leads to crop failure, hence resulting in food insecurity. At most, we have a 30 year window of opportunity to deal with the threats of climate change because most of the farmers in Nigeria already experience a variety of stress and shocks on a regular basis. In this sense, the impacts of climate change are nothing new but the scale and in some situations, the nature of the impacts will change dramatically as the pace of change increases. Small-scale farmers who are dependent on low-input and low output rain-fed mixed farming with traditional technologies dominate the agricultural sector. The government has given top priority to this sector and has taken steps to increase crop productivity. However, various problems are encountered such as poor crop production due to declining farm size; subsistence farming because of population growth; land degradation due to inappropriate use of land, such as cultivation of steep slopes; over cultivation and overgrazing; and inappropriate farming polices. Other problems include land tenure insecurity; weak agricultural research and extension services; lack of agricultural marketing; an inadequate transportation network; low-in use of fertilizers, inadequate improved seeds varieties and pesticides; and the use of traditional farm implements. However, the major cause of under production is drought, which often causes famine, flood and total crop failure (Fakihri & Sombroek, 2010).

IMPACT OF CLIMATE CHANGE ON ROOT AND TUBER CROPS

Root and tuber crops especially cassava, yam, cocoyam, potato and sweet potatoes are among major food crops grown in Nigeria. Statistics has shown that those who engaged in arable crop production suffer from higher rainfall variability than other agricultural sectors (Rikko, 2010). There are uncertainties in the onset of the farming season due to changes in pattern of rainfall, early rain may not be sustained and crop planted at their instance may become smothered by heat waves that can lead to unusual sequence of crop planting and replanting which may result to food shortage due to harvest crop failure (Mendelsohn et al., 2000). Crops like yam may survive heavy sunshine in the field but may be highly affected during harvesting and storage, whereas crops like cocoyam, potato would be both highly affected in the field and in the storage as a result of heavy sunshine and increased temperature. Cassava as a root crop does not like much flood which can easily result to root rot diseases caused by fungi. However crop yield projection from the second half of the year shows much lower yield due to high temperature. Rural and peasant farmers are particularly at this risk from failing crop yield (Adejumo, 2005).Floods across Africa are reported to be the worst in decades in some places and extend in an arc from Mauritania in the west to Kenya in the east. They have affected over 17 countries in Africa including Uganda, Kenya, Ghana, Sudan, Zambia, Nigeria, Burkina Faso, Togo, Rwanda, Egypt (in the coastal zone especially on the Red Sea and Sinai coasts) and Ethiopia among others. Floods displaced hundreds of people, e.g. in Nigeria in 2011, it destroyed infrastructures such as roads, bridges, thus limiting access of agricultural produce to markets, wide spread crop failure, food insecurity and famine (AU, 2008; Sidi, 2012). The population of Nigeria is projected to increase by more than 50 percent in the coming two decades (FAO 2001). During this 20 year period, the rural population is projected to increase by more than 25 percent, and the agricultural component is expected to grow by a slightly lower proportion, moderated by climate change and under capitalization of the smallholder farmers. Perhaps it is with this projected scenario in mind that Davidson et al. (2003) noted that the food security threat posed by climate change is greatest for Africa, where agricultural yields and per capita food production have been steadily declining, and where population growth will double the demand for food, water and forage in the next 30 years.

CLIMATE CHANGE AND POTATO PRODUCTION IN NIGERIA

Potato (*Solanumtu berosum* L) is one of the seasonal tuber crops grown in the temperate zones all over the world, but primarily in the northern hemisphere. The crop originated in the tropical highlands of the Andes of South America (Horton and Anderson, 1992) and was introduced into Nigeria by the European tin miners in Jos Plateau in the later part of the 19th and 20th century (Ifenkwe, 1981). Jos and Mambilla plateaus are the major areas known for potato production because of the relatively cool temperature of the environment. About 85% of the potato grown in Nigeria comes from Jos plateau state (Okonkwo et al., 1995). Potato is by far the most efficient tuber crop in term of tuber yields and days to maturity. It matures in about 80-90 days (3 months) as to compare to other tuber crops like cassava and yam (Amadi, 2011). Thus potato becomes available at a critical time in the food supply cycle of the populace. Optimum temperatures for tuber initiation and root growth is 15°C – 20°C while halm growth is at optimum when temperatures are between 20°C-25°C . High temperature subjects potato plants to heat stress which include changes in physiological and biochemical processes including photosynthesis, dark respiration, enzyme activities, membrane stability, and ultimately affects their productivity (Amadi, 2011). Temperatures above 30°C can have a range of negative effects on potato, including:

- Slowing tuber growth and initiation.
- Less partitioning of starch to the tubers.
- Physiological damage to tubers (e.g. brown spots).
- Shortened/non-existent tuber dormancy, making tubers sprout too early.

Rainfall is another climatic element that affects crop yield. Jos plateau an annual rainfall of 146 mm.The rainfall pattern of Jos plateau state showed that rains generally start late April or early May each year and stop in October. Rainfall is heavy between July and early September and in this period soil is frequently waterlogged and unsuitable for potato production. As a result of climate change there has been variability in the pattern of rainfall. There was erratic change in rainfall pattern between 1996 and 2000, there was gradual decline in amount of rainfall from the year 2000, and also from 1996 the relative humidity moves downward and progressively continued to decline. Another declining trends start again in march 2004. All these changes have negative impact on potato production. For example high rainfall, relative humidity and temperature encourage the spread of late blight and bacteria wilt disease of potato (Lenka et.al., 2008) this has the tendency to decrease potato production.

Potatoes are more sensitive to soil water deficits compared to other crops such as wheat, and need frequent irrigation especially while the tubers are growing. Reduced rainfall in many areas is predicted to increase the need for irrigation of potato crops. Apart from reduced volume of rainfall, potato crops also face challenges from changing seasonal rainfall patterns. Shortage of rainfall in recent decades has resulted in a shorter potato growing season. There are early maturing varieties of potato which are able to escape late blight which attacks most potato varieties from mid-July each year; these varieties include Br 63-18., Roslin Ruaka and Nicola (Okonkwo et al., 2009). There are dry season varieties which are susceptible to late blight and bacterial wilt which occur more in rainy season, these include Bertitia, Cardinal, Rc767-2, and Diamant. As a result of climate change and variability in the pattern of rainfall in Nigeria, there is problem in potato production because rainy and dry season cannot be predicted again; there had been increase in temperature over the years which is not favorable for potato production. Climate variability has contributed to low output of potato in recent time as a result of late blight disease, bacteria wilt, over flooding and high temperature. Potato production must be adapted to climate change to avoid reductions in crop yields.

Table 1 , show the impact of climate change on individual farming activities among farmers. The result indicates that climate change affected root and tuber crops production activities right from the period of planting to the periods of harvesting and storage. Majority of root and tuber crops were affected most during storage, followed by planting period. It was observed that yam, cocoyam and sweet potato were highly affected by climate change at storage level whereas cassava was affected mainly at the planting stage of operation.

Table 1. Impact of climate change on root and tuber crops production activities

	Crops							
Activity	**Cassava**	**Yam**	**Sweetpotato**	**Ginger**	**Cocoyam**	**Cassava itc**	**Yam ITC***	**Mean**
Planting	166 (40.3)	16 (5.5)	24(8.3)	12 (4.2)	08(2.8)	96 (33.3)	80 (27.8)	1.2
Weeding	36(12.5)	72(25)	- -	- -	20(6.9)	32 (11.1)	48 (16.7)	0.7
Fert. Application	16(5.5)	16 (5.5)	- -	- -	24(8.3)	24 (8.3)	12(4.2)	0.2
Harvesting	64(22.2)	48 (16.7)	- -	04 (1.4)	40(13.9)	60 (20.8)	48 (16.7)	0.9
Processing	24(8.3)	8(2.8)	12 (4.2)	52 (18.0)	72(25.0)	108 (37.5)	84(29.5)	1.0
Storage	56(19.4)	140 (48.6)	100 (34.7)	60 (20.8)	140 (48.6)	48 (16.7)	24 (8.5)	**2.0**

*Itc= intercrops. Figures in brackets are percentages
Source: Nwakor et al., 2013

CLIMATE CHANGE ADAPTATION IN NIGERIA AGRICULTURE

Adaptation is the adjustment of natural or human systems in response to current or expected climate change (or to its effects), to moderate negative consequences and take advantage of any opportunities (IPCC, 2001).Adaptation to climate change includes activities that are taken before impacts are observed (anticipatory) and after impacts have been felt (reactive).Adaptation can also be planned or autonomous. As Ifeanyi-obi et al. (2012) rightly explained, autonomous adaptation refers to reaction of farmers to changing precipitation patterns, in that he/she changes crops, uses different harvest and planting/sowing dates while planned adaptation measures are conscious policy options or response strategies, often multi-sectorial in nature and aimed at altering the adaptive capacity of the agricultural system of facilitating specific adaptations. Most agricultural systems have a measure of in-built adaptation capacity ("autonomous adaptation") but the current rapid rate of climate change will impose new and potentially overwhelming pressures on existing adaptation capacity(Agoumi,2003). This is particularly true given that the secondary changes induced by climate change are expected to undermine the ability of people and ecosystems to cope with, and recover from, extreme climate events and other natural hazards. It is for this reason that the IPCC encourages "planned adaptation", that is deliberate steps aimed at creating the capacity to cope with climate change impacts (IPCC, 2007). Effective adaptation strategies and actions should aim to secure well-being in the face of climate variability, climate change and a wide variety of difficult to predict biophysical and social contingencies.

Adapting potato production to climate change, shifting growing areas, improving water capture, use of irrigation and use of improved potato cultivars should be adopted by farmers. In pursuing this aim, climate change adaptation should focus on support for the decision-making and capacity building processes that shape social learning, technology transfer, innovation and development pathways. Adaptation is most relevant when it influences decisions that exist irrespective of climate change, but which have longer-term consequences (Stainforth et al., 2007). A key component of climate change adaptation involves building resilience, where resilience is the capacity of a system to tolerate disturbance without collapsing into a qualitatively different state that is controlled by a different set of processes: a resilient system can withstand shocks and rebuild itself when necessary. Over sixty per cent of Africans remain directly dependent on agriculture and natural resources for their well-being (FAO, 2003). Agriculture is highly dependent on climate variability (Salinger et al., 2005) which is why the threat of climate change is particularly urgent in Africa (Boko et al, 2007).Despite the reliance on large proportions of the population on agriculture, agricultural development has historically not been a priority of governments, with 1% or less of the average national budgets going to agriculture (FAO, 2003). However, many donors and NGOs have supported agriculture across the continent because of this reliance on agriculture and the potential to improve yields. Flooding is a major climate stressor which leads to secondary impacts of health and destruction of livelihoods. Communities are resilient if the household capacity, neighborhood setting and livelihood strategies would permit adjustment to flooding shock and hazards. Adaptation measures that can reduce poverty and impacts of climate change which have been tested include: adoption of neat crops and clean technologies among farmers. There is a need for better storage and use of water in dams to stabilize agricultural production for perennial irrigation. Other approaches are needed to prevent sea encroachment and soil erosion. Of critical importance is to improve awareness of the challenges among coastal communities. Ensuring decisive but gradual responses to uncertainty and providing quality information for people to assist them to adapt. Giving farmers the means to adapt, for example access to credit and insurance, and involving people by enabling them to participate. Adapting plant to high temperature enables their cultivation to be sustained with increasing global warming and in addition may help extended production beyond the traditional growing areas. For instance in Nigeria, Amadi (2011), suggested that adapting potato to heat stress prevalent in warm environment through the development of heat tolerant genotype will help not only to sustain but will lead to expand rainy season production beyond the highlands.

Table 2. Farmers adoption of climatic change adaptation strategies in Nigeria

Strategy	Frequency	Percentage
Adoption of agricultural innovation	108	37.5
Application of different family systems	128	44.4
Good cropping system	184	63.9
Mixed cropping	136	47.2
Crop rotation	200	69.4
Use of cover crops	102	35.4
Change of planting time	140	48.6
Use of organic manure	112	38.9
Land rotation	160	55.5
Bush fallow	120	41.6
Erosion control measure	164	56.9
Mulching	136	47.2
Planting of shed trees	192	68.0
Communicate community effort	116	40.2
Government assistance	56	19.4
Traditional measures	82	28.4
Soil preservation	104	36.1
Extension export advice	140	48.6

Source: Nwakor et al., 2013

Adapting to climate change will require that individuals change their practices, which, in turn, are likely to require changes in the rules under which agriculture industry operates, the lack of enforcement of such rules, and the lack of participatory and accountable decision-making mechanisms are likely to increase socioeconomic vulnerabilities and limit the adaptive capacity of communities and societies. Fortunately farmers already have experience of many innovative technologies which have been adapted over the years that can be shared both within and outside countries. As shown in Table 2, farmers are already familiar with and have even adopted some climate adaptation strategies.

As shown in Table 2 crop rotation, planting of shed trees, good cropping systems, and good erosion control measures were the major strategies adopted or suggested by the farmers in order to adapt to climate change impact on root and tuber crops production.

CHALLENGES TO CLIMATE CHANGE ADAPTATION IN NIGERIA

The following are challenges to climate change adaptation experienced in Nigeria:

- Lack of funds for dealing with the problems wrought by climate change like increased flood and drought.
- Uncertainty about the actual impacts of climate change, rendering it difficult to make informed decisions on agriculture.
- Lack of political will to take measures to respond to climate change.
- Mobilization of cross-sector collaboration to respond to climate change.
- Changing the mindset, e.g., moving towards nuclear power.
- Mistrust among countries across the region or within sub regions, which may hinder a common approach to dealing with the issues of climate change.
- Provision of enough resources to maintain a system of information distribution to farmers.
- Newness of the concept, and the long-term nature of the projections, which means that it is not easy to integrate into development processes.
- Lack of appreciation of future risks, despite the widespread publication of the projections.
- High cost of technology to adapt.
- Low stock of knowledge and inadequate research and capacity to develop viable clean development mechanisms (CDMs).
- The need to reconcile poverty reduction with sustainable development.

CONCLUSION

Climate change is already a reality in all agro-ecological zones of Nigeria, with sharp reduction and uneven distribution of rainfall, prolonged drought in some parts and very heavy rains leading to floods in others, shifts in precipitation, high evaporation, and so on. These have had mostly negative impacts on agricultural production especially in the area of root and tuber crops. Crop yields have reduced due to increased incidence of pests and diseases thereby threatening food security and livelihoods of the people. Though farmers have developed autonomous measures in response to climate change, planned proactive measures such as development of improved varieties of root and tuber crops tolerant to heat, drought, waterlogging and other abiotic stresses are of utmost importance to mitigate the impact of climate change. In addition, improved training and general education on climatic change impact on agriculture and adaptation strategies should be extended to the farmers in rural communities. Agricultural research centers and experimental stations can examine the present farming systems and their resilience to extreme heat, floods, water shortage, pest damage and other factors and also try to develop new farming strategies to meet changes in climate, technology and other factors.

REFERENCES

Adejuwon, J. O. (2005). Food crop production in Nigeria. I: Present effects of climate change. *Climate Research*, *32*, 229–245. doi:10.3354/cr032229

Adejuwon, J. O. (2006). Food crop production in Nigeria. II: Potential effects of climate variabilbity. *Climate Research*, *30*, 53–60. doi:10.3354/cr030053

Agbola, T., & Ijeleye, D. (2007). Climatic change and food crop production in Ibadan, Nigeria. *African Crop Science Conference Proceedings, 8,* 1423-1433.

Agoumi, A. (2003). Vulnerability of North African countries to climatic changes: Adaptation and implementation strategies for climatic change. *Developing Perspectives on Climate Change: Issues and Analysis from Developing Countries and Countries with Economies in Transition.* IISD/Climate Change Knowledge Network. Retrieved July 31, 2016, from http://danida.vnu.edu.vn/cpis/files/Papers_on_CC/Vulnerability/Vulnerability%20of%20North%20African%20Countries%20to%20Climatic%20Changes.pdf

Amadi, C. O. (2011). *Breeding potato for Adaptation to Heat stress: Implication for Food Security in Nigeria. Root and Tuber Crops Research for Food Security and Empowerment.* Retrieved January 6, 2016, from http://www.nrcri gov.ng

AU (African Union) Summit. (2008). *Ban Ki-Moon. Secretary-General's address to the Summit of the African Union. Addis Ababa, Ethiopia.* Retrieved 31 January 2008, from http://www.un.org/apps/sg/sgstats.asp?nid=2978

Boko, M., Niang, I., Nnong, A., Vogel, C., Githeko, A., Medany, B., … Yanda, P. (2007). Africa. In M. L. Parry, O. F. Canziani, J. P. Palutikof, P. J. Van der Linden & C. E. Hanson (Eds.), *Impact, Adaptation and Vulnerability contribution of working group II to the fourth assessment report of the inter-governmental panel on climate change* (pp. 433-467). Cambridge University Press UK. Retrieved July 31, 2016, from https://www.ipcc.ch/publications_and_data/ar4/wg2/en/ch9.html

Davidson, D. J., Williamson, T., & Parkins, J. R. (2003). Understanding climate change risk and vulnerability in northern forest-based communities. *Canadian Journal of Forest Research, 33*(11), 2252–2261. doi:10.1139/x03-138

Enete, A. A., & Amusa, T. A. (2010). Challenges of Agricultural Adaptation to Climate Change in Nigeria: a Synthesis from the Literature. *Field Actions Science Reports, 4*, 1-11. Retrieved 31 July 2016, from http://factsreports.revues.org/678

Fakihri, A. B., & Sombroek, W. G. (2010). Global climate change and agricultural production: An assessment of current knowledge and critical gaps. *FAO Corporate Document Repository*, 1-17.

FAO (Food and Agricultural Organisation of the Unites Nations). (2001). *Farming systems and Poverty: Improving farmer's livelihoods in a Challenging world.* FAO Rome.

FAO (Food and Agricultural Organization of the Unites Nations). (2003). Responding to Agricultural and Food Insecurity Challenges Mobilizing Africa to Implement NEPAD Programmes. Conference of ministers of agriculture of the African union, Maputo, Mozambique.

Fischer, G. S., Shah, M., & Uelthuizen, H. Z. (2002). Climate change and agriculture vulnerability: international institute for applied systems analysis. Report prepared under UN institutional contract agreement 1113 for world summit on sustainable development, Luxemburg.

Heagle, A. S., Heck, W. W., Rawlings, J. O., & Philbeck, R. B. (1983). Effects of Chronic Doses of O_3 and Sulfur Dioxide on Injury and Yield of Soybeans in Open-Top Field Chambers. *Crop Science, 23*(6), 1184. doi:10.2135/cropsci1983.0011183X002300060037x

Horton, D. E., & Anderson, J. L. (1992). Potato production in the context of the world and farm economy. In P. M. Harris (Ed.), *The potato crop* (pp. 794–15). Chapman and Hall London. doi:10.1007/978-94-011-2340-2_16

Ifeanyi-obi, C. C., Etuk, U. R. & Jike-wai, O. (2012). Climate Change, Effects and Adaptation Strategies; Implication for Agricultural Extension System in Nigeria. *Greener Journal of Agricultural Sciences, 2*(2), 53-60.

Ifenkwe, O. P. (1981). A review of the status of potato production in Plateau State Nigeria. Academic Press.

IPCC. (2001). *Third Assessment Report.* Cambridge, UK: Intergovernmental Panel on Climate Change Cambridge University Press.

IPCC. (2007), Contribution of working group 11. Intergovernmental Panel on climate change. Cambridge University Press.

IPCC. (2014). Impacts, Adaptation and Vulnerability. *Summary for Policy Makers.* Working Group II Contribution to the Fifth Assessment Report of the Intergovernmental Panel on Climate Change.

Jones, P. G., & Thornton, P. K. (2002). *Croppers to livestock keepers: Livelihood transition to 2010 in Africa due to climate change. Global Environmental Change.* Geneva, Switzerland: World Health Organization.

Lenka, D. M., Okonkwo, J. C., Amadi, C. O., Asumgha, G. N., & Nwosu, K. I. (2008). *Potato production, storage and marketing in Nigeria. In Farmer's perspectives and practice* (p. 33). National Root Crops Research Institute Umudike.

Ludlow, L., & Smit, B. (1987). Assessing the Implications of Environmental Change for Agricultural Production: The Case of Acid Rain in Ontario, Canada. *Journal of Environmental Management, 25*, 27–44.

Mendelsohn, R., Nordhaus, W. D., & Shaw, D. (1994). The Impact of Climate Change on Agriculture. A Ricardian Analysis. *The American Economic Review, 84*(4), 753–771. Retrieved from http://www.jstor.org/stable/2118029?origin=JSTOR-pdf&seq=1#page_scan_tab_contents

Ngaira, J. K. W. (2009). Challenges of water resource management and food production in a changing climate in Kenya. *Journal of Geography and Regional Planning, 2*(4), 97-103.

Nwakor, F. N., Anyaegbunam, H. N., & Ekedo, T. O. (2013). Effects of Climate change on root and tuber crop production among farmers in southeastern Nigeria. *International Journal of Applied Research and Technology*, 2(6), 42–47. Retrieved from http://www.esxpublishers.com/images/ijrt-0613-0220.pdf

Obioha, I. B. (2002). The Late Onset of the Rains in the Coastal Area of Nigeria. *Weather*, *5*(6), 222–227.

Odjugo, P. A. O. (2010). General Overview of Climate Change Impacts in Nigeria. *Journal of Human Ecology (Delhi, India)*, *29*(1), 47–55.

Okonkwo, J. C. Amadi, C. O. & Nwosu, K.I. (2009). *Potato production, storage, processing and utilization in Nigeria.* National Root Crop Research Institute, Umudike, Nigeria.

Okonkwo, J. C., Ene, L. S. O., & Okoli, O. O. (1995). Potato production in Nigeria. National Root Crops Research Institute, Umudike, Nigeria.

Rikko, L.S. (2010). Emerging issues from world climate trends and technologies for climate change adaptation. *Journal of Sustainable Development, 7*(2), 27-33.

SEI. (2008). *Climate change and adaptation in African agriculture*. Rockefeller Foundation. Retrieved July 31, 2016 from https://www.sei-international.org/mediamanager/documents/Publications/Climate/Using_Science_To_Improve_Adaptation_Africa.pdf

Sidi, M. S. (2012). *The Impact of the 2012 Floods on Agriculture and Food Security in Nigeria Using GIS*. United Nations International Conference on Space-based Technologies for Disaster Management -Risk Assessment in the Context of Global Climate Change, Beijing, China.

Smit, B., Ludlow, L., & Brklacich, M. (1988). Implications of a Global Climatic Warming for Agriculture: A Review and Appraisal. *Journal of Environmental Quality*, *17*(4), 517–527. doi:10.2134/jeq1988.00472425001700040001x

Stainforth, D. A., Thomas, E. D., Richard, W., Ana, L., & Mark, N. (2007). Issues in the interpretation of climate model ensembles to inform decisions. *Philosophical Transactions of the Royal Society A: Mathematical, Physical and Engineering Sciences*, *365*(1857), 2163–2177. doi:10.1098/rsta.2007.2073 PMID:17569657

Watson, R. T., Patz, J., Gubler, D. J., Parson, E. A., & Vincent, J. H. (2005). Environmental health implications of global climate change. *Journal of Environmental Monitoring*, *7*(9), 834–843. doi:10.1039/b504683a PMID:16121261

Wikipedia. (2010). Climate change and agriculture. Author.

Wikipedia. (2016). *Climate change and potato*. Author.

Zoellick, R. B. (2009). *A Climate Smart Future. In The Nation Newspapers*. Lagos, Nigeria: Vintage Press Limited.

Chapter 10
Impact of Global Climate Change on Potato Diseases and Strategies for Their Mitigation

Mehi Lal
CPRIC, India

Vijay Kumar Dua
CPRI, India

Saurabh Yadav
CPRIC, India

B. P. Singh
CPRI, India

Rajendra Prasad Pant
CPRIC, India

Surinder Kumar Kaushik
NBPGR, India

ABSTRACT

The quality and quantity of the potatoes produced is directly affected by the climatic factors that prevailed during the crop season. It is also well established that abiotic and biotic stresses cause tremendous losses to the crop. Host plants and their pathogens are prone to various climatic factors like temperature, relative humidity, rainfall and CO2 which are behaving in erratic manner. Phytophthora infestans has adapted itself at higher temperature so there are chances to spread at a larger area. The other potato diseases like early blight, bacterial wilt, soft rot and viral diseases may also behave differently at elevated temperature and high rainfall. Viral diseases of potato are serious threat to potato industry as most of the viruses are transmitted by vectors and vector populations are bound to increase with these changed climatic conditions. Therefore, potato researchers need to simulate these conditions and devise mitigation strategies for sustained potato production.

DOI: 10.4018/978-1-5225-1715-3.ch010

INTRODUCTION

Potatoes are staple diet of millions throughout the world. It can be grown in temperate, tropical and subtropical regions of the world. Over the years, potato production has increased substantially in developing countries. Today, China is the main producer of potato producing about 20% of global production (Staubli et al., 2008). It has been predicted that potato yield will decrease by 18-32% due to climate change. Recent years have witnessed a steady increase in national and international concern over the sustainability of the global environment. Climate change has emerged as the most prominent of the global environment issues. Global climate has changed ever since industrial revolution. The atmospheric concentrations of the greenhouse gases carbon dioxide (CO_2), methane (CH_4), and nitrous oxide (N_2O) have increased since 1750 due to human activity. In 2011, the concentrations of these greenhouse gases were 391 ppm, 1803 ppb, and 324 ppb, exceeding the pre-industrial levels by about 40%, 150%, and 20%, respectively (Intergovernmental Panel on climate Change [IPCC], 2013). Global surface temperature change by the end of the 21st century is likely to exceed 1.5°C relative to 1850 to 1900 for all RCP (Representative Concentration Pathways) scenarios except RCP2.6. It is likely to exceed 2°C for RCP6.0 and RCP8.5, and more likely than not to exceed 2°C for RCP4.5. Warming will continue beyond 2100 under all RCP scenarios except RCP2.6. Warming will continue to exhibit inter annual-to-decadal variability and will not be regionally uniform (IPCC, 2013).This is expected to have grave consequences for mankind and the environment. The critical threshold is said to be around a temperature increase of 2°C (IPCC, 2007). CO_2 and temperature interactions are recognized as a key factor in determining plant damage from pests in future decade; likewise CO_2 and precipitation interaction will also be important. Climate change implies that the average conditions (mean and/or variability) are changing over time and may never return to those previously experienced (Coakley, 1988).

Pace of climate change and the unpredictability of its characteristics are of great concern with respect to the pathogens, insect pests and weeds that reduce crop yield. The classic disease triangle recognizes the role of climate in plant diseases, as no virulent pathogen can induce disease on a highly susceptible host if prevailed climatic conditions are not favorable. Climate change could "alter stages and rate of development of the pathogen, modify host resistance, and result in changes in physiology of host pathogen interaction" (Garrett et al., 2006). The climate change and global warming with increases in temperature, moisture and CO_2 levels can impact all three legs of the plant disease triangle in various ways. Climate influences all stages of host and pathogen life cycles as well as development of disease. Climate change and global warming will allow survival of plant and pathogens outside their existing geographical range. "The lack of action on climate change not only risks putting prosperity out of reach of millions of people in the developing world, it also threatens to roll back decades of sustainable development" (Kim,19th November, 2012; a foreword to the report, Turn Down the Heat: Why a 4°C Warmer World Must be Avoided). The climatic factors including changes in temperature, rainfall and other atmospheric composition along with predominantly elevated CO_2 levels would accelerate the reproduction time of many plant pathogens and pests, thereby increasing their infection pressure on crop plants (Boonekamp, 2012).

The impact of climate change on the occurrence and activity of pathogens, pests and diseases of agricultural crops have recently been considered more seriously. This is documented by the fact that renowned scientific journals in plant pathology (e.g. *Annual Review of Phytopathology*) published the papers focused on the implications of climate change for plant disease occurrence and management (Garrett et al., 2006). The range of plant pathogens and insect pests are mainly constrained by temperature, and the frequency and severity of weather events affecting the timing, intensity and nature of outbreaks of most organisms (Yang & Scherm, 1997). It is well established that short- and long-term changes in climate, will impact on which strains of pathogens prove most aggressive via their ability to grow and sporulate under particular environmental conditions. Change in environmental factors, particularly temperature and leaf wetness, influence the rate of progress at any stage of plant disease development (Agrios, 1997; Coakley et al., 1999).

Potato crop is affected by a number of diseases. Amongst these, Late blight of potato caused by *Phytophthora infestans* still remains the leading threat to potato cultivation globally. This notorious pathogen is known to break host resistances across *Solanum species* and has the ability to develop resistance to chemical fungicides in a very short span of time. The pathogen possesses this quality owing to very large genome size of about 240 Mbps (Hass et al., 2009) that is largely composed of transposons. Consequently, it is also expected that this pathogen will quickly adapt itself to changed climatic conditions and will continue to play havoc in potato growing regions across the globe. Some of the other diseases include early blight, black scurf, black leg/soft rot, bacterial wilt and viral diseases etc. Practically, the impact of a changed climate cannot be predicted with certainty on diseases affecting potato crop. However, it can be safely postulated that there may be several diseases that are not causing epidemics at present may be due to non-availability of favorable environmental conditions. A changed climate, if favourable, will certainly lead to flare up of these diseases. In another such scenario, there are diseases that cannot establish or cause epidemics upon their introduction in new potato growing areas because of unfavourable weather conditions. However, a changed climate may provide excellent opportunity to such diseases, not only to establish but also to cause severe damage to the potato production. Pamella et al. (2004) reported that the introduction (40%), weather (41%) and farming techniques (19%) are important drive in fungi for spreading emerging infectious disease. Fungal diseases such as those caused by *Ustilago* and *Sclerotia* species are not expected to be affected by climate change because of their monocyclic nature. In polycyclic diseases such as those caused by *Colletotrichum, Peronospora, Phytophthora* and *Puccinia* species, each additional disease cycle multiplies inoculums many fold, so an increased duration of growing season in Ontario (Canada) would be expected to results in an increased number of disease cycles and inoculums. Primary inoculums will be increased for disease established in tomato late blight whereas potential duration for epidemic of potato late blight will be increased in Ontario (Boland et al., 2004). These experiences of potato researchers needs to be simulated for other diseases and potato growing regions and suitable mitigation strategies needs to be devised for sustained potato production globally.

FUNGAL DISEASES

Effect of Major Components of Climate Change on Potato Late Blight *(Phytophthora Infestans)*

Elevated Temperature

Temperature and its duration of exposure are important in determining the effect of climate change on disease intensity. Temperature change might lead to appearance of new races of the pathogens hitherto not active but might cause sudden epidemic. Change in temperature will directly influence infection, reproduction, dispersal, and survival between seasons and other critical stages in the life cycle of a pathogen. Some of the soil-borne diseases may increase at the rise of soil temperature. If climate change causes a gradual shift of cropping regions, pathogens will follow their host. The susceptible host has important role in epidemiology of the disease. Epidemiological components are having significant role in climate change aspects. Incubation period of *P.infestans* was short at higher temperature and shortest Incubation Period (IP) occurred at 28°C (Becktell et al., 2005). Mizubuti & Fry (1998) also reported that IP declined exponentially and Lesion Area (LA) increased exponentially with increasing temperature, in general. Less than 20°C temperature with optimum 12-13°C for indirect germination (5-10 zoospore/sporangium) whereas more than 20°C with optimum at 24°C favours for direct germination (one sporangium gives rise to a germ tube). Even at optimum temperature, the percentage of direct germination is usually much lower than the percentage of indirect germination (Crosier, 1934; Goodwin et al., 1994).

A positive co-relation was observed between incubation period and lesion area development (Central Potato Research Institute [CPRI], 2011-12). It was observed that up to 15°C incubation period was more and lesion area was less, while from 18 to 25°C incubation period was less and lesion area was more and an incremental increase in temperature thereafter resulted in increased incubation period with decrease in lesion area (CPRI, 2012-13). Considering an increase in temperature in West Bengal during the potato growing season, the late blight, disease severity is expected to be reduced by 5-7% from 1981-2010 levels to 2031-40 (APN Project Report, 2011). Becktell et al. (2005) studied sporulation of US-17, a new tomato adapted lineage, at 28°C and found that sporulation was almost nil, whereas the sporulation of US-1 isolates at 27°C was found similar to that recorded at its optimum temperature. It might have happened due to the location of tomato producing areas at lower altitudes, that may have resulted in the higher fitness of US-1 on tomato over the longer time of co-existence of US-1 with tomato plants and the subsequent adaptation of US-1 isolates to higher temperatures. Studies conducted at CIP, Peru to work out the risk of late blight (expressed as number of pesticide sprays) at global level in climate change scenario revealed that with rise in global temperature of 2°C, there will be lower risk of late blight in warmer areas (<22°C) and higher risk in cooler areas (>13°C). Earlier onset of warm temperatures could result in an early appearance of late blight disease in temperate regions with the potential for more severe epidemics and increased number of fungicide applications needed for its control. Studies carried out in Finland predicted that for each 1°C warming, late blight would occur four to seven days earlier, and the susceptibility period extended by 10 to 20 days (Kaukoranta, 1996). This would result in 1 to 4 additional fungicide applications, increasing both cost of cultivation and environmental risk. Analysis of historical data from 1948 to 1999 indicated that late blight risk over a standardized growing season from 1 May to 30 September increased in the Upper Great Lakes region of the US. Predominant genotypes of *Phytophthora infestans* (e.g. US8) in the US appear more tolerant to temperatures close to 0°C and their survival in warming conditions may explain their supremacy (Baker, 2004).

In India also, the late blight scenario may change drastically with climate change. Earlier, late blight was not a serious problem in the states of Punjab, Haryana and parts of Uttar Pradesh, primarily due to sub-optimal temperature regimes during December-January. However, disease outbreaks will become more intense with increase in ambient temperature coupled with high RH. Such scenarios have been witnessed during warmer years *i.e.* 1997-98 and 2006-07, when average crop losses in this region exceeded 40%. During recent past also severity of late blight increased due to occurrence of rainfall at advanced stage of crop growth. States like Madhya Pradesh, Gujarat and Central Uttar Pradesh which are comparatively less affected by late blight may also witness frequent outbreaks of the disease under the climate change scenario (Singh & Bhat, 2010). Increase in both, temperature and RH has added new dimensions to late blight disease cycle across the world. Under such a situation, *P. infestans* attacks potato stems more often than foliage. In fact, in recent years it has been observed that 'stem blight' is more common than the foliar blight. This phase of the disease is more serious than the foliar stage as it affects stem and subsequently tubers also. In a recent study conducted using JHULSACAST model, it was found that if the host physiology does not change than under the favourable relative humidity in Punjab, late blight is expected to be delayed by 0 to 6 days in 2020 and 12 to 14 days in 2055 scenarios, over the baseline of 2000. In western Uttar Pradesh, earliest late blight appearance during the potato crop season is expected to be delayed by 0 to 8 days in 2020 and 10-21 days in 2055. In Punjab, there were average 105 late blight favourable days out of 182 suitable potato growth days in year 2000; the number is likely to be increased to 135 and 140 days in 2020 and 2055 respectively in Punjab. In western Uttar Pradesh, potato growing season was warmer which would decrease late blight favorable days by 7 and 27 in 2020 and 2055, respectively. The number of sprays required to control late blight in potato seed crop would be 7.3 and 8 in future scenario (2020 and 2055) in comparison to 6.5 in baseline (2000) in Punjab. In contrast, there would be no change in number of sprays in year 2020 (over baseline year 2000) in western Uttar Pradesh, however, due to further increase in temperature in year 2055, it could be reduced by 2 numbers of sprays (Dua et al., 2015).

Rainfall/High Humidity

An increase in rainfall and high humidity will result in elevated threat of potato diseases, such as late blight (*P. infestans*), especially when combined with longer growing seasons. In Upper Great Lakes region of the US, increase in annual precipitation and increase in number of days with precipitation over the years is supposed to be the reason for the increased risk of potato late blight infection and subsequent yield and economic losses. In India, in Lahaul valley of Himachal Pradesh, which was earlier free from late blight because of lack of precipitation, has now experienced attack of late blight due to occurrence of rainfall (Singh et al., 2013) However, hotter and drier summers which are likely in UK may reduce the importance of late blight, although earlier disease onset may prevent this advantage. An empirical climate disease model has suggested that under the climate change scenario of 1°C increase with 30% reduction in precipitation in Germany will decrease potato late blight to a mere 16% of its current level. In India, late blight attacks are likely to increase in North-Western plains with the increase in temperature. The disease is likely to be reduced in Eastern plains including West Bengal. Hilly regions may get less prolonged attack of late blight with the increase in temperature regime and decrease in precipitation.

Elevated CO_2

Elevated CO_2 levels can impact both the host and pathogen in various ways. It was reported that the higher growth rates of leaves and stems observed for plants grown under high CO_2 concentrations may result in denser canopies with higher humidity that favour pathogens (www. Climate and forming.org). Osswald et al. (2006) investigated that elevation of CO_2 (400 up to 700 ppm) and/or ozone (ambient or two-fold ambient) resulted in a change in susceptibility of potato plants infected with *P. infestans*. The main result was that a rise in CO_2 caused a significantly enhanced resistance of the susceptible potato cultivar 'Indira' towards *P. infestans*, whereas ozone had no significant effect. These authors also investigated the effect of N-fertilization in combination with CO_2-treatment on the resistance of potato to *P. infestans*, because CO_2 induced an increase in resistance correlated with an increased C/N-ratio in potato leaves. The lower C/N-ratio, due to higher N-concentrations, decreased in resistance to *P. infestans*. Similarly, Plessl et al. (2007) reported that the potato cultivar Indira, which under normal conditions showed a high degree of susceptibility to *P. infestans*, developed resistance after exposure to 700 ppm CO_2. Ywa et al. (1995) reported that an increased tolerance of tomato plants to *Phytophthora* root rot when grown at elevated CO_2. Mitchell et al. (2003) reported that elevated CO_2 increased the pathogen load of C_3 grass, perhaps due to increased leaf longevity and photosynthetic rate.

IMPACT OF CLIMATE CHANGE ON LATE BLIGHT MANAGEMENT

Generally, fungicides, host resistance and bio-control strategies are adopted for late blight management. These strategies will also be affected due to variation in climate. Efficacy of fungicides will be mainly affected by elevated temperature and rain fall while host resistance and bio-agents will also be affected by increasing temperature.

Temperature

Temperature and precipitation can alter fungicides residues dynamics in the foliage and degradation of product can be modified. Penetration, translocation and mode of systemic fungicides could affect with CO_2 enriched atmosphere or from different temperature and precipitation conditions (Ghini et al., 2008). Systemic fungicides could have been affected negatively by physiological changes that slow uptake rates, such as smaller stomata opening or thicker epicuticular waxes in crop plants grown under higher temperature. These fungicides could affect positively by increasing plant metabolites rates that could be increase fungicides up take (Petzoldt & Seman. www. climate farming). Fungicide, cymoxanil is known to degrade rapidly (Klopping & Delp, 1980) and high temperature may accelerate the degradation. Day time temperatures in field plots were regularly above 24°C, even reaching above 30°C on some occasions. They believe that high temperatures limited the efficacy of cymoxanil in their field experiments (Mayton et al., 2001). In one of our experiments to study the efficacy of fungicides at elevated temperatures, potted plants were sprayed with different fungicides at different concentration and exposed to 25°C and 30°C. Leaves were detached at regular interval and challenge inoculated with *P. infestans*. At 25°C, metalaxyl 8% +mancozeb 64% WP fungicide showed degradation after 14 days of spraying though the rate of degradation was very low compared to control while at 30°C degradation was more. At 25°C cymoxanil 8% + mancozeb 64% WP showed no degradation even after 14 days while at 30°C degradation was found at 14 days after spraying. No degradation was recorded in dimethomorph 50% WP fungicide, even after 14 days of spraying at 25°C and 30°C (Lal & Yadav, 2014).

The effect of temperature on host resistance was studied at 20°C and 25°C· Two late blight differentials (R1 & R1.2.3) and two cultivars (Kufri Girdhari and K. Jyoti) were grown at above temperatures and then detached leaves were challenge inoculated with *P. infestans*. Results revealed that at higher temperature, more lesions size was produced (Indian Institute of Spices Research [IISR], 2012). Some studies revealed that in spite of a delay in the initial development and a reduction in host penetration, the established colonies develop faster under increased CO_2 concentration (Ghini et al., 2008). The most likely impact of climate change will be felt in three areas: in losses from plant diseases, in the efficacy of disease management strategies and in the geographical distribution of plant diseases (Chakraborty et al., 2000). Increase of 3°C temperature from ambient temperature had increased potato-yield loss caused by late blight in the 3-year-long controlled-environment study, reducing potential benefits from yield increases due to warmer temperatures (Kaukoranta, 1996). De Jong et al. (2002) observed that R genes cf4 and cf9, mediated HR in tomato against fungal pathogen *Cladosporium fulvum* can be suppressed at 33°C. Probably, it is a general phenomenon that disease resistance to biotrophic pathogens is generally inhibited by a mild elevation of the normal growth temperature. Wang et al. (2009) also reported that higher temperature inhibits the hypersensitive (HR) against *Pseudomonas syringae* pv tomato.

Rainfall

Frequent rainfall may affect the efficacy of contact fungicides due to washing off from leaf surface (Kocmankova et al., 2009). Effect of rain on the efficacy of fungicide upon its deposition on leaves, the interaction of precipitation frequency, intensity, and fungicides dynamics are complex, and for certain fungicides precipitation following application may result in enhance diseases control because of a redistribution of the active ingredient on the foliage (Schepers, 1996). Chen & McCarl (2001) performed a regression analysis between pesticide usage and climate variations in different locations of USA and reported that average cost of pesticides application per acre of different crops (cotton, corn, potato, soybean and wheat) was found to increase as precipitation increases. Similarly, the pesticides usage average cost for cotton, corn, soybean and potato also increase as temperature increases, while the pesticides usage cost for wheat decreased.

Early Blight of Potato

The early blight disease initially affects lower leaves of the plant. However, if proper doses of fertilizer is not applied (N) and prevailed of favorable condition; in such scenarios, the disease may cause significant losses. The early blight disease is favoured by temperature ranging from 25-30°C with an optimum of 26°C (Dutt, 1979) or when relative humidity is more than 90% (Sherf & MacNab, 1986). The host resistance to early blight of tomato and potato caused by *Alternaria solani* / *A. alternata* is likely to be enhanced by the elevated CO_2 effect along with enhanced sporulation and infection through increased summer rainfall. Therefore, under climate change, damage due to early blight will depend on the balance between enhanced host plant resistance and increased pathogen sporulation. Although, potato researchers' efforts are directed mainly for the management of *A. solani* than *A. alternata*. Different temperature requirements for spore germination and subsequent infection of the leaves by *A. solani* have been reported by different researchers but it is clear that the maximum temperature for growth of these fungi is < 40°C (Waggoner & Parlange, 1975; Rotem, 1994; Chaerani & Voorrips, 2006). Bashi & Rotem (1974) reported that under favorable relative humidity, infection can be caused at a temperature range of 10°C - 35°C, and 25°C is optimum for infection. Since rainfall pattern is erratic in most of the Asian regions, therefore, it may be conducive for early blight disease. Interruption in the rainfall will affect the wetting period and subsequently lead to the more sporulation and dispersal. Van der Waals et al. (2003) also opined that interrupted wetting periods are more important than high leaf wetness for spore germination and dispersal of *A. solani*. Therefore, it is required that forecasting models should be developed with keeping changed pattern of pathogen behavior.

Sclerotium Wilt

Sclerotium wilt caused by *Sclerotium rolfsi* is soil and tuber borne in nature. It favours high moisture and temperature regimes. The fungus can grow in wide range of temperatures 8-37°C with an optimum between 30-35°C (Dutt, 1979). In India, sclerotium wilt was reported mainly in plateau region, however, due to effect of erratic behavior of climatic factors, it may spread to other regions of the country wherever potato is being grown in late season and temperature is likely to increase from normal level.

BACTERIAL DISEASES

There are two major bacterial (wilt & soft rot) diseases in potato. Elevated temperature and irregular rainfall with increased risk of higher intensity rainfall and occasional flooding increase the risks of bacterial infection. Temperature is one of the most important factors affecting the occurrence of bacterial diseases such as *Ralstonia solanacearum*, *Acidovorax avenae* and *Burkholderia glumea*. Therefore, bacteria could proliferate in areas where diseases have not been previously observed due to non availability of favorable climatic condition (Kudela, 2009).

Bacterial Wilt

Bacterial wilt caused by *R. solanacearum* is both soil and seed borne in nature. Generally, bacterial wilt multiples in hot climate but more recently it has been observed that it can also multiply in cooler climates. Bacterial wilt may also increase as the climate becomes warmer and wetter; and potato pests, including disease-carrying aphids, will survive at higher altitudes (http://www.new-ag.info/en/focus/focusItem.php?a=532). In cool temperate climates, this disease drew attention in the 1990s by outbreaks of potato brown rot disease in some European countries (Elphinstone, 1996). There are a number of quarantine diseases, which might become more serious under global warming, including brown rot (*R. solanacearum*) and sudden oak death (*P. ramorum*). The pathogen is disseminated through latent infected potato tubers and in contaminated irrigation water and overwintering in aquatic roots of the weed host *Solanum dulcamara*. The *R. solanacearum* strains RS-2 (phylotype II bv 2) pathogen could survive for one month at 40°C, two months at 37°C, 4 months at 32°C and six months at 20-28°C in sterilized soil. However in un-sterilized soil, the pathogen could survive only for one month at 37°C and 4 months at 20-28°C, while no growth of *R. solanacearum* was observed at 40°C after one month and at 32°C and above after two months (CPRI, 2014-15). It seems that *R. solanacearum* can grow wide range of temperature subsequently it will be spread at large areas where this is not reported till date.

Black Leg of Potato

Black leg or bacterial soft rot is caused by *Pectobacterium* spp (Erwinia spp). Incidence of black leg of potato increased over the years in Europe due to effect of climatic variation. Similarly in India, due to fluctuation in climatic condition its incidence also increased in western Uttar Pradesh. The black leg bacteria are opportunistic pathogens that infect plants when conditions become favorable for their multiplication (Pérombelon, 1992). This is a seed-borne disease that survives in lenticels and wounds during storage. The contaminated mother tuber is the main source of progeny tuber contamination and not the spread of bacteria through the soil/water. Most seed tubers are contaminated but black leg incidence is related to the seed contamination level that increases strongly with excess soil water. *Pectobacterium* spp dominate in the rotting mother tuber depends on the prevalent temperatures, for example *P. carotovora* favours temperatures (below 25°C) and *P. chrysanthemi* higher temperatures.

VIRAL DISEASES

Virus diseases will pose greater threat to potato cultivation at global level, particularly in climate change scenario since most of the viral diseases spread through insect vectors. Vectors will also be affected due to change in climatic condition. An increase in atmospheric temperature will increase the rate of multiplication of potato viruses. In subtropical plains, where most of the potato varieties are grown in India, global warming may not affect potato viruses directly but may have a serious repercussion through the altered biology of insect vectors. The increase in temperature will enhance vector population thereby increasing insecticide sprays for keeping the insect population under control. The rate of multiplication of virus in host tissue will also increase substantially leading to early expression of virus symptoms. During the last 12 years (1994-2008), it has been reported that new strains of the Potato Virus Y (PVY) i.e. PVY^{NTN} and PVY^{NW} has been introduced in Holland due to the impact of climate change (Singh et al., 2013).

Population sizes of *M. persicae* at elevated CO_2 generally increases but not the number of alatae, which are considered to be more important in spreading PVY and PLRV than apterae (Hughes & Bazzaz, 2001). In a study in Sweden, it had been predicted that aphid population will increase substantially thereby problem with PVY in seed production may increase as a result of larger aphid population

(SJV, 2007). The increase in soil moisture and temperature in northern Europe are expected to increase the activities of zoospores and nematodes that transmit viruses (Jones, 2009). Potato mop top virus (PMTV) is transmitted by plasmodiophorids, *Spongospora subterranea*, has been recently introduced in Sweden and is spreading (Ryden et al., 1986; Sandgren, 1995; Lennefors et al., 2000; Santala et al., 2010). The most important factor of this spread is probably movement of infected material and changed soil and climatic conditions. In Canada, it has been predicted that a net positive increase in prevalence of potato leaf roll virus due to increased vector population is resulting from warmer winter temperatures (Boland et al., 2004).

In South Africa, the relative aphid population growth will increase the infection of PVY and PLRV transmission in summer and winter crops, the increase will be relatively higher in winter crop than the summer crop. Due to changed climatic conditions the potato industry may experience greater problems with PVY and PLRV due to an increase in aphid population. The expected increase in aphid number is of particular concern with regard to Potato Virus Y strains PVY ^{NTN} and PVY $^{N\text{-}Wilga}$ that are classified as emerging problems in Europe (Verbeek et al., 2010) and North America (Mello et al., 2011). The PVY strains are also on the increase in South Africa (Visser & Bellstedt, 2009). PVY and PLRV are two important viral diseases of potato globally (Robert et al., 2000; Redcliffe & Ragsdale, 2002). The infection caused by PVY and PLRV in tubers leads to downgrading of seed lot because of the low tolerance allowed by seed certification programme for high quality seed. The virus is generally transmitted to new crop primarily by winged (alatae) aphids (Radcliffe, 1982). PVY is transmitted nonpersistently while PLRV is transmitted by persistent circulative non propagative manner by several aphid species that colonize potato. *M. persicae* is considered the most efficient vector of PVY and PLRV. This species has been recorded from seed potato growing regions of South Africa (Daiber, 1965). The aphid multiplies holocyclic (sexual) and unholocyclic (asexual) life cycle in South Africa depending on climate of the region (Daiber & Scholl, 1959). *M. persicae* reproduces sexually when monthly temperatures fall below 20°C (Blackman, 1974).

Aphids generally have low developmental temperature threshold and a short generation time and reproduce parthenogenitically 18 generation in a year under British conditions (Harrington, 1994; Harrington et al., 2007). Under warming condition with an increase of 2°C, aphids multiply extra 5 generations in a year. In Sweden, it has been observed that Increase in CO_2 concentration, soil nitrogen content and host population density increases aphid populations (Newman et al., 2003) and expected to increase in importance as pests (Fagelfors et al., 2009). Aphid show considerable variation in their life cycle traits. Some species are holocyclic often placing eggs on woody plants where as other species are anholocyclic which do not go through sexual phase and continue with parthenogenetic and viviparious reproduction throughout the year. In India, the whitefly (*Bemisia tabaci*) was a minor pest till recently and not affecting potato crop. It has been noticed that the population buildup of whitefly has increased from 11 whitefly/100 leaves to 24.24 in the last two decades. The ambient temperature has also increased by 1.07°C during this period. This increased temperature is responsible for increase in whitefly population in Indo-Gangetic plains and led to the outbreak of potato apical leaf curl disease, a whitefly transmitted *Begomovirus* which was not reported earlier from potato (Singh et al., 2013). The virus had been characterized and identified as strain of Tomato Leaf curl New Delhi virus and named as *Tomato Leaf Curl New Delhi Virus* – potato (*ToLCNDV*-pot) (Usharani et al., 2004). Therefore, a new dimension has been added to the seed potato production in the subtropics. In addition to the increased population of whitefly in the subtropical regions of India, the population of *M. persicae*, an important vector of PVY and PLRV has increased remarkably. The mean aphid population/100 compound leave was recorded 567 during 1984-85 has increased to 653 after two decades. The appearance of aphid population has also been advanced 5 days during the last two decades, reducing low aphid pressure window from 80 to 75 days. The population of another important aphid, *Aphis gossypii* has increased three folds during the last 20 years. Although *A. gossypii* has not been reported efficient vector of potato viruses but its presence throughout the crop season is posing a serious threat to potato seed production in the subtropical plains. Another serious pest of potato seed crop in India is leaf hopper (*Empoasca fabae*) in the subtropical plains. It has been reported that its population has increased from 16.6 in 1984 to 23.8 in 2004. Their infestation is responsible for causing hopper burn which has increased from 45% to 65% in early sown crop during the last two decades. The mite infestation has also increased from 80% to 100% in early sown crop during the last two decades (Singh et al., 2013).

MITIGATION STRATEGIES

Following mitigation strategies can be applied to minimize the impact of climate change:

- Selection of new resistant/tolerant genotypes in climate change scenario. For example Kufri Surya is tolerant to early heat stress at the time of planting (Minhas et al., 2006).
- Use of new molecules with higher efficacy at elevated temperature for diseases management. New molecule (rain fastness) with higher efficacy in high rainfall regions.

- New forecasting models for prediction of appearance of diseases at spatial scale *i.e.* region, state, or country. For example *Indo-blightcast* for forecasting first appearance of late blight in India (http://cpri.ernet.in)
- Adjustment of date of sowing to avoid favourable environmental conditions of the insects/vector, which are responsible for transmission of various viral diseases.
- Selection of bio-agents having wide range of adaptability (elevated temperature, CO_2, rainfall, humidity etc) in climate change scenario.
- Management of diseases through integration of all the existing strategies.

FUTURE IMPACT

Climate change is a reality now and here to stay. Every organism needs to adjust to the changed environmental conditions in order to survive and perpetuate. These conditions would have a direct bearing on pathogen characteristics such as frequency of generations, proportion of sexual reproduction, and rate of adaptation. Similarly host characteristics i.e host anatomy, host physiology and life span would also be affected. The invasive plant species may be better adapted to climate change and move to new areas rapidly, leaving pathogens behind or at least limiting their evolutionary options through bottlenecks (Mitchell & Power, 2003). Various researchers reported that changes in the environmental conditions are known to exacerbate plant diseases symptoms (Boyer, 1995; McElrone et al., 2001) and are implicated in 44% of new diseases emergence (Anderson et al., 2004). Potato industry may incrementally experience greater problems with viruses particularly PVY and PLRV due to an increase in aphid population size with climate change. The expected increase in aphid numbers is of particular concern with regard to Potato virus Y strains PVY ^{NTN} and PVY $^{N\text{-}Wilga}$ that are classified as emerging pathogens in Europe (Verbeek et al., 2010) and North America (Mello et al., 2011). In India, the whitefly (*Bemisia tabaci*) was a minor pest till recently and not affecting potato crop, but due to change in cropping pattern and climate change, the whitefly has become a major pest of potato and causing apical leaf curl disease in Indo-Gangetic plains of India. The population has increased from an average 11 to 24.24 whitefly/plant during the last two decades which is a serious threat to potato seed crop production. Therefore, it is the need of the hour that we collect more genetic resources from their wider areas of adaptation and diversity besides their centres of origin and conserve in different forms for the benefit of our future generations.

FUTURE RESEARCH DIRECTIONS

- Development and deployment of higher temperature tolerant varieties possessing resistance/tolerance against various pests and diseases.
- Forecasting models for precise prediction of pests and diseases in future climate scenarios.
- Investigation of host-pathogen interactions at molecular level in climate change scenarios.
- Virus-vector relationships in changed climatic conditions.
- Appearance of new pathogens/ hitherto lesser important pathogen having potential to flare up in the new environmental conditions and development of their management schedules in advance.

CONCLUSION

Systematic research on climate change particularly involving plant diseases is still in nascent stage. Most of the efforts on how climate change may affect plant diseases had been concentrated on the effects of a single atmospheric constituent or meteorological variable on the host, pathogen, or the interaction of the two under controlled conditions. However, interactions are more complex in the real situation, where multiple climatological and biological factors are varying simultaneously in a dynamic environment. Climate change has the potential to modify host physiology and resistances and to alter the stages and rates of development of the pathogen. The most likely impacts would be the shift in the geographical distribution of the host and pathogens, change in the physiology of host-pathogen interactions and change in crop losses. New disease complexes may arise and some diseases may cease to be economically important if warming causes a pole ward shift of agro climatic zones and host plant migrate into new regions. Pathogen would be following the migrating hosts and may infect vegetation of natural plant communities not previously exposed to the often more aggressive strains from agricultural crops.

REFERENCES

Agrios, G. N. (1997). Plant pathology. London: Elsevier.

Anderson, P. K., Cunningham, A. A., Patel, N. G., Morales, F. J., Epstein, P. R., & Daszak, P. (2004). Emerging infectious diseases of plants: Pathogen pollution, climate change and agrotechnology drivers. *Trends in Ecology & Evolution*, *19*(10), 535–544. doi:10.1016/j.tree.2004.07.021 PMID:16701319

APN Project Report. (2011). *The Effects of Climate Change on Pests and Diseases of Major Food Crops in the Asia Pacific Region*. Author.

Baker, K. M., Kirk, W. W., Andresen, J., & Stein, J. M. (2004). A problem case study: influence of climatic trends on late blight epidemiology in potatoes. Acta Horticulturae, 638, 37-42. doi:10.17660/ActaHortic.2004.638.3

Bashi, E., & Rotem, J. (1974). Adaptation of four pathogens to semi-arid habitats as conditioned by penetration rate and germinating spore survival. *Phytopathology*, *64*(7), 1035–1039. doi:10.1094/Phyto-64-1035

Becktell, M. C., Daughtrey, M. L., & Fry, W. E. (2005). Temperature and leaf wetness requirements for pathogen establishment, incubation period, and sporulation of *Phytophthora infestans* on *Petunia × hybrida. Plant Disease*, *89*(9), 975–979. doi:10.1094/PD-89-0975

Blackman, R. L. (1974). Life-cycle variation of Myzus persicae (Sulz.) (Hom., Aphididae) in different parts of the world, in relation to genotype and environment. *Bulletin of Entomological Research*, *63*(04), 595–607. doi:10.1017/S0007485300047830

Boland, G. J., Melzer, M. S., Hopkin, A., Higgins, V., & Nassuth, A. (2004). Climate change and plant diseases in Ontario. *Canadian Journal of Plant Pathology*, *26*(3), 335–350. doi:10.1080/07060660409507151

Boonekamp, P. M. (2012). Are plant diseases too much ignored in the climate change debate? *European Journal of Plant Pathology*, *133*(1), 291–294. doi:10.1007/s10658-011-9934-8

Boyer, J. S. (1995). Biochemical and biophysical aspects of water deficits and the predisposition to disease. *Annual Review of Phytopathology*, *33*(1), 251–274. doi:10.1146/annurev.py.33.090195.001343 PMID:18999961

Central Potato Research Institute. (2011-12). *Annual Report 1998-1999*. Central Potato Research Institute, Shimla, India.

Central Potato Research Institute. (2012-13). *Annual Report 1998-1999*. Central Potato Research Institute, Shimla, India.

Central Potato Research Institute. (2014-15). *Annual Report 1998-1999*. Central Potato Research Institute, Shimla, India.

Chaerani, R., & Voorrips, R. E. (2006). Tomato early blight (*Alternaria solani*): The pathogen, genetics, and breeding for resistance. *Journal of General Plant Pathology*, *72*(6), 335–347. doi:10.1007/s10327-006-0299-3

Chakraborty, S., Tiedemann, A. V., & Teng, P. S. (2000). Climate change: Potential impact on plant diseases. *Environmental Pollution*, *108*(3), 317–326. doi:10.1016/S0269-7491(99)00210-9 PMID:15092926

Chen, C. C., & McCarl, B. A. (2001). An investigation of the relationship between pesticide usage and climate change. *Climatic Change*, *50*(4), 475–487. doi:10.1023/A:1010655503471

Intergovernmental Panel on climate Change. (2013). Summary for Policymakers. In Climate Change2013: The Physical Science Basis. Contribution of Working Group I to the Fifth Assessment Report of the Intergovernmental Panel on Climate Change. Cambridge University Press.

Coakley, S. M. (1988). Variation in Climate and prediction of disease in Plants. *Annual Review of Phytopathology*, *26*(1), 163–181. doi:10.1146/annurev.py.26.090188.001115

Coakley, S. M., Scherm, H., & Chakraborty, S. (1999). Climate change and plant disease management. *Annual Review of Phytopathology*, *37*(1), 399–426. doi:10.1146/annurev.phyto.37.1.399 PMID:11701829

Crosier, W. (1934). Studies in the biology of *Phytophthora infestans* (Mont.) De Bary. *Cornell Univ. Exp. Stn. Memoir, 155*.

Daiber, C. C. (1965). Notes on potato aphids and leaf roll spread at different South African localities. *Journal of the Entomological Society of Southern Africa*, *27*, 191–215.

Daiber, C. C., & Schöll, S. E. (1959). Further notes on the overwintering of the green peach aphid, *Myzus persicae* (Sulzer), in South Africa. *Journal of the Entomological Society of Southern Africa*, *22*, 494–520.

De Jong, C. F., Takken, F. L. W., Cai, X., Dewit, P. J. G. M., & Joosten, M. H. A. J. (2002). Attenuation of Cf-mediated defense responses at elevated temperatures correlates with a decrease in elicitor-binding sites. *Molecular Plant-Microbe Interactions*, *15*(10), 1040–1049. doi:10.1094/MPMI.2002.15.10.1040 PMID:12437302

Dua, V. K., Singh, B. P., Ahmad, I., & Sharma, S. (2015). Potential impact of climate change on late blight outbreak in western Uttar Pradesh and Punjab using JHULSACAST model. *Potato Journal*, *42*(1), 58–71.

Dutt, B. L. (1979). *Bacterial and Fungal Diseases of potato*. ICAR.

Elphinstone, J. G. (1996). Survival and possibilities for extinction of *Pseudomonas solanacearum* (Smith) Smith in cool climates. *Potato Research*, *39*(3), 403–410. doi:10.1007/BF02357946

Fagelfors, H., Wivstad, M., Eckersten, H., Holstein, F., Johansson, S., & Verwijst, T. (2009). *Strategic analysis of Swedish agriculture. Production systems and agricultural landscapes in a time of change*. Swedish University of Agricultural Sciences. Retrieved September 5, 2010, from http://pub-epsilon.slu.se:8080/1626

Garrett, K. A., Dendy, S. P., Frank, E. E., Rouse, M. N., & Travers, S. E. (2006). Climate change effects on plant disease: Genomes to ecosystems. *Annual Review of Phytopathology*, *44*(1), 489–509. doi:10.1146/annurev.phyto.44.070505.143420 PMID:16722808

Ghini, R., Hamada, E., & Bettiol, W. (2008). Climate change and plant diseases. *Scientia Agricola (Piracicaba, Braz.)*, *65*, 98-107.

Goodwin, S. B., Cohen, B. A., & Fry, W. E. (1994). Panglobal distribution of a single clonal lineage of the Irish potato famine fungus. *Proceedings of the National Academy of Sciences of the United States of America*, *91*(24), 11591–11595. doi:10.1073/pnas.91.24.11591 PMID:7972108

Haas, B. J., Kamoun, S., Zody, M. C., Jiang, R. H., Handsaker, R. E., Cano, L. M., & Nusbaum, C. et al. (2009). Genome. sequence and analysis of the Irish potato famine pathogen *Phytophthora infestans Letters*. *Nature*, *461*(7262), 393–398. doi:10.1038/nature08358 PMID:19741609

Harrington, R. (1994). Aphid Layer. *Antenna*, *18*, 50–51.

Harrington, R., Clark, S. J., Welham, S. J., Verrier, P. J., Denholm, C. H., Hullé, M., & Cocu, N. et al. (2007). Environmental change and the phenology of European aphids. *Global Change Biology*, *13*(8), 1550–1564. doi:10.1111/j.1365-2486.2007.01394.x

Hughes, L., & Bazzaz, F. A. (2001). Effects of elevated CO_2 on five plant-aphid interactions. *Entomologia Experimentalis et Applicata, 99*, 87–96.

Indian Institute of Spices Research. (2012). *PhytoFura consolidated report 2009-12*. Kozhikode, Kerala: Indian Institute of Spices Research.

Intergovernmental Panel on Climate Change. (2007). *Climate change: Impact, adaptation and vulnerability, contribution of WG II to the third assessment report of the Intergovernmental Panel on climate Change*. Cambridge Univ. Press. Retrieved from www.ipcc-wg2.org

Jones, R. A. C. (2009). Plant virus emergence and evolution: Origins, new encounter scenarios, factors driving emergence, effects of changing world conditions, and prospects for control. *Virus Research, 141*(2), 113–130. doi:10.1016/j.virusres.2008.07.028 PMID:19159652

Kaukoranta, T. (1996). Impact of global warming on potato late blight: Risk, yield loss, and control. *Agriculture Food Science Finland, 5*, 311–327.

Klopping, H. L., & Delp, C. J. (1980). 2-cyano-N [(ethylamino) carbonyl]-2-(methoxylmino) acetamide, a new fungicide. *Journal of Agricultural and Food Chemistry, 28*(2), 467–468. doi:10.1021/jf60228a010 PMID:7391384

Kocmánková, E., Miroslav, T., Jan, J., Martin, D. D. S., Martin, M., & Zdeněk, Ž. (2009). Impact of climate change on the occurrence and activity of harmful organisms. *Plant Protection Science, 45*(Special Issue), S48–S52.

Kudela, V. (2009). Potential impact of climate change on geographic distribution of plant pathogenic bacteria in Central Europe. *Plant Protection Science, 45*, S27–S32.

Lal, M., & Yadav, S. (2014). *Efficacy of fungicides against Phytophthora infestans in the era of global climate change*. Global conference on technological challenges and human resources for climate smart horticulture-issues and strategies, NAU Navsari.

Lennefors, B. L., Lindsten, K., & Koenig, R. (2000). First record of A and B type *Beet necrotic yellow vein virus* in sugar beets in Sweden. *European Journal of Plant Pathology, 106*(2), 199–201. doi:10.1023/A:1008780812215

Mayton, H., Forbes, G. A., Mizubuti, E. S. G., & Fry, W. E. (2001). The roles of three fungicides in the epidemiology of potato late blight. *Plant Disease, 85*(9), 1006–1012. doi:10.1094/PDIS.2001.85.9.1006

McElrone, A. J., Sherald, J. L., & Forseth, I. N. (2001). Effects of water stress on symptomatology and growth of *Parthenocissus quinquefolia* infected by *Xylella fastidiosa*. *Plant Disease, 85*(11), 1160–1204. doi:10.1094/PDIS.2001.85.11.1160

Mello, A. F. S., Olarte, R. A., Gray, S. M., & Perry, K. L. (2011). Transmission efficiency of Potato virus Y strains $PVY^{O}$and $PVY^{N\text{-}Wi}$ by five aphid species. *Plant Disease*, *95*(10), 1279–1283. doi:10.1094/PDIS-11-10-0855

Minhas, J. S., Kumar, D., Joseph, T. A., Raj, B. T., Paul Khurana, S. M., Pandey, S. K., & Naik, P. S. et al. (2006). Variety suitable for early planting in north-western plains, peninsular India and processing into french fries and chips. *Potato Journal*, *33*(1-2), 35–43.

Mitchell, C. E., & Power, A. G. (2003). Release of invasive plants from fungal and viral pathogens. *Nature*, *24*, 421. PMID:12571594

Mitchell, C. E., Reich, P. B., Tilman, D., & Groth, J. V. (2003). Effects of elevated CO_2, nitrogen deposition, and decreased species diversity on foliar fungal plant disease. *Global Change Biology*, *9*(3), 438–451. doi:10.1046/j.1365-2486.2003.00602.x

Mizubuti, E. S. G., & Fry, W. E. (1998). Temperature effects on developmental stages of isolates from three clonal lineages of *Phytophthora infestans*. *Phytopathology*, *88*(8), 837–843. doi:10.1094/PHYTO.1998.88.8.837 PMID:18944891

Newman, J. A., Gibson, D. J., Parsons, A. J., & Thornley, J. H. M. (2003). Hoe predictable are aphid population responses to elevated CO_2? *Journal of Animal Ecology*, *72*(4), 556–566. doi:10.1046/j.1365-2656.2003.00725.x

Osswald, W. F., Fleischmann, F., & Heiser, I. (2006). Investigations on the effect of ozone, elevated CO_2 and nitrogen fertilization on host- parasite interactions. *Summa Phytopathologica*, *32S*, S111–S113.

Pamella, K. (2004). Emerging infectious diseases of plant: Pathogen pollution, climate change and agrotechnology drivers. *Trends in Ecology & Evolution*, *19*(10), 535–544. doi:10.1016/j.tree.2004.07.021 PMID:16701319

Pérombelon, M. C. M. (1992). Potato blackleg: Epidemiology, host–pathogen interaction and control. *European Journal of Plant Pathology*, *98*, 135–146.

Plessl, M., Elstner, E. F., Rennenberg, H., Habermeyer, J., & Heiser, I. (2007). Influence of elevated CO_2 and ozone concentrations on late blight resistance and growth of potato plants. *Environmental and Experimental Botany*, *60*(3), 447–457. doi:10.1016/j.envexpbot.2007.01.003

Radcliffe, E. B. (1982). Insect pests of potato. *Annual Review of Entomology*, *27*(1), 173–204. doi:10.1146/annurev.en.27.010182.001133

Radcliffe, E. B., & Ragsdale, D. W. (2002). Aphid-transmitted potato viruses: The importance of understanding vector biology. *American Journal of Potato Research, 79*(5), 353–386. doi:10.1007/BF02870173

Robert, Y., Woodford, J. A. T., & Ducray-Bourdin, D. G. (2000). Some epidemiological approaches to the control of aphid-borne virus diseases in seed potato crops in northern Europe. *Virus Research, 71*(1-2), 33–47. doi:10.1016/S0168-1702(00)00186-6 PMID:11137160

Rotem, J. (1994). The Genus Alternaria: Biology, Epidemiology, and Pathogenicity. APS Press.

Rydén, K., Eriksson, B., & Insunza, V. (1986). Rostringar hos potatis orsakade av potatismopptoppvirus (PMTV). *Växtskyddsnotiser, 50*, 97–102.

S. J. V. (2007). *Swedish Board of Agriculture. En meter i timmen—klimatförändringarnas påverkan på jordbruket i Sverige*. (English summary). Report no. 16. ISSN 1102–3007.

Sandgren, M. (1995). Potato mop-top virus (PMTV): Distribution in Sweden, development of symptoms during storage and cultivar trials in field and glasshouse. *Potato Research, 38*(4), 387–397. doi:10.1007/BF02357743

Santala, J., Samuilova, O., Hannukkala, A., Latvala, S., Kortemaa, H., Beuch, U., & Valkonen, J. P. T. et al. (2010). Detection, distribution and control of *Potato mop-top virus*, a soil-borne virus, in northern Europe. *Annals of Applied Biology, 157*(2), 163–178. doi:10.1111/j.1744-7348.2010.00423.x

Scehepers, H. T. A. M. (1996). Effect of rain on efficacy of fungicides deposits potato against *Phytophthora infestans. Potato Research, 39*(4), 541–550. doi:10.1007/BF02358473

Sherf, A. F., & MacNab, A. A. (1986). Vegetable Disease and Their Control. John Wiley and Sons.

Singh, B. P., & Bhat, M. N. (2010). Impact assessment of climate change on potato diseases and insect pests. In H. P. Singh, J. P. Singh, & S. S. Lal (Eds.), *Challenges of climate change Indian horticulture* (pp. 178–184). New Delhi: Westville Publishing house.

Singh, B. P., Dua, V. K., Govindakrishnan, P. M., & Sharma, S. (2013). Impact on climate change on potato. In H. P. Singh et al. (Eds.), Climate-Resilient Horticulture: Adaptation and Mitigation Studies (pp. 125-135). . doi:10.1007/978-81-322-0974-4_12

Staubli, B., Wenger, R., & Wymann Von Dach, S. (2008). *Potatoes and climate change*. Retrieved March 6, 2016 from http://www.inforesources.ch/pdf/focus08_1_e.pdf

Usharani, K. S., Surendernath, B., Paul-Khurana, S. M., Garg, I. D., & Malathi, V. G. (2004). Potato leaf curl – a new disease of potato in northern India caused by a strain of Tomato leaf curl New Delhi virus. *Plant Pathology*, *53*(2), 235. doi:10.1111/j.0032-0862.2004.00959.x

Van der Waals, J. E., Korsten, L., Aveling, T. A. S., & Denner, F. D. N. (2003). Influence of environmental factors on field concentrations of *Alternaria solani* conidia above a South African potato crop. *Phytoparasitica*, *31*(4), 353–364. doi:10.1007/BF02979806

Verbeek, M., Piron, P. G. M., Dullemans, A. M., Cuperus, C., & Van der Vlugt, R. A. A. (2010). Determination of aphid transmission efficiencies for N, NTN and Wilga strains of Potato virus Y. *Annals of Applied Biology*, *156*(1), 39–49. doi:10.1111/j.1744-7348.2009.00359.x

Visser, J. C., & Bellstedt, D. U. (2009). An assessment of molecular variability and recombination patterns in South African isolates of Potato virus Y. *Archives of Virology*, *154*(12), 1891–1900. doi:10.1007/s00705-009-0525-3 PMID:19862472

Waggoner, P. E., & Parlange, J. Y. (1975). Slowing of spore germination with changes between moderately warm and cool temperatures. *Phytopathology*, *65*(5), 551–553. doi:10.1094/Phyto-65-551

Wang, Y., Bao, Z., Zhu, Y., & Hua, J. (2009). Analysis of temperature modulation of plant defense against biotrophic microbes. *Molecular Plant-Microbe Interactions*, *22*(5), 498–506. doi:10.1094/MPMI-22-5-0498 PMID:19348568

Yang, X. B., & Scherm, H. (1997). El Niño and infectious disease. *Science*, *275*(5301), 7399. doi:10.1126/science.275.5301.737c PMID:9036530

Ywa, N. S., Walling, L., & McCool, P. M. (1995). Influence of elevated CO_2 on disease development and induction of PR proteins in tomato roots by *Phytophthora parasitica. Plant Physiololgy, 85*(Supplement), 113. Retrieved November 4, 2015, from http://www.climateandforming.org

KEY TERMS AND DEFINITIONS

Fungicide: Any chemicals or molecules which have the ability to inhibit or kill the growth of pathogens.

IPCC: Intergovernmental Panel on Climate Change. It is an organization which works on issues related to climate change and informing the information to policy makers and public.

Late Blight: Late blight is a disease of potato/tomato, which is caused by a fungus like organism.

Pathogen: Any organisms which cause the economic loss are called pathogen.

Resistance: it is ability of an organism to overcome or reduced the effect of a pathogen.

Seed Borne: A pathogen which survives on/in the seed is called seed borne pathogen.

Soil Borne: A pathogen which survives on/in the soil is called seed borne pathogen.

Wilt: Any loss of turgidity and dropping of plant, which is caused by lack of water or infection in vascular system by organism.

Chapter 11
Diseases of Potato:
A Major Constraint to Potato Production

Baby Summuna
Sher-e-Kashmir University of Agricultural Sciences and Technology of Kashmir, India

Sachin Gupta
Sher-e-Kashmir University of Agricultural Sciences and Technology of Jammu, India

Moni Gupta
Sher-e-Kashmir University of Agricultural Sciences and Technology of Jammu, India

ABSTRACT

Potato production is seriously compromised due to prevalence of a number of diseases and they are the major constraints in potato production resulting in significant yield reduction. Integrated disease management of potato includes regular inspection for healthy seed or nursery, crop production, correct identification of the problem, cultural practices (crop rotation, sanitation etc.), biological control, soil fumigation (if necessary), seed or nursery stock treatment and disinfestations of cutting tools. Due to the ever increasing number of new fungicide resistant fungal pathogens, proper and timely diagnosis of potato diseases is becoming paramount to effective disease management, and growers need up-to-date information to help make important decisions on optimal use and timing of pesticides and other control options.

DOI: 10.4018/978-1-5225-1715-3.ch011

INTRODUCTION

Potato (*Solanum tuberosum* L.) is a herbaceous perennial plant in the family Solanaceae which is grown for its edible tubers. The importance of potato as a food crop was duly recognized soon after its introduction in Europe in the 16th century. Potato (*Solanum tuberosum* L.) is one of mankind's most valuable food crops (FAO, 2004). It is the most important vegetable crop in terms of quantities produced and consumed worldwide (FAO, 2005). In volume of production it ranks fourth in the world after wheat (*Triticum aestivum* L.), rice (*Oryza sativa* L.), and maize (*Zea mays* L.) (Bowen, 2003). It remains an essential crop in Europe where per capita production is still the highest in the world, but the most rapid expansion over the past few decades has occurred in Southern and Eastern Asia. However, the local importance of the potato is variable and changing rapidly. Potato has become an integral part of much of the world's cuisine providing a balanced source of starch, vitamins and minerals to many communities across the globe. Potato can produce more food per unit area per unit time than the conventional cereal crops. Potatoes have the potential to relieve the pressure of increasing cereal prices on the poorest people and contribute significantly to food security particularly in the under developed and developing countries. Potato production and consumption is accelerating in most of the developing countries including India primarily because of increasing industrialization.

Potatoes in India are grown under varied climatic conditions ranging from tropics, subtropics to temperate highlands. Consequently, the spectrum of insect-pests and diseases is very large. All of them put together have the potential to limit potato production upto 85% depending upon the weather/region. Potato production is seriously compromised due to prevalence of a number of diseases and they are the major constraints in potato production resulting in significant yield reduction. Full potential of potato crop can only be realized if the diseases of potato are kept under control. Successful production of quality potato can be undertaken only in those areas and fields which are free from serious soil borne pathogens like wart, bacterial wilt, black scurf and common scab. Commercial production of most potatoes is primarily through vegetative propagation by means of lateral buds formed on the tuber, a modified stem. Through such vegetative propagation, many diseases are transmitted from generation to generation. Suppression of such diseases and reduction of yield losses due to disease are a necessary part of increasing the food supply. Diseases of potato assume significance for the fact that late blight of potato laid the foundation of plant pathology. Late blight is the disease that triggered the Irish potato famine of the 1840s. Between 1800 and 1845, severe food shortages had occurred in various parts of Ireland. However, during the famine, the crop failure became national for the first time, affecting the entire country at once. It also was the first plant disease for which a microorganism was proved to be the causal agent, leading to the birth of plant pathology as a science.

Potato is a vegetative propagated crop and therefore the availability of good quality planting material is a major constraint in successful production of potatoes as seed tubers suffer from many diseases. Potato diseases can be classified according to the class of organisms which are responsible for causing them. Primarily bacteria, fungi, virus and viroids are responsible for causing potato diseases. The principles, strategies, and tactics of plant disease management are important to preventing yield losses. Integrated disease management (IDM) may supply effective control of the potato diseases e.g., verticillium wilt, blackleg, bacterial ring rot, late blight, early blight, scab etc. It includes regular inspection for healthy seed or nursery, crop production, correct identification of the problem, cultural practices (crop rotation, sanitation etc.), biological control, soil fumigation, seed or nursery stock treatment and disinfestations of cutting tools. Due to the ever increasing number of new fungicide resistant fungal pathogens, proper and timely diagnosis of potato diseases is becoming paramount to effective disease management, and growers need up-to-date information to help make important decisions on optimal use and timing of pesticides and other control options.

A brief account of diseases which infect potato and reduces their production is presented as under:

BACTERIAL DISEASES

Potato is affected mainly by six bacterial diseases *viz.,* bacterial wilt or brown rot (*Ralstonia solanacearum*), soft rot of stem and tubers (*Erwinia carotovora, Bacillus* spp., *Pseudomonas* spp.), ring rot (*Clavibacter michiganensis*), common scab (*Streptomyces* spp.), pink eye (*Pseudomonas* spp.) and leaf spot (*Xanthomonas vesicatoria*). Bacterial wilt/brown rot is the most destructive followed by common scab and soft rot.

Bacterial Wilt or Brown Rot

Bacterial wilt or brown rot is caused by *Ralstonia solanacearum.* In all potato growing areas, bacterial wilt caused by *R. solanacearum* is regarded as an important disease contributing to yield reduction. It is considered more problematic than late blight since it has no known chemical control procedures and many farmers do not know how to control it. It has been estimated to affect potato crop in 3.75 million acres in approximately 80 countries with global damage estimates exceeding $950 million per year (Shekhawat *et al.,* 2000). In India, the disease is prevalent in all potato growing areas except north western plains comprising of Rajasthan, Punjab and Haryana and north central part of Uttar Pradesh. Losses up to 75 per cent have been recorded under extreme conditions. With increase in global temperature, the disease is likely to spread to new areas and affect potato cultivation. Bacterial wilt is a primary factor restricting seed production in states of West Bengal, Karnataka, Maharashtra, Orissa, NEH Region, and Bihar that constitutes about 50-55% potato area. The disease causes damage at two stages; (1) killing the standing plants by causing wilt and (2) causing rot of infected tubers in storage and transit. Another indirect loss is spread of the disease through planting of healthy looking tubers harvested from infested fields. Bacterial wilt poses a serious restriction to seed and processing potato production. Potato breeder seed production cannot be undertaken in those fields having even slightest bacterial wilt incidence. The tolerance limit in Foundation seed I is zero and in Foundation seed II and certified seed, it is only 3 wilted plants per hectare. The infected tubers are not suitable for processing purpose also.

Symptoms

The earliest symptom of the disease is slight wilting in leaves of top branches during hot sunny clear days. The leaves show drooping due to loss of turgidity followed by total unrecoverable wilt. In advanced stages of wilt, cut end of base of the stem may show dull white ooze on squeezing. Bacterial wilt in field can be distinguished from any fungal wilt by carrying out ooze test. In tubers, two types of symptoms are produced; they are vascular rot and pitted lesion on surface. In vascular rot, the vascular tissues of transversely cut tuber show water soaked brown circles and in about 2-3 minutes, dirty white sticky drops appear in the circle. In advanced stages of wilt, bacterial mass may ooze out from eyes. Such eyes may carry soil glued with the bacterial ooze. Second kind of symptom is the lesions on tubers. The lesions are produced due to infection through lenticels (skin pore). Initially, water soaked spot develop which enlarges in the form of pitted lesion. The tubers may not rot in storage and also may not show vascular browning. These symptoms on tuber surface are more common in north eastern regions of India.

Survival and Spread

The pathogen survives through infected seed tubers and in plant debris in soil. Symptomless plants may harbour the bacterium and transmit it to progeny tubers as latent infection. This could lead to severe disease outbreaks when the tubers are grown at disease free sites. High soil moisture, temperature, oxygen stress and soil type affect the survival of the pathogen. The pathogen population decline gradually in soil devoid of host plants and their debris. Transmission of *R. solanacearum* from one area to another occurs through infected seed, irrigation water and farm implements.

Management

Absolute control of the disease is difficult to achieve due to seed and soil borne nature of the pathogen, however economic losses certainly can be brought down considerably using the following eco-friendly means:

- **Healthy Seed:** Use of healthy planting material can take care of almost 80% of bacterial wilt problem. Western and Central Indo-Gangetic plains have been identified as bacterial wilt free areas, so potato seeds obtained from these areas can be used for growing disease free crop. Infected tubers should not be cut as the cutting knife may spread the disease.
- **Field Sanitation:** The disease can be minimized by adopting the following agronomic practices:
    - **Crop Rotation:** 2-3 years crop rotation using non solanaceous crops like maize, cereals, garlic, onion, cabbage etc. Paddy and sugarcane although are not host, still they carry pathogen and contribute to the disease perpetuation.
    - **Avoid Tillage Operations:** Pathogen enters in plant through root or stolen injuries. Such injuries cannot be avoided during intercultural operations. Therefore, tillage should be restricted to the minimum and it is advisable to follow full earthing up at planting.
    - **Off-Season Management of Field:** The pathogen perpetuates in the root system of many weeds and crops. Weeds and root/foliage remnants should be destroyed and burnt. The pathogen in remnants can be exposed to high temperature above 40ºC in summer in plains and plateau and low temperature below 5ºC in hills by giving deep ploughing. This may cause extinction of pathogen from the field.
    - **Chemical Control:** Application of stable bleaching powder @ 12 kg/ha at the time of potato planting in furrow along with fertilizer reduces pathogen population from field and gives effective control.

Soft Rot or Black Leg

Potato blackleg is caused by *P. atrosepticum*, *P. carotovorum*, *P. carotovorum*, *P. wasabiae* and *Dickeya* spp. wherever potato is cultivated. Disease symptoms caused by these different pathogens are indistinguishable. Bacterial species belonging to different genera like *Pectobacterium*, *Dickeya*, *Pseudomonas*, *Bacillus*, *Clostridium*,

Aerobacter, *Flavobacterium* and *Rhodococcus* are able to cause tuber rot. All these bacteria possess the ability to produce plant tissue macerating enzymes. Of these, soft rot and blackleg causing *Pectobacterium* and *Dickeya* spp. are regarded as the most important. *Pectobacterium* and *Dickeya* spp. are primary pathogens whereas species belonging to the other genera are in general only able to enhance decay after rot has been initiated by *Dickeya* and/or *Pectobacterium* spp. All species easily cause soft rot of tubers during storage if environmental conditions favour disease progression, but the bacteria differ in their relative contribution to blackleg incidences (Perombelon, 2002). For a long time, *P. carotovorum* was considered to play a minor role in potato blackleg, but recently it has been proven that *P. carotovorum* infections can result in typical blackleg in Europe. This pathogen is more virulent than *P. atrosepticum* under warm climate conditions. Bacterial soft rot can cause significant loss of potato tubers at harvest, transit and storage. Losses under bad handling of the produce, poorly ventilated storage or transit may go up to 100 per cent. Soft rot bacteria usually infect potato tubers which have been damaged by mechanical injury or in the presence of other tuber borne pathogens. Bacterial soft rot develops much faster under warm and humid conditions. The disease also results in blackleg of foliage during the crop growing season. Soft rot is mainly caused by coliform bacteria called *Pectobacterium atrosepticum*. Excessive moisture in field predisposes the tubers to soft rot. It leads to the proliferation of lenticels and create anaerobic conditions. Water film and injury on tubers is essential for soft rot bacteria to infect and proliferate. High temperature during harvest predisposes tubers to soft rot. Tubers which are immature, large, damaged by hail, exposed to late blight or dry rot are more prone to soft rot. Nitrogen fertilization particularly ammonium chloride application during cultivation makes tuber more prone to soft rot. High nitrogen dose delays the maturity while chloride ions increase the water content and decrease dry matter in tubers leading to increased soft rot.

Symptoms

Initially a small area of tuber tissue around lenticels or stolon attachment point becomes water soaked and soft. Under low humidity, the initial soft rot lesions may become dry and sunken. Under high humidity, the lesions may enlarge and spread to larger area. Tubers in advanced stages of decay are usually invaded by other organisms and the decaying tissue becomes slimy with foul smell and brown liquid ooze. The tuber skin remains intact and sometimes the rotten tubers are swollen due to gas formation. At harvest, many small rotten tubers with intact skin can be seen. The infected seed tubers rot before emergence resulting in poor stand of the crop. Black-leg phase of the disease is not common in India. However, the disease infests in the form of soft rot infected seed tubers and the affected haulms become black at collar region just above the ground. Infected plants develop yellowing, start wilting and die early without producing any tubers. Water soaked lesions develop on succulent stems, petioles, and leaves. On stem and petioles, the lesions first enlarge into stripes, turn black and then invade the affected parts causing soft rot and toppling of the stem and leaves. Veins on lower surface of the leaves become water-soaked and gradually turn necrotic. The entire leaf or leaflets curl downwards. The apical buds become bunchy and stunted. The growth and unfolding of the leaves get delayed. The partly expanded leaves are distorted with necrotic margins.

Survival

Soft rot bacteria may survive in soil, on tuber surface, lenticels, periderm, cortex, ground tissue and vascular tissue. Rotting and decay of infected tubers in fields or stores may cause extensive contamination of adjacent healthy tubers, which serves as the most important source of primary inoculums. Contaminated irrigation water, aerosols of rains, farm implements, soil micro-fauna, nematodes, earthworms, larvae and adults of some insects etc. also help in secondary spread of the disease. Excessive moisture creating anaerobic condition, high temperature, excess nitrogen, tuber injuries and poor ventilation during storage are the important factors helping in disease development.

Spread

In warm climates, where one potato crop follows another or where only short rotation cycles are applied, the bacteria can pass easily from one crop to the next, especially in poorly drained soil. The bacteria can be disseminated in the potato fields by irrigation water, insects, rain or bacterial aerosols. The pathogen may also spread through water during washing of the produce with contaminated water. Soft rot causing bacteria can also spread easily from diseased to healthy tubers during storage, handling and grading. Insects especially maggots of anthomyiid fly (*Hylemyia* spp.) may also transmit the bacteria from one tuber to another.

Management

Pectolytic bacteria causing soft rot are present in soil, water and tubers. Soft rot bacteria are carried deep inside the tuber, in lenticels and surface wounds making it difficult to eradicate. These quiescent bacteria proliferate in high moisture condition and require water film that cause anaerobic conditions leading to disease development. Surface injury predisposes the tubers to soft rot infection. Avoidance of excess irrigation and proper drainage coupled with restricted nitrogen dose helps to minimize the disease incidence. Adjustment of planting time so as to avoid hot weather during plant emergence and harvesting the crop before soil temperature rises above 28^{0}C is effective in reducing disease incidence. Sorting out of injured and bruised tubers, treatment of tubers with 3% boric acid for 30 min. and storing the produce either in well-ventilated cool stores or cold stores reduces the losses caused by disease during storage.

Common Scab of Potato

Common scab of potato has become endemic in various potato growing states. The disease doesn't cause a detectable yield loss but does cause economic losses to the growers as the affected tubers fetch low prices in market due to bad look (pit, lesion and russeting) and is out rightly rejected for seed purposes by seed agencies when incidence is 3-5% or by processing industries as peeling loss is very high in such tubers.

Symptoms

Scab begins as small reddish or brownish spot on the surface of the potato tubers and its initial infection take place during juvenile period of tuber. Infection takes place mainly through lenticels and surrounding periderm turns brown and rough. Lesion becomes corky due to elongation and division of invaded cells. Under Indian conditions, multiple kinds of symptoms have been recorded and they are grouped as:

1. A mere brownish roughening or abrasion of tuber skin.
2. Proliferated lenticels with hard corky deposition that might lead to star shaped lesion.
3. Raised rough and corky pustules.
4. 3-4 mm deep pits surrounded by hard corky tissue.
5. Concentric series of wrinkled layers of cork around central black core.

The last type of symptoms has been attributed to mix infection of scab and *Fusarium oxysporum*.

Survival and Spread

The organism is tuber-borne and is well-adapted saprophyte that persists in soil on decaying organic matter and manure for several years. Infected tubers serve as inoculum foci in the field giving rise to infected progeny tubers. The pathogenic

Streptomyces spp. are both soil as well as tuber borne. Tuber-borne inoculum is likely to be involved in the distribution of new strains or species.

Management

The pathogen is difficult to eradicate because of long survival both on seed tubers and in soils. Therefore, practices to minimize the inoculum like use of disease free seed tubers and tuber treatment with boric acid (3% for 30 min.) before or after cold storage and creating adverse condition for pathogen spread/disease development like irrigating the crop repeatedly to keep the moisture near to field capacity right from tuber initiation until the tubers measure 1 cm in diameter, maintaining high moisture in ridges for at least a few weeks during the initial tuberization phase and crop rotation with non solanaceous crops are recommended.

FUNGAL DISEASES OF POTATO

Late Blight

Late blight is probably the single most important disease of potatoes and tomatoes worldwide (Son *et al.* 2008). Worldwide losses due to late blight are estimated to exceed $5 billion annually and thus the pathogen is regarded as a threat to global food security (Latijnhouwers *et al.* 2004). Late blight was responsible for the Irish potato famine in the 1840s (Mercure 1998). Ireland's potato crop failures in the past had always been regional and short-lived with modest loss of life. Between 1800 and 1845, sixteen food shortages had occurred in various parts of Ireland. However, during the Famine the crop failure assumed national importance for the first time, affecting the entire country at once. It also was the first plant disease for which a microorganism was proved to be the causal agent, leading to the birth of plant pathology as a science. It is caused by the fungus *Phytopthora infestans*.

Symptoms

The disease affects all plant parts *viz.* leave, stems and tubers. It appears on leaves as small pale green spots, which enlarge into large water soaked lesions. A white mildew (cottony growth) ring forms around the dead areas on the lower side of leaves. In dry weather, water soaked areas turn necrotic brown. On stems, light brown elongated lesions are formed which may encircle the stem. Tubers develop reddish brown, shallow to deep, dry rot lesions. The affected tuber flesh becomes 'caramalised' with a sugary texture. Frequently metallic tinge develops on the margins of the affected tissue. Tubers carrying the pathogen are the real carriers and serve as the source of the disease in the subsequent season. Infected seed tubers grow into healthy plants but under favourable conditions for the disease (10-12^{0}C and RH>80%) development, the disease infects the stem and lower leaves.

Management

Seed potatoes should be checked thoroughly before storage. All blighted tubers must be removed and buried deep in the soil. Ridges should be made high enough to cover all daughter tubers and reduce chance of their infection upon exposure. If the weather conditions (temperature 10-20ºC, RH>80%) are favourable for the disease development irrigation should be stopped immediately. When the disease affects 75% crop foliage, the haulms should be cut, removed from the field, and buried deep. Protective sprays with a contact fungicide, *viz.,* Mancozeb (0.2%) before appearance of the disease is effective. Subsequent sprays if necessary should be repeated at 8 to 10 days interval. In case of severe blight attack, one or two sprays of Metalaxyl (0.25%) are given to check the further spread of the disease. Mancozeb is applied at an interval of 15 days after the Metalaxyl application.

Early Blight

Early blight, that is caused *Alternaria* (*A.solani* and *A.alternata*) occurs worldwide on potato crop particularly in the regions with high temperature and alternating periods of dry weather and high humidity and/or irrigated potato soils, light-textured, sandy, low in organic matter. Early blight occurs in all potato production areas, but there is a significant impact on the tuber yield and the quality only in warm, wet conditions in the early season, which favour a rapid disease development. Disease symptoms are characteristic dark brown to black lesions with concentric rings, which produce a 'target spot' effect. Symptoms are initially observed on older, senescing leaves. The causal organism of early blight was first described by Ellis & Martin (1882) as *Macrosporium solani*. The first reference to the fungus as a parasite and its association with potato leaf blight was by Galloway (1891) in Australia. In the USA, Chester (1892) noted the disease on potatoes and other cultivated plants of the Solanaceae, particularly tomato (*Lycopersicon esculentum* Mill.) and eggplant (*Solanum melongena* L.). He described the symptoms and observed that the progress of early blight was slower than that of late blight, caused by *Phytophthora infestans* (Mont.) de Bary. He also noticed that potato plants severely affected by early blight had lower yields and produced smaller tubers. The disease mainly infects leaves and tubers. Initially the symptoms occur on the lower and older leaves in the form of small (1-2 mm) circular to oval brown spots. These lesions have the tendency to become large and angular at later stage. Mature lesions on foliage look dry and papery, and often have the concentric rings, looking like bulls eye. The symptoms on the tuber comprise of brown, circular to irregular and depressed lesions with underneath flesh turning dry, brown and corky. Lesions tend to enlarge during storage and affected tubers later become shriveled. Jones (1893) was the first to suggest the name early blight to distinguish the disease from late blight. The name stems from the fact that early blight attacks early maturing cultivars more severely than medium or late maturing ones, whereas late blight is more severe on medium or late-maturing cultivars.

Epidemiology

The fungus can survive in soil and plant debris particularly in temperate climate. The infected tubers form the primary source of inoculum. Moderate temperature (17-25^{0}C), high humidity, intermittent dry and wet weather is more conducive for early blight. Early blight is considered polycyclic with repeating cycles of new infection. This is the period when the disease has the potential to spread rapidly and build up to damaging levels in the crop.

Dispersal

Conidia are spread by wind and splashing water. Wind, rain and insects are the principle methods of dissemination of *A. solani*. The primary inoculum produces conidia in the spring, which are then dispersed to the lower leaves of the plant where they germinate and penetrate the epidermis directly, through stomata or wounds. With the use of spore traps, it has been found that peak spore dispersal precedes the hottest and driest hour of the day by two hours, and the time of maximum wind velocity by four hours. A rapid rate of dispersal does not set in until infection has reached the stage at which whole leaves dry up and plants begin to die. The curve of wind velocity resembles that of spore dispersal, however, there appears to be no significant correlation between spore dispersal and other climatological factors (Van der Waals, 2002); (Figure 1).

Management

Use of Disease Free Seed Tubers for Raising the Crop

The crop must be given balanced doses of fertilizers, especially nitrogen. Spraying the crop with urea (1.0%) at 45 days after sowing and giving subsequent sprays 8-10 days after the first spray helps the crop to easily escape the severe on-slaught of early blight disease. In the hilly regions, spraying of copper oxychloride (0.3%)

Figure 1. Disease cycle of early blight pathogen, Alternaria solani
Source: (Wharton and Kirk, 2012)

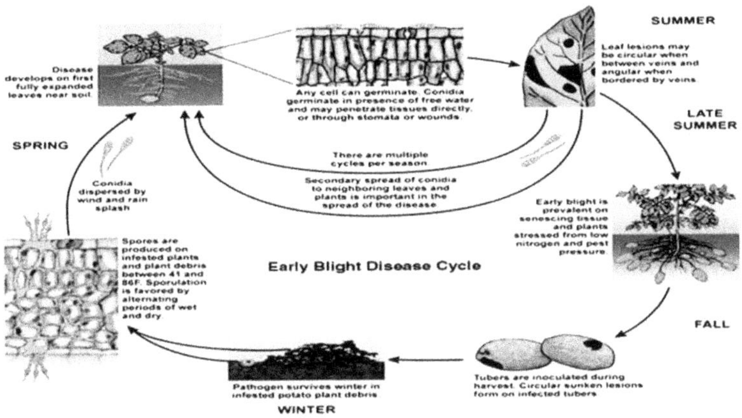

and bordeaux mixture (1.0%) is recommended for control of early blight disease. Solanaceous crops act as the collateral hosts for the disease organism, hence their cultivation nearby potato fields must be avoided.

Early blight can be controlled by efficient use of cultural practices, such as a 3–5-year crop rotation with non-host crops, site selection, sanitation of fields, providing proper plant nutrition, avoiding water stress and planting disease-free seed (Madden *et al.* 1978). Generally, the best crops for rotation are forage crops and grains, including maize (*Zea mays* L.). A high frequency of potato or tomato cropping in one field, as well as consecutive plantings of potatoes or tomatoes, are associated with an earlier appearance of initial early blight lesions (Shtienberg & Fry, 1990). Planting cultivars that are less susceptible to early blight may also reduce disease severity. However, Shtienberg & Fry (1990) showed that host resistance has no significant effect on the initial appearance of early blight. Tuber infection can be decreased by allowing tubers to mature before harvesting, avoiding excessive wounding at harvest and providing storage conditions conducive to wound healing (Venette and Harrison 1973; Workman *et al.* 1983).

These various cultural practices can reduce the severity of early blight, but under situations of sufficient inoculum and environmental conditions favourable for disease, complete control will not be achieved.The most effective control method is a protectant fungicide spray programme used from early in the growing season to vine kill (Jones 1912; Harrison *et al.* 1965 a,b; Harrison & Venette 1970; Douglas & Groskopp 1974). In Colorado field trials, yields in plots treated with fungicide were approximately 20–40% higher than in untreated plots (Harrison & Venette 1970), while in Minnesota, chemical control of early blight resulted in yield increases of upto 90% compared to unsprayed controls (Teng & Bissonnette, 1985). Proper timing of initial and subsequent fungicide applications can reduce the overall number of sprays with no significant loss in yield. The most important consideration in the use of fungicides to control early blight is coverage. With aerial application of fungicides, it is important to ensure that the lower, senescing leaves (where most of the early blight lesions occur) receive fungicide to prevent spread of the disease.

Verticillium Wilt

Verticillium wilt of potatoes can be caused by two different soil-borne fungi, *Verticillium alboatrum* or *Verticillium dahliae*. In addition to tomatoes and potatoes, these fungi can infect cucumber, eggplant, pepper, rhubarb, watermelon, artichoke, beet, broad bean, strawberries, raspberries and a number of weedy plants.

Symptoms

The infection starts from the roots and the fungus grows into the stem and colonizes the xylem vessels thereby disrupting the water and mineral supply to the aerial parts that results in plants remain stunted, lack vigour, lower leaves, tend to droop and there is loss of turgidity. Vascular bundles of stem and tuber become brown. In tuber, initial infection is seen as yellowish discolouration at the stolon end. Wilting is the most characteristic symptom of infection by *Verticillium* spp. In addition, infected plants often have a characteristic V-shaped lesion at the edge of the leaf occurring in a fan pattern. These foliar lesions can enlarge, resulting in complete browning and death of the leaves.

Management

Management of this disease is difficult since the pathogen survives in the soil and can infect many species of plants. A combination of methods may decrease its effect. Planting of resistant varieties and removal followed by destruction of infected plant material is effective in disease spread. Susceptible crops can be rotated with non-hosts such as cereals and grasses.

Powdery Mildew

The fungus *Erysiphe cichoracearum* is responsible for this disease. It first appears on potatoes as brown lesions of various sizes on stems and petioles. Lesions coalesce to form short streaks or stippled patches. If the air is moist, diseased leaves eventually may be covered by it. Leaves and stems are killed and only the tip of the plant may remain green. Late in the season, small black specks may develop in the powdery growth.

Management

Planting of resistant cultivars helps to manage the disease incidence. Wider spacing between the hills and the rows to increase air flow among plants reduces disease incidence. Crop rotation and removal of infected plant materials and alternative hosts, pruning of overcrowded plants to increase air circulation reduces the relative humidity and increases the light penetration thereby reducing the disease infection.

Wart

Potato wart is an important disease of the cultivated potato (*Solanum tuberosum* L.) known by various names like black wart, black scab, potato tumor, potato cancer or canker, cauliflower disease, warty disease and many other descriptive terms in diverse languages and cultural backgrounds where potato is grown and the disease is present (Frank, 2007). The causative agent of wart is the obligate biotrophic, soil-borne fungus *Synchytrium endobioticum* (Schilb.) Perc. The most favourable conditions for the development of the disease are periodic flooding followed by lack of proper drainage and aeration. The disease is characterized by 'cauliflower-like' warty growths on tubers, stolons and stem bases but not roots. Under wet conditions, it may be seen in the form of greenish-yellow crust on the stems and leaves at or near the soil level. All the tubers on diseased plant do not necessarily develop warts. Diseased tubers may show formation of either one or more tumours. Such tubers sometimes are completely transformed into warty mass. The tumors may turn brown to black with age.

Management

Wart affected tubers used as seed are the chief means of the disease spread. The disease may also spread through seed of wart immune varieties grown in wart infested land, contaminated soil carried on the feet of men, animals or farm implements and manure containing diseased material. Control of the disease is possible only by cultivation of resistant varieties. However preventive measures like practicing long crop rotation (5 years or more), using disease free potatoes as seed material and burning of wart affected lumps and potato peelings are effective in checking the spread of the disease.

Charcoal Rot

The disease is caused by the fungi *Macrophomina phaseolina* and infects the tubers in the soil through proliferated lenticels and injuries. Black spots appear around the lenticels and eyes which enlarge into patches extending deep into the tuber flesh. The pathogen infects through lenticels, eyes, stolons and wounds made by larvae of the tuber moth to cause black sunken lesions and later blackening of internal tissues.

Management

Harvesting the crop early before the soil temperature reaches 28^0C can check the disease.

Black Scurf and Stem Canker

The disease caused by *Rhizoctonia solani* commonly affects the tubers, sprouts, stems and stolons. The most common symptom is black scurf comprising of dark brown to black irregular lumps sticking on the surface of tubers. These irregular lumps are closely adhered to the tuber surface and do not wash off easily. Other symptoms on the tuber include skin cracks, crater like depressions, pitting, stem-end necrosis and shape deformity. The disease often causes sprout injury both in storage and in fields after planting. The affected sprouts show discoloration of tissue. The heavily infected sprouts cannot emerge from soil leading to loss of germination. The emerging sprouts when infected later develop cankers causing girdling of stem bases. Such affected plants show upward rolling of leaves with pinkish or purplish margin. Often small green or reddish aerial tubers are also formed in the axils. The infection also spreads to roots and developing stolons resulting in rotting of cortical tissues. Such infected roots later shed away, hence infected plants have poor root system. Infected stolons give rise to deformed tubers.

Management

Combination of tuber disinfection and improved cultural practices successfully checks the incidence and severity of black scurf. In the hills tuber treatment with an organomercurial compound and soil application of pentachloronitrobenzene (PCNB) @ 30kg/ha is most effective. In the North India plains, treatment of the diseased seed with thiabendazole, thiabendazole (TBZ) + 8 hydroxyquinoline, acetic acid + zinc sulphate, carbendazim and boric acid effectively controls the disease. The progeny tubers of such treated seeds are usually free from black scurf. A continuous use of treated seed for 2-3 crop seasons is found to completely check the disease.

Crop rotation with maize (*Zea mays*) or *Sesbania aegyptiaca* (dhaincha) for green manure also checks the disease build up.

Fusarium Dry Rot

The dry rot is an important disease of storage caused by *Fusarium* spp. The skin of dry rot infected tubers first becomes brown then turns darker and develops wrinkles. These wrinkles are often arranged in irregular concentric circles. In the later stage of infection, a hole may be observed in the center of the concentric ring with whitish or pinkish growth of fungal mycelium. On cutting these affected tubers, whitish or brownish tissues are seen with one or more cavities. Eventually the infected tubers loose water and become dry, hard and shriveled.

Management

Use of clean and healthy seed tubers for planting and tuber washing followed by drying under shade substantially reduces the infection. Dipping the tubers in organo mercurial compounds (0.2%) for 30 minutes is effective. Tuber damage and injury must be avoided during harvest and storage. Tubers should be stored in cold stores in plains.

POTATO VIRUSES

Viruses disseminated in tubers can cut yields by 50 per cent. The major virus affecting the yield of potato and causing diseases are:

- **Potato Leaf Roll Virus (PLRV):** Causes an important disease of potatoes affecting the quantity and quality of production and may cause a crop to be ineligible for certification. Foliar symptoms of PLRV can be divided into primary and secondary infections. Primary infection results when an initially healthy plant is inoculated via aphids during the current season. Symptoms first appear where inoculation occurs. The upper leaves become pale, upright, rolled and show some reddening of the tissue around the leaf edges. The lower leaves may or may not have symptoms. Secondary infection occurs when an infected tuber is planted, giving rise to an infected plant. The lower leaves are severely rolled and leathery to the touch. The plant frequently has an overall stunted, upright, chlorotic appearance. The oldest leaves may show reddening on the margins or chlorosis. The upper leaves may not have obvious symptoms. PLRV is transmitted in a persistent manner by several aphid species, the most important being the green peach aphid *(Myzas persicae)*. In addition to infecting potato, the virus infects other solanaceous crops (tomato, tobacco) and weeds (jimsonweed). Control consists of suppressing aphid populations with systemic and foliar insecticides and planting certified seed.

- **Potato Virus Y (PVY):** One of the most important viruses infecting potatoes. It is readily spread by aphids in a non-persistent manner as well as mechanically by human activity and may result in severely depressed yields. PVY is tuber borne and can interact with other viruses such as PVX and PVA to result in heavier losses. Control depends on the use of disease-free seed, insecticides to reduce aphid populations, and mineral oil sprays to interfere with the aphid transmission process.
- **Potato Virus X (PVX):** One of the most widely distributed viruses of potatoes because no symptoms develop in some varieties (latent mosaic), the full extent of damage with PVX is not recognized. Mixed infections of PVX with other viruses like PVY and PVA cause more damage than PVX alone. PVX is tuber borne and is readily mechanically transmitted by human activities. Tobacco, pepper, and tomato are additional hosts for this virus.
- **Potato Spindle Tuber Viroid (PSTV):** PSTV consists of a small RNA molecule lacking the protein coat of viruses. It is often transmitted through breeders' progenies mechanically, as well as through pollen and true seed.

The leaflets of infected plants may be smaller and curve inwardly giving a stiff upward growth habit. Use of these tests and the selection of certified seed are important steps in eliminating this disease from potato stocks.

Thus, it is evident that potato diseases are of significant economic importance leading to huge economic losses and if managed timely can prevent national losses ultimately leading to food security.

REFERENCES

Bowen, W. T. (2003). Water productivity and potato cultivation. In J. W. Kijne, R. Barker & D. Molden (Eds.), *Water Productivity in Agriculture: Limits and Opportunities for Improvement* (pp. 229-238). CAB International. Retrieved April 10, 2016, from http://www.iwmi.cgiar.org/pubs/Book/CA_CABISeries/Water_Productivity/Protected/0 851996698ch14.pdf/

Chester, F. D. (1892). Diseases of the round potato and their treatment. *Annual Report of the Delaware College Agricultural Experimental Station*, *5*, 67–70.

Douglas, D. R., & Groskopp, M. D. (1974). Control of early blight in eastern and south central Idaho. *American Potato Journal, 51*(11), 361–368. doi:10.1007/BF02852082

Ellis, J. B., & Martin, G. B. (1882). Macrosporium solani E&M. *American Naturalist, 16*, 1003.

FAO. (2004). *Agricultural data. Production and Indices Data Crop Primary*. Retrieved February 11, 2016, from http://www.fao.org/

FAO. (2005). *FAOSTAT Agricultural Data. Agricultural production, crops, primary*. Retrieved February 10, 2016, from http://faostat.fao.org/faostat/collections?subset=agriculture

Frank, G. (2007). *Potato Wart. Online*. APSnet Features. doi:PSnetFeature-2007-060710.1094/A

Galloway, B. T. (1891). The new potato disease. *Garden and Field, Adelaide, Australia, 16*, 158.

Harrison, M. D., Livingston, C. H., & Oshima, N. (1965a). Control of early blight in Colorado. I. Fungicidal spray schedules in relation to the epidemiology of the disease. *American Potato Journal, 42*(11), 319–327. doi:10.1007/BF02869697

Harrison, M. D., Livingston, C. H., & Oshima, N. (1965b). Control of early blight in Colorado. II. Spore traps as a guide for initiating applications of fungicides. *American Potato Journal, 42*(11), 333–340. doi:10.1007/BF02869699

Harrison, M. D., & Venette, J. R. (1970). Control of potato early blight and its effect on potato yield. *American Potato Journal, 47*(3), 81–86. doi:10.1007/BF02864808

Jones, L. R. (1893). The new potato disease or early blight. *Vermont Agricultural Experimental Station Bulletin, 6*, 66–70.

Jones, L. R. (1912). *Potato diseases in Wisconsin and their control.* University ofWisconsin Experimental Station Circular of Information 36.

Latijnhouwers, M., Ligterink, W., Vleeshouwers, V. G., VanWest, P., & Govers, F. (2004). A Gα subunit controls zoospore mobility and virulence in the potato late blight pathogen Phytophthora infestans. *Molecular Microbiology, 51*(4), 925–936. doi:10.1046/j.1365-2958.2003.03893.x PMID:14763970

Madden, L., Pennypacker, S. P., & MacNab, A. A. (1978). FAST, a forecasting system Alternaria solani on tomato. *Phytopathology, 68*(9), 1354–1358. doi:10.1094/Phyto-68-1354

Mercure, P. (1998). *Early Blight and Late Blight of Potato*. University of Connecticut, Integrated Pest Management. Retrieved December 15, 2015, from www.hort.uconn.edu/IPM/VEG/HTMS/BLTPOT.HTM

Shekhawat, G. S., Gadewar, A. V., & Chakrabarti, S. K. (2000). *Potato bacterial wilt in India*. Technical Bulletin No. 38 (Revised). C.P.R.I.

Shtienberg, D., & Fry, W. E. (1990). Influence of host resistance and crop rotation on initial appearance of potato early blight. *Plant Disease*, *74*(11), 849–852. doi:10.1094/PD-74-0849

Son, S. W., Kim, H. Y., Choi, G. J., Lim, H. K., Jang, K. S., Lee, S. O., & Kim, J. C. et al. (2008). Bikaver.in and fusaric acid from Fusarium oxysporum show antioomycete activity against P infesta.ns. *Journal of Applied Microbiology*, *104*(3), 692–698. doi:10.1111/j.1365-2672.2007.03581.x PMID:17927749

Teng, P. S., & Bissonnette, H. L. (1985). Estimating potato yield responses for chemical control of early blight in Minnesota. *American Potato Journal*, *62*(11), 595–606. doi:10.1007/BF02854434

Van der Waals, J. E. (2002). *A review of early blight on potatoes*. University of Pretoria.

Venette, J. R., & Harrison, M. D. (1973). Factors affecting infection of potato tubers by Alternaria solani in Colorado. *American Potato Journal*, *59*, 313–325.

Wharton, P., & Kirk, W. (2012). *Early Blight. Potato Disease*. Michigan State University. Retrieved January, 05, 2016, from http://www.potatodiseases.org/earlyblight.html

Workman, M., Harrison, M. D. & Franc, G. D. (1983). *Effects of nitrogen fertilization and fungicide treatments on the Monona potato cultivar*. Colorado State University Experimental Station Bulletin.

Chapter 12
Agro-Geoinformatics, Potato Cultivation, and Climate Change

Upasana Dutta
C-DAC, India

ABSTRACT

The agriculture sector is reeling under the pressures of population, land and water scarcity, diseases, disasters and the most challenging of them all, climate change. Although climate change is yet to be charged with affecting agriculture, but in recent years trends of change have been witnessed in various crop production, with a hint of climate's role in it. With the advent of technology, these trends have become easier to analyse and in certain cases predict too. Information Technology (ICT) tools like Geoinformatics are playing a profound role in the agriculture sector and is helping to understand and assess the various factors affecting the growth of crops along with finding out the alternative suitability parameters for better production and distribution. The main aim of this chapter on agro-geoinformatics is to look into this linkage between technology usage and better potato production during adverse conditions.

DOI: 10.4018/978-1-5225-1715-3.ch012

INTRODUCTION

Agro-Geoinformation

Management of agricultural resources comprises of myriad activities in conservation practices and land/water resources aimed at increasing the food production. Substantial increase in crop production can also be achieved by bringing additional land under cultivation, improved crop management technology through use of high yielding, input responsive and stress tolerant crop varieties, improved pest control as well as by sustainably practicing irrigation and fertilizer inputs. These inputs together with reliable information on:

1. Existing land use and acreage under various crops;
2. Soil types and extent of problem soils;
3. Monitoring of surface water bodies (to determine water availability in irrigation systems) for ground water development; and
4. Management of natural calamities etc., will enable formulation of appropriate strategies to sustain the pace of agricultural development.

This in turn calls for a holistic approach, which must combine short-term management of agricultural resources at micro-level with long-term global perspectives, keeping in view the socio-economic and cultural environment of the people (Rai et al., 2008).

Growth rates of crops are of utmost importance to planners and policy makers. These show the past trend of the variables and enable us to forecast the near future trend and to study the growth behavior in area, production and yield a sound technique is required. With the advent of information technology into the agriculture sector, these effects and trends can help in joining the dots and provide a holistic view of the entire scenario.

Agriculture-related geoinformation has become one of the key information sources in agricultural decision-making and policy formulation process. Recent advances in geoinformatics have created new opportunities and challenges in applying geoinformatics to agriculture. The issues related to handling and applying agro-geoinformation, such as collecting (including field visits and remote sensing), processing, storing, archiving, preserving, retrieving, transmitting, accessing, visualizing, analyzing, synthesizing, presenting, disseminating have been addressed actively in the past several years.

According to Wikipedia, Geoinformatics is an amalgamation of geography science and information science. It develops and uses information science to address the problems of geography, cartography, geosciences and related branches of science and engineering. It is an integrated technology for collection, transformation and generation of information from spatial and non-spatial databases. Geo-informatics constitutes of Remote sensing, Geographical Information Sciences (GIS), Global Positioning Systems (GPS), Relational Data Base Management Systems (RDBMS), etc.

Agro-Geoinformatics uses the power of Geoinformatics in bettering the agriculture sector. It uses it as a tool for assessment, monitoring, planning and management of agricultural research and development. Agro-geoinformation is therefore gaining more importance and prominence in the agricultural decision-making process. It is a powerful tool and is critical for agricultural sustainability, food security, environmental research, bio-energy, natural resource conservation, land use management, carbon accounting, global climate change, health research, agricultural industry, commodity trading, economy research, education, agricultural decision-making and policy formulation, etc. Concisely, Agro-geoinformation will help farmers better their produce in a much more efficient way in the face of today's climate and industry challenges.

Agricultural remote sensing involving crops and soils are quite complex. These complexities are due to the dynamic nature and inherent complexity of biological materials. In order to handle these complex problems, remote sensing technology offers numerous advantages over traditional methods of conducting agricultural and other resource surveys. Advantages include the potential for accelerated surveys, capability to achieve a synoptic view under relatively uniform lighting conditions, availability of multispectral data for providing intense information, capability of repetitive coverage to depict seasonal and long-term changes and availability of imagery with minimum distortion etc. Therefore, it permits direct measurement of important agro-physical parameters.

As is known, remote sensing of earth resources utilizes electromagnetic waves, which ranges from short wavelength ultraviolet to visible near infrared and thermal infrared in the longer wavelength, active radar and passive microwave systems. A great advancement in applications of computers to this science is the development of capability of storing vast and varied information, ranging from historical information and aerial photography to spacecraft data, ground reference, and other forms of ancillary data. All these information is stored in the form of highly useful database/ information system. Thus, remotely sensed data and its derived information have become an integral component of agricultural management system in the country (Rai et al., 2008). In addition, it is also imperative to investigate how far the latest developments in agro-geoinformatics and remote and proximal sensing can be applied for agricultural monitoring, management, and decision-making.

This chapter tries to highlight these growing concerns and looks into few of the trends emerging in the direction of how best agro-geoinformatics can help ensure the safety of the same. The following paragraphs show that agro-geoinformatics is not only an active and vibrant research field, but also the research in that direction is beginning to make a real impact in agricultural decision-making and food security. This chapter aims to provide a beneficial summary of recent progress and useful references on agro-geoinformatics research and its scope for the effective management of potato cultivation in climate change scenario.

CLIMATE CHANGE

The agriculture industry in the recent years has undergone many changes owing to population pressure, land pressure, etc. The population is growing at approximately 77.6 million per year, and is expected to reach nearly 10 billion by 2050. At the same time, the middle class is expected to double by 2030. And as Engel's law states, as incomes rise people spend more on food and eat more animal protein. It has also been observed that the rate of yield of major crops has been trending negatively on a 10-year curve. Experts, therefore predict that to meet the demand for food, fuel and fiber from a growing and increasingly affluent population we will need to double the global crop production over the next 35 years (Leclerc & Tilney, 2015).

The recent years have also seen climate change and extreme weather patterns (Figure 1) as the new game-changers in agriculture or so it is believed. The stage is however, too nascent to conclude climate change's direct impact on agriculture. But, there have been many reports of impact on the surrounding abiotic factors leading to the effect on the quality and quantity of production, like yield reliability and increased susceptibility to weeds and diseases, to name a few.

One such crop, whose growth has reportedly been affected, even when we are debating on climate change, is the Potato. It is a temperate crop but grown in a wide range of climatic conditions and is highly vulnerable in higher temperatures, which can lead to severe yield loss. Scientists report that the potato, around the world, is especially vulnerable to heat stress and further high temperature reduces growth and starch formation. The potato is more sensitive for change in climatic conditions; hence, it makes potato an almost ideal crop to be studied for signs of climate change effects.

CLIMATE CHANGE AND POTATO CULTIVATION

POTATO (*Solanum tuberosum L.*) is one of the important food crops of the world, a dietary staple for millions of people. Potatoes provide a source of low cost energy to the human diet. It is a rich source of starch, vitamins C and B and minerals. It is perennial but as a crop, treated as an annual and mostly vegetatively propagated by

Figure 1. Extreme Weather and Climate Change
Source: IPCC, 2012

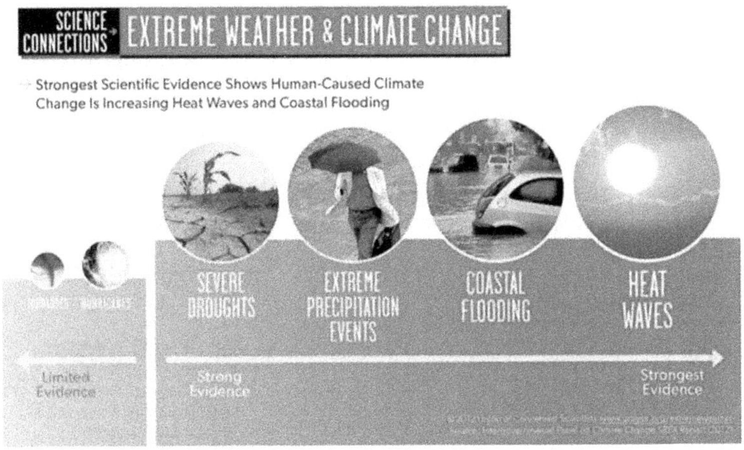

means of tubers (Figure 2). Potato occupies 1.33% of the cropped area in the world. Potato is cultivated in almost all the states in India. Nearly 80% of the crop grown in the Indo-Gangetic plains comprising of Punjab, Haryana, Uttar Pradesh, Bihar and West Bengal. (NBSS Manual, 2006)

During 2006-07, the country produced 22.09 million tonnes of potato from 1.48 million ha of land with the yield level increasing to 14.90 t/ha from a meager 6.58 t/ha during 1949-50. Presently, India is the third largest potato producer in the world (Pandey et.al, 2007). Moreover, India has one of the most organized potato seed production programme in the world.

Let us also take a global example. The agricultural sector in Flevoland, which is a province of the Netherlands, both nationally and internationally famous for its cultivation of (seed)-potatoes. The province houses Emmeloord, also known as the "World Potato City", is the second largest exporter in the world, and generates the highest yield per acre worldwide.

With the earth becoming warmer, with higher minimum temperatures these potato belts are now feeling the 'heat'. Even though the evidences are circumstantial, one report lists out the climate change effects on the quality of potatoes in northern Europe.

- Increased tuber size;
- Increased Vitamin C;
- Increased dry matter concentration;
- More hollow hearts;

Figure 2. The potato
Source: www.technologyreview.com

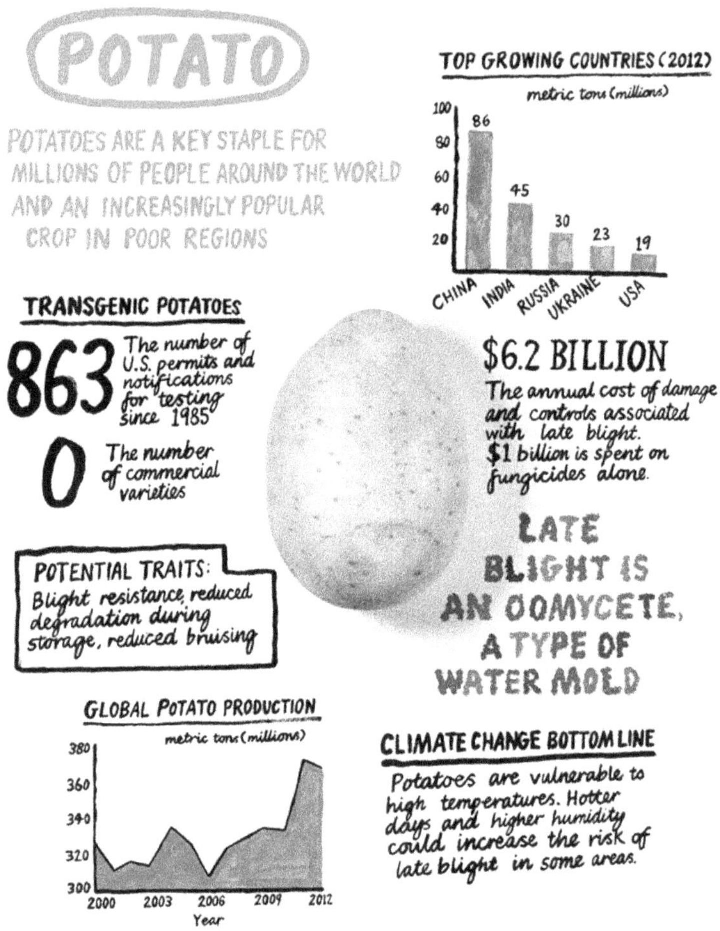

- More secondary growth and sugar ends;
- Winter temperatures up: reduced storability; and
- More rotten tubers due to water logging.

There are also evidences of climate change affecting the spread of disease causing agents like the potato tuber moth (Figure 3). Studies also show that, because of increases in temperature, future potato yields could decrease in many regions. In some regions, mainly in temperate regions, yield decline can partly be avoided through adaptation. Yields may even go up at high latitudes because of a lengthening of the growing season. In some regions, such as in parts of Algeria, Morocco, China, and

Figure 3. Climate change scenarios for the potato tuber moth, Phthorimaea operculella
Source: Kroschel et al, 2012

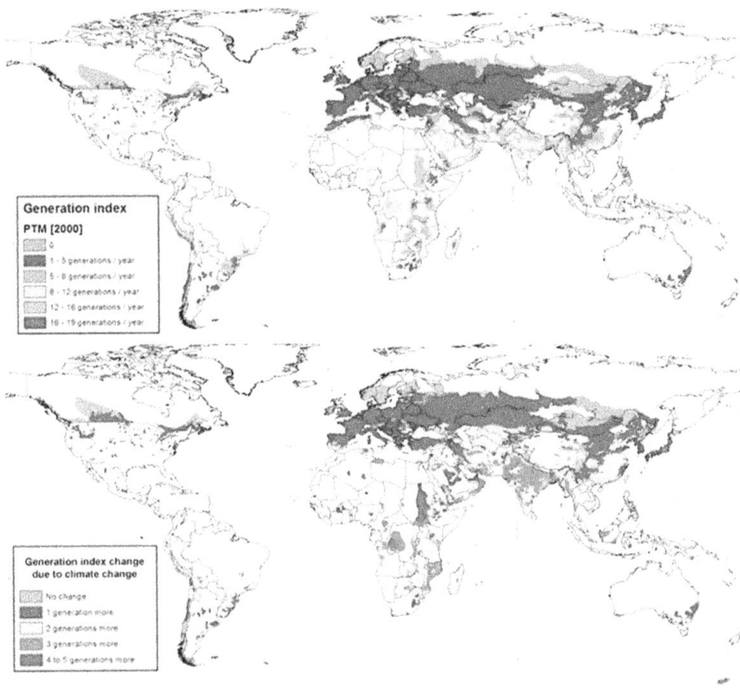

South Africa, yield may increase because a warmer climate would allow growing a winter crop (instead of an autumn or spring crop) (Hijmans, 2003).

Many such examples can be found in the global scenario, which depicts the facts surrounding climate changes effects on Potato yield, disease and storage issues. Now the question remains as to how such problems can be effectively managed or predicted, or a method or technology found that will minimize the losses as well as improve the surrounding environment for sustainable cultivation and production. Development in latest information technology based advancements can be used for management of future potato cultivation.

METHODOLOGIES AND RECOMMENDATIONS

Agro-Geoinformatics

Technocrats' job is to dream up the future and make that dream a reality. In agriculture, with the latest advancements in technology, small harvesting robots, skyscraper

Figure 4. Smart farming
Source: www.techcrunch.com

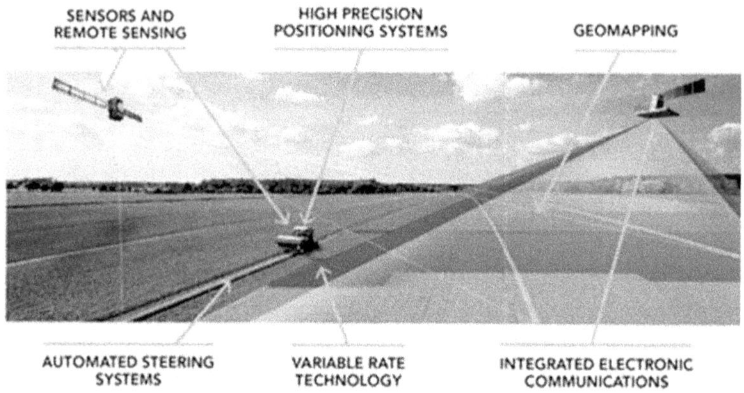

vertical farms in every city and sensor-studded fields overseen by drones may not be too far from reality (Figure 4). Yet, as fun as it is to imagine this kind of future for agriculture, it is imperative to find the challenges and work towards it. In addition, this has led to a confluence of hardware and software technology advances, which is creating opportunities to address the agriculture sector. Inexpensive and infinitely configurable mobile devices have liberated technology from the office desktop. At the same time, inexpensive but sophisticated hardware sensors have emerged to automate the collection of massive data sets. These were earlier unheard of in the agriculture sector, but with more and more successful demonstrations, the cultivators, farmers and planners are opening up to it.

With these technology shifts, exciting technologies like drones, Artificial Intelligence (AI), satellite mapping, robotics, and the Internet of Things (IOT), have quickly realized that the agriculture value chain provides fertile first market opportunities for many technologies that are not advanced enough or have not yet found solutions in the consumer space. Many investors looking at these technologies are going to need to get smart on agriculture if they are going to make informed investment decisions (Leclerc & Tilney, 2015).

A great example of such advancement is Blue River Technology, a company that is using computer vision and machine learning to build a future in which 'every plant counts.' Blue River's Lettuce Bot is an impressive piece of hardware. The mechanical engineering behind the robot is not trivial, as it gets pulled through fields by large tractors. It is able to visually characterize each plant through real-time image capture and processing, use algorithms to determine which portions of the plant to keep and precisely eliminate the portions of the plants that are unwanted. Moreover, their Zea product enables high-throughput, field-based phenotyping. Using computer vision,

Zea counts plants, measures plant spacing, builds canopy height distributions and measures key physiological parameters — all based on imagery.

CropX another such product uses three simple, low-cost soil sensors to collect data, cloud analytics to run algorithms and generate adaptive irrigation maps so farmers have a better handle on where their water is going. The accuracy generated with data from only three simple sensors is revolutionary. The CropX system also can link into irrigation systems to provide a more end-to-end solution for farmers.

Blue River and CropX are among a large number of startups tackling the opportunities in precision agriculture and data management — from Granular to OnFarm to TerrAvion. Whether it is IoT or Remote Sensing, these companies are working toward making the farming industry more efficient and effective (www.cropx.com).

Application of Space Borne Remote Sensing Data in the Agriculture Sector

In 1971, large area crop survey was explored in USA under Corn Blight Watch Experiment (CBWE), which was followed by 80 experiments/large-scale remote sensing program. In a country like India, with vast geographic spread and great diversity in its set up, the need to apply remote sensing technology for national development was recognized during the early 70's. The pioneering experiment was of coconut root–wilt disease using Color-infrared aerial photography. Numbers of studies conducted for methodological development in this area. Remote sensing activities in India received a tremendous boost with the launch of Indian Remote Sensing Satellite-1A (IRS 1A) in March 1988. In addition, with the development of new satellite systems, India is moving fast in this direction and recently successfully launched number of satellites dedicated to specific area of applications such as OceanSat, CartoSat, ResourceSat etc. Radar Imaging Satellite (RISAT) (Figure 5), a microwave remote sensing mission with Synthetic Aperture Radar (SAR) operating in C-band and having a 6x2 meter planar active array antenna based on trans-receiver module architecture has also been launched recently. Many such technological developments occurred in the 20th century, which contributed largely to the development of the concept of precision farming and agro-geoinformatics, which includes GPS, GIS and high resolution remote sensing (Rai et al., 2008). All these advancements in data computation and collection are enabling farms to run efficiently-with precision, data-driven decisions and automation.

Agro-Modelling using Geoinformatics

According to Granell et.al (2015), due to the ever-increasing availability of remotely sensed data, and accompanying advances in image analysis, data storage and process-

Figure 5. Indian remote sensing satellite series
Source:www.indiandefensenew.in

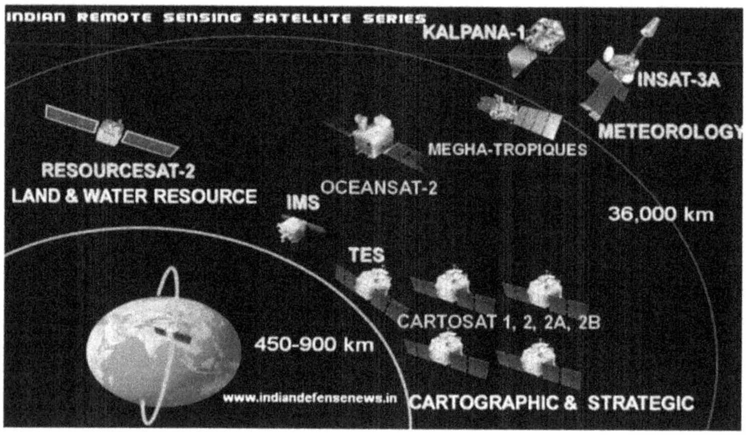

ing solutions, a wealth of actionable data is now available for various stakeholders, such as farmers, authorities and agro-service providers. In Potato farming, such geodata may deployed for applications at various spatial and temporal scales such as:

- Site/Land Suitability Analysis;
- Potato Growth and Productivity Monitoring and Simulation;
- Precision Farming;
- GIS-Agro-Modeling- Data Handling;
- Predictive Landcover Modeling;
- Sustainable Resource Protection;
- Disease onset and Spread Analysis;
- Transfer and Traceability of Products; and
- Storage Management.

Let us elucidate some of these applications for a better understanding:

Site/Land Suitability Analysis

Land suitability evaluation is a prerequisite for land-use planning and development (Sys 1985; Van Ranst et al., 1996) and is one of the most useful applications of GIS for spatial planning and management (Collins et al., 2001; Malczewski, 2004). Characterizing the edaphic (soil related) and climatic regions suitable for the production of a specific crop type generally requires a long time frame, coupled with extensive experimentation and experience, and significant resources (Siddons et

Figure 6. Schematic framework for assessing potato land suitability
Source: Daccache et al., 2011

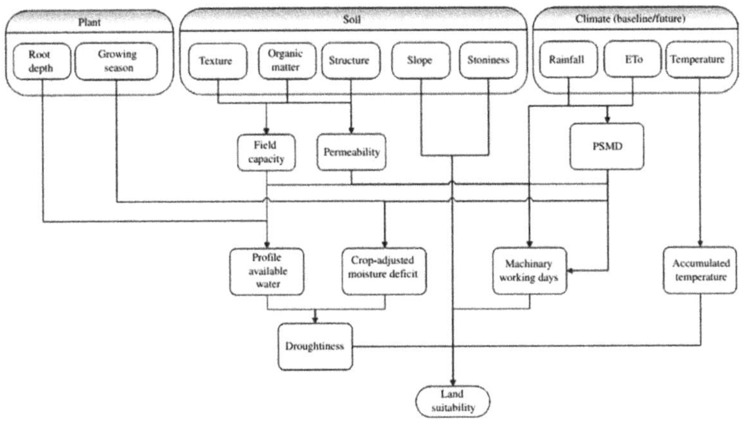

al., 1994). However, if land types and local climates can be adequately specified and sufficient knowledge is available regarding crop responses to soil and weather factors, then land suitability models (Figure 6) offer an alternative and rapid means of producing maps and data sets showing suitable areas for a particular crop, in our case, for the potato. It aims at identifying the most appropriate spatial pattern for future land uses according to specific requirements, preferences, or predictors of some activity (Hopkins, 1977; Collins et al., 2001). It provides information on the constraints and opportunities for the use of the land and therefore guides decisions on optimal utilization of land resources (FAO 1983 and Daccache et al., 2011).

GIS serves the multi-criteria evaluation function of suitability assessment well, providing the attribute values for each location and both the arithmetic and logical operators for combining attributes (Jiang and Eastman, 2000). Furthermore multi-criteria evaluation may be used to develop and evaluate alternative plans that may facilitate compromise among interested parties (Malczewski, 1996). In general, the GIS-based land suitability analysis assumes that a given study area is subdivided into a set of basic unit of observations such as polygons or raster. Then, the land-use suitability problem involves evaluation and classification of the areal units according to their suitability for a particular activity. Over the last 10 years or so, land-use suitability problems increasingly been conceptualized in terms of the GIS-based multi-criteria evaluation procedures (e.g. Banai, 1993; Jankowski and Richard, 1994; Joerin, 1995; Barredo, 1996; Antonie et al., 1997; Lin et al., 1997; Beedasy and Whyatt, 1999; Malczewski, 1999; Barredo et al., 2000; Mohamed et al., 2000; Bojorquez-Tapia et al., 2001; Dai et al., 2001; Joerin et al., 2001). There are two fundamental classes of multi-criteria evaluation methods in GIS: The Boolean overlay operations (non-compensatory combination rules) and the weighted linear combination (WLC) methods (compensatory combination rules). These are the most often used approaches for land use suitability analysis (Heywood et al., 1995; Jankowski, 1995; Barredo, 1996; Beedasy and Whyatt, 1999; Malczewski, 2004). These approaches can be generalize within the framework of OWA (Asproth et al., 1999; Jiang and Eastman, 2000; Makropoulos et al., 2003; Malczewski et al., 2003; Malczewski and Rinner, 2005; Malczewski, 2006 and Mokarram & Aminzadeh, 2009).

The aim in integrating Multi-Criteria Decision Analysis (MCDA) with Geographical Information Systems (GIS) is to provide more flexible and more accurate decisions to the decision makers in order to evaluate the effective factors. Furthermore, by changing the parameters in this type of method, a wide range of decision strategies or scenarios can generated. It also takes the advantage of incorporation of fuzzy quantifiers into GIS-based land suitability analysis by ordered weighted averaging (OWA). OWA is a multi-criteria evaluation procedure (or combination operator) (Yager, 1988). Conventional OWA can utilize the qualitative statements in the form of fuzzy quantifiers (Yager, 1988, 1996).

The optimal use of reserved land resources for agriculture is a complex problem that involves subjective assessments with multiple criteria. GIS-based multi-criteria land suitability evaluation using Ordered Weight Averaging with fuzzy quantifier approach has been found to be effective for solving this problem. An empirical study in Shavour, Iran has conducted using this approach. The fuzzy-quantifier-based OWA approach (Figure 7) is capable of capturing qualitative information, the decision maker or analyst may have regarding their perceived relationship between the different evaluation criteria. It is in this effort one can see the benefit of the fuzzy quantifier approach to GIS-based multi-criteria analysis. This is especially true in situations involving a large number of criterion maps. In such situations, it is impractical or even impossible to specify the exact relationships between the evaluation criteria. The OWA approach provides a mechanism for guiding the decision maker/analysis through the multi-criteria combination procedures and allows exploration of different decision strategies and scenarios. Consequently, the approach facilitates a better understanding of the alternative land-use suitability patterns (Mokarram & Aminzadeh, 2009).

Potato Growth and Productivity Monitoring and Simulation Using Remote Sensing and GIS

Crop growth and productivity are determined by a large number of factors like weather, soil and management, which vary significantly across space. Remote Sensing (RS) data, acquired repetitively over agricultural land helps in identifica-

Figure 7. Suitability map based on OWA approach
Source www.agronomy.it

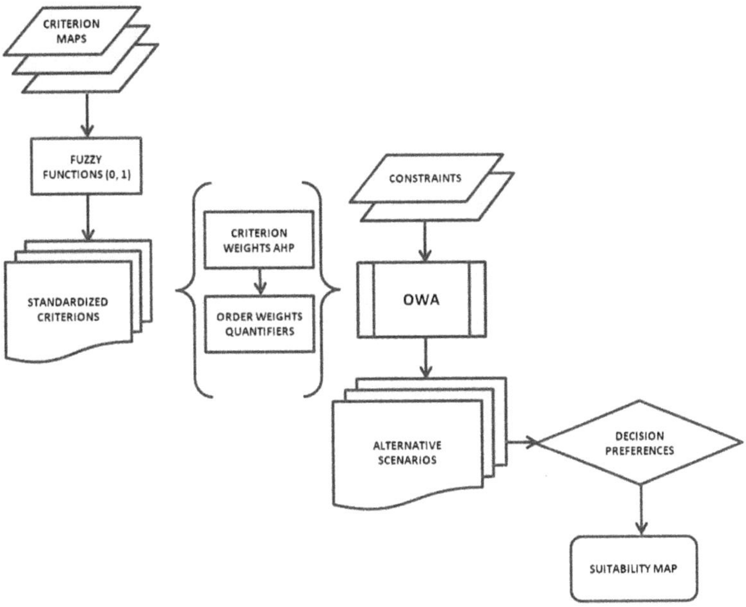

tion and mapping of crops and in assessing crop vigor. As RS data and techniques have improved, the initial efforts that directly related RS-derived vegetation indices (VI) to crop yield have replaced by approaches that involve retrieved biophysical quantities from RS data. Thus, crop simulation models (CSM) that have been successful in field-scale applications are being adapted in a GIS framework to model and monitor crop growth with remote sensing inputs making assessments sensitive to seasonal weather factors, local variability and crop management signals. The RS data can provide information of crop environment, crop distribution, leaf area index (LAI), and crop phenology. Such information integrated in CSM, in a number of ways such as use as direct forcing variable, use for re-calibrating specific parameters, or use simulation-observation differences in a variable to correct yield prediction (Dadhwal, 2003).

The use of GIS along with RS data for crop monitoring is an established approach in all phases of the activity, namely preparatory, analysis and output. In the preparatory phase, GIS is used for stratification/zonation using one or more input layers (climate, soil, physiography, crop dominance etc.), or preparing input data (weather, soil and collateral data) available in different formats to a common format. In the analysis phase, use of GIS is mainly through operations on raster layers of NDVI or computing vegetative index profiles within specified administrative boundaries. The final output phase involves GIS for aggregation and display of outputs for defined regions (e.g., administrative regions) and creating map output products with required data integration through overlays. Wade et al.(1994) described efforts within National Agricultural Statistics Service (NASS) of U.S. Department of Agriculture (USDA) of using NOAA AVHRR NDVI for crop monitoring and assessment of damage due to flood and drought by providing analysts a set of map products. Combining satellite data in a GIS can enhance the AVHRR NDVI composite imagery by overlaying State and District boundaries. The use of raster-based (grid-cell) capabilities of ARC/INFO (GRID) for the generation of difference image helps compare a season with previous year or average of a number of years. Overlaying a crop mask helps in highlighting only effects on crops. Generation and overlay of contours of precipitation data generated using TIN function of ARC/INFO also is an aid to interpreting NDVI difference image.

Applications of CSM interfaced with GIS covers a wide range of applications like spatial yield calculation (regional and global), precision farming, climate change studies, and agro-ecological zonation, etc.

Crop Growth Monitoring System (CGMS) of Monitoring Agriculture with Remote Sensing (MARS) is a project of the European Union, which uses WOFOST model and Arc/Info for operational yield forecasting of important crops (Meyer-Roux & Vossen, 1994). The CGMS of MARS integrates crop growth modeling (WOFOST), relational database ORACLE and GIS (ARC/INFO) with system analytical part for yield forecasting (Bouman et al., 1997). There are databases on soil, weather, crop, and yield statistics that cover the whole of EU. The system-analytical part consists of three modules: agro-meteorological module, a crop growth module and a statistical module. The meteorological module takes care of the processing of daily meteorological data that received in real time to a regular grid of 50x50 km for use as input by crop growth model or for assessment of 'alarm' conditions. The crop growth module consists of the dynamic simulation model WOFOST in which crop growth is calculated and crop indicators generated for two production levels: potential and water-limited. In CGMS, WOFOST is run on a daily basis for each so-called 'simulation unit', i.e. a unique combination of weather, soil, and crop (mapping) units. In the statistical module, crop indicators (total above ground dry weight and dry weight storage organs) calculated with WOFOST are related to historical yield statistics through regression analysis in combination with a time-trend, for at least 15 years of simulated and historical data (Vossen, 1995). The resulting regression equations per crop per region are used to make actual yield forecasts. CGMS generates on a 10 day and monthly basis three types of output on current cropping season:

1. Maps of accumulated daily weather variables on 50x50 km grid to detect any abnormalities, e.g. drought, frost,

2. Maps of agricultural quality indicators based on comparison of simulated crop indicators with their long-term means, and
3. Maps and tables of yield forecasts.

Precision Farming

Han et al. (1995) developed an interface between PC ARC/INFO GIS and SIMPOTATO simulation model to study potato yield and N leaching distribution for site-specific crop management (precision farming) in a 50 ha field. The GIS input layers, corresponding to important distributed input parameters for the model, were irrigated water/N layer, soil texture layers and initial soil N layers. For each unique sub-area stored in the GIS database, the interface program extracts the attribute codes of that sub-area from the GIS database, converts the attribute codes to the input parameters of the SIMPOTATO and sends them to the model. After running the model, the interface program retrieves the output data (potato yield and N leaching), converts them to the attribute codes and stores the output data in the GIS database.

GIS-Agro-Modeling: Data Handling

With the advancements the challenge of making sense of the flood of data being produced via weather stations, in-field sensors, satellites and GPS monitoring and mapping tools has however cropped up. This has been a humongous task for which many software and hardware are present in the market. The USDA is making much of this data, part of its Agricultural Statistical Service (NASS) and Economic Research Service (ERS), available to the public for the first time. NASS, for example, has opened up access to its CropScape and VegScape APIs. CropScape provides geospatial data through annual land cover data sets, while VegScape is a vegetative condition database compiled from satellite data (Leclerc & Tilney, 2015).

Many spatial data models, as a result have also developed to understand the scenario and make the data more understandable and usable for the purpose intended. Gupta (2007) attempted to develop spatial prediction model under different situations. He grouped the situation in to four groups viz:

1. Using prior information about parameters, which is non-informative for known and unknown variance;
2. Using prior information about parameters which is informative for known and unknown variance;
3. Using prior information about parameters as natural conjugate prior for known and unknown variance; and

4. Using fuzzy approach for linear interval model for vague characters under study.

The study showed through simulation that Bayesian regression analysis is always better than simple regression analysis. This may be due to the fact information contained in the sample as well as about the parameter of the model has been utilized in the estimation procedure. Further, there is significant gain in the precision in case of geographical variables when spatial effects were taken in to account in the estimation procedure under Bayesian framework. Rai et al. (2008) further reported that the variogram models plays significant role in capturing the spatial effect. Spatial Bayesian regression model performs better when spatial effects incorporated through variogram models. The results obtained through exponential and spherical variogram models found to be encouraging as compared to other models.

Land suitability models have long used in agriculture for increasing farming efficiency. With the advent of Geoinformatics in the field multi-criteria analysis has become much easier and such models easier to apply. There is great scope for effective management of potato crop in climate change situation in different parts of the globe.

Disease Onset and Spread Analysis

Spatial and temporal dynamics of diseases in Potato crops be investigated using RS. There are many incidences when disease detection and spread using aerial photographs. Aerial photography coupled with spatial analyses of late blight-infected plants is an effective technique to quantitatively assess disease patterns in relatively large fields and is useful in quantifying an intensification of aggregation during the epidemic process on a large scale (Dennis et al., 2003). Further, the authors reported that for Columbia Basin during epidemics in 1993, 1995, and 1998, the photographs scanned and exported in TIF format and further separated into the RGB data layers (Figure 8). Unsupervised clustering used to create 10 classes for each field. An indicator transformation was used to classify each pixel as diseased or non-diseased. This transformation allowed comparison of disease occurrence patterns between fields with differing levels of disease assigned to the 10 classes for each field. A rectangular section within each field was then used for statistical analyses. Runs analysis and indicator variograms were used to examine the spatial association of late blight between nearby pixels and for pixels more separated in space. Each analysis was carried out in four directions (north, northeast, east, and southeast). A runs analysis was then used to examine aggregation of disease pixels compared with the mixing of diseased and healthy pixels expected for randomly distributed occurrence of disease. Spatial autocorrelation was examined in four directions with indicator variograms. Variograms related to correlograms display the degree of correlation of disease from spatially contiguous pixels at successively larger distances. The minimum lag distance used to plot variograms was approximately 1m (Dennis et. al, 2003). (Figure 8)

Figure 8. Color, infrared aerial photograph of cv. Norkotah Russet in 1998 (field V) on A, 29 June, B, 6 July and C, 27 July. Healthy foliage appears red; diseased appears tan to dark. Dry soil without vegetation (plants previous blighted and killed) appears white to blue)
Source: Dennis et al., 2003

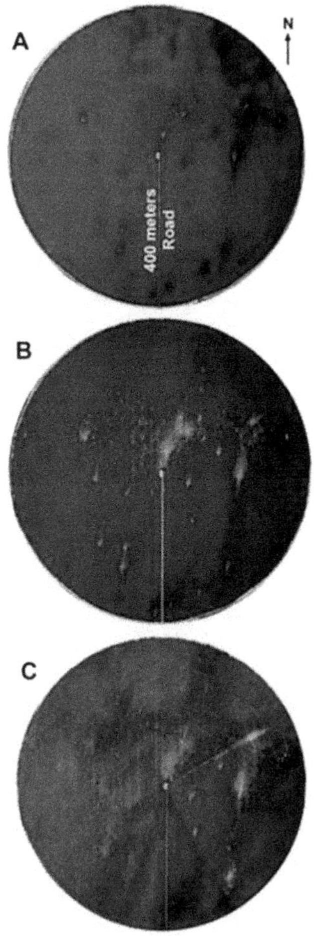

The above example has been cited to show the capability of geoinformatics as an information technology tool to determine the spread of the potato disease. The importance of this study was to demonstrate the potential of using aerial photography in combination with geostatistical analyses and further quantitatively describe spatial patterns of a plant disease during an epidemic.

Storage Management

Storage is a huge problem as regards to Potato in particular or any crop in general is concerned. In India, post-harvest loses of horticultural crops are estimated to be in the range of 20-40% due to weak infrastructure. The situation of storage is more or less similar in all under developed and developing countries. A scientific approach is therefore required to evolve a methodology to locate sites for storage that would optimally utilized and maintained by the growers. RS and GIS can be

Figure 9. Flowchart representing methodology involved in site selection of cold storage
Source: Hiremath et al., 2013

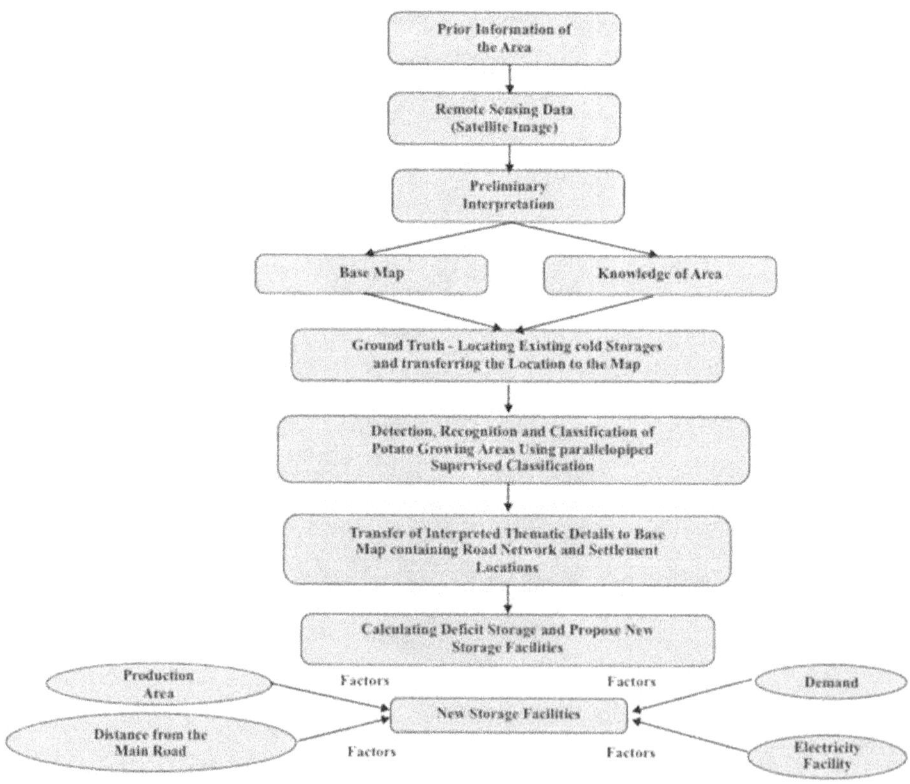

utilized to solve this problem. According to one case study from Gujarat, India where authors used GIS to create the spatial database for road network, settlement and cold storage locations (Figure 9). Buffer and location-allocation analysis was performed for locating optimum cold storage sites. Using the already existing cold storages and the new probable sites location-allocation analysis was carried out to locate the required number of optimum new sites and allocate potato growing area to respective cold storage.

In another case study conducted in Banaskantha district of Gujarat to analyse the price and storages issue related to potato in the State. Banaskantha district is the major potato growing area, produces around 50 percent of the state's total potato production. It was observed that 80 percent of the potato in Gujarat is grown in winter and sold throughout the year. Hence, emphasis was given to develop post-harvest infrastructure like cold storage. A study was taken up to analyse the demand and supply situation, the effect on price variations and to evolve an optimum plan to locate cold storages using satellite remote sensing (RS) data and Geographic Information System (GIS). Resourcesat -1 and IRS 1C LISS-III (23m resolution) images were acquired for the study. The major factors, which account for economics in potato pricing includes weather, production, infrastructural such as service area and location of cold storages. All these factors directly or indirectly affect the prices.

Farmers have to face huge losses due to inadequate storages facilities. The authors reported that the analysis of data showed that there is a requirement to increase the number of cold storages and these need to be located over a larger area nearer to farms rather than to concentrate them in a particular town. The study helped in generating revenue for both the government and private sectors as well as circumvents losses due to transportation (Hiremath et al., 2013).

There are also other ways of ensuring proper management of the produce in the cold storages. GIS enabled comprehensive inventory management can be an answer for storage management of potato. GIS information would not only help in correctly locating the potato storage but also will give a real time mechanism to monitor the flow within the storage unit from the time it registered in to its sale. This system will keep track of the conditions of the potato inside the storage units. Invoice and payment monitoring will also be carried out to ensure that taxes and fees related to potato production and trade are collected. The use of ICT and GIS would allow a more comprehensive overview of the entire storage and management scenario.

These are few of many such examples showing the utility of Geoinformatics technology for effective management of potato crop from cultivation to marketing and prove to be powerful analysis and assessment tool in the changing climatic situation for various regions across the globe.

FUTURE RESEARCH DIRECTIONS

The era of climate-driven agriculture has arrived. Agriculture is at the crossroads of some of the world's most critical challenges: climate change, sustainable food production, providing clean and abundant water, renewable energy, and improving human health. Those challenges are immense, especially when you consider that we must meet the food needs of a world population that is expected to top 9 billion people by mid-century. To meet these challenges, we must take advantage of knowledge from all sectors, and encourage new and innovative thinking (Milne, 2015).

Potato is temperate crop and farmers are growing it in tropical and subtropical climatic situation at various parts of the globe. The existing potato cultivation practices may not be efficient in climate change situation for the potato cultivation. Hence, thinking in scientific community is changing and the possibility of effective utilization of geoinformatics from data collection to decision making is getting momentum. Such ICT based tools can help to understand and assess the various factors involved in a potato crop cycle along with finding out the alternative suitability parameters for better production.

Further, refinement in the research is required for production and prediction of diseases and disease agents of potatoes that spread to storage, transportation and distribution of the same. Research on effective ICT based potato storage is equally important as increasing production because much potato crops go waste because of inadequate storage conditions, transportation issues and distribution failures. A completely new avenue can open up if these issues are tackled properly and Geoinformatics is one of the few technologies that have the capability to address such issues in climate change situation for better adoption and mitigation.

CONCLUSION

Climate change may prove to be a villain in matters of potato production and might show a very different picture of the future than we had envisaged. However, ICT based technologies will enhance farmers' resilience in climate change situation for the sustainable potato production. The evolution of ubiquitous computing in the form of cheap sensors now allows getting high-resolution, real-time data from the field. Technologies like cloud computing allows to use this massive amount of data to make smarter real-time decisions that improve yield and reduce costs. Such technologies integrated with Geoinformatics and result in efficient tools and systems for management of potato crop. All is needed now is more innovations and creative ideas from technocrats and their dissemination to the concerned stockholders. Then only cultivators will able to use such information technologies to increase yields and incomes while creating resilience to climate change.

REFERENCES

Antonie, J., Fischer, G., & Makowski, M. (1997). Multiple criteria land use analysis. *Applied Mathematics and Computation*, *85*(2-3), 195–215. doi:10.1016/S0096-3003(96)00190-7

Banai, R. (1993). Fuzziness in geographical information systems: Contributions from an analytic hierarchy process. *International Journal of Geographical Information Systems*, *7*(4), 315–329. doi:10.1080/02693799308901964

Barredo, J. I. (1996). *Sistemas de Informacion Geografica y Evaluacio Multicriterio en la Ordenacion del Territorio*. Madrid: Ra-Ma.

Barredo, J. I., Benavidesz, A., Hervhl, J., & van Westen, C. J. (2000). Comparing heuristic landslide hazard assessment techniques using GIS in the Tirajana basin, Gran Canaria Island, Spain. *International Journal of Applied Earth Observation and Geoinformation*, *2*(1), 9–23. doi:10.1016/S0303-2434(00)85022-9

Beedasy, J., & Whyatt, D. (1999). Diverting the tourists: A spatial decision support system for tourism planning on a developing island. *J. Appl. Earth Observ. Geoinform.*, *3/4*(3-4), 163–174. doi:10.1016/S0303-2434(99)85009-0

Bojorquez-Tapia, L. A., Diaz-Mondragon, S., & Ezcurra, E. (2001). GIS-based approach for participatory decision making and land suitability assessment. *International Journal of Geographical Information Science*, *15*(2), 129–151. doi:10.1080/13658810010005534

Bouman, B. A. M., van Dipen, C. A., Vossen, P., & van Der Wal, T. (1997). Simulation and systems analysis tools for crop yield forecasting. In P. S. Teng, M. J. Kropff, H. F. M. ten Berge, J. B. Dent, F. P. Lansigan, & H. H. van Laar (Eds.), *Applications of Systems Approaches at the Farm and Regional Levels* (pp. 325–340). Dordrecht, The Netherlands: Kluwer Academic Publishers. doi:10.1007/978-94-011-5416-1_24

Collins, M. G., Steiner, F. R., & Rushman, M. J. (2001). Landuse suitability analysis in the United States: Historical development and promising technological achievements. *Environmental Management*, *28*(5), 611–621. doi:10.1007/s002670010247 PMID:11568842

Daccache, A., Keay, C., Jones, R. J. A., Weatherhead, E. K. M., Stalham, A., & Knox, J. W. (2011). Climate change and land suitability for potato production in England and Wales: Impacts and adaptation. *The Journal of Agricultural Science, 150*(2), 1–17. Retrieved from https://dspace.lib.cranfield.ac.uk/bitstream/1826/8188/1/Land%20suitability.pdf

Dadhwal, V. K. (2003). *Crop Growth and Productivity Monitoring and Simulation using Remote Sensing and GIS*. Satellite Remote Sensing and GIS Applications in Agricultural Meteorology.

Dai, F. C., Lee, C. F., & Zhang, X. H. (2001). GIS-based geo-environmental evaluation for urban land-use planning: A case study. *Engineering Geology, 61*(4), 257–271. doi:10.1016/S0013-7952(01)00028-X

Dennis, A., Johnson, J., Alldredge, R., Hamm, P. B. & Frazier, B. E. (2003). Aerial Photography Used for Spatial Pattern Analysis of Late Blight Infection in Irrigated Potato Circles. *Epidemiology- The American Phytopathological Society, 93*(7), 805-812.

FAO. (1983). *Guidelines: Land evaluation for rainfed agriculture. Soils Bulletin 52*. Rome: FAO.

Granell, C., Casteleyn, S., & Atzberger, C. (2015). Geospatial Data Capturing, Processing, Analysis, and Visualization in Agro-Geoinformatics. *Frontiers in Environmental Science*. Retrieved 10 December 2015 from http://journal.frontiersin.org/article/10.3389/fenvs.2015.00076/full

Gupta, N. K. (2007). *On spatial prediction modeling* (Unpublished Ph.D. thesis). P. G. School IARI, New Delhi, India.

Han, S., Evans, R. G., Hodges, T., & Rawlins, S. L. (1995). Linking geographic information system with a potato simulation model for site-specific crop management. *Journal of Environmental Quality, 24*(4), 772–777. doi:10.2134/jeq1995.00472425002400040031x

Heywood, I., Oliver, J., & Tomlinson, S. (1995). Building an exploratory multi-criteria modelling environment for spatial decision support. In P. Fisher (Ed.), *Innovations in GIS* (vol. 2, pp. 127–136). London: Taylor & Francis.

Hijmans, R. J. (2003). The Effect of Climate Change on Global Potato Production. *American Journal of Potato Research, 80*(4), 271–280. doi:10.1007/BF02855363

Hiremath, D. B., Patil, N. R., & Dasgupta, A. (2013). Geospatial technique for potato cold storage allocation. *Journal of Geomatics, 7*(1), 13–17.

Hopkins, L. (1977). Methods for generating land suitability maps: A comparative evaluation. *Journal of the American Institute of Planners*, *34*(1), 19–29.

IPCC. (2012). *Managing the risks of extreme events and disasters to advance climate change adaptation IPCC 2012*. Retrieved October 10, 2016, from https://www.ipcc.ch/pdf/special-reports/srex/SREX_Full_Report.pdf

Jankowski, P. (1995). Integrating geographical information systems and multiple criteria decision-making methods. *International Journal of Geographical Information Science*, *9*, 251–273.

Jankowski, P., & Richard, L. (1994). Integration of GIS-based suitability analysis and multicriteria evaluation in a spatial decision support system for route selection. *Environment & Planning B*, *21*(3), 326–339.

Jiang, H., & Eastman, J. R. (2000). Application of fuzzy measures in multi-criteria evaluation in GIS. *International Journal of Geographical Information Systems*, *14*(2), 173–184. doi:10.1080/136588100240903

Joerin, F. (1995). Me ́thode multicrite ̀re d'aide a ̀ la de ́cision et SIG pour la recherche d'un site. Rev. *Int. Ge ́omatique*, *5*(1), 37–51.

Joerin, F., Thériault, M., & Musy, A. (2001). Using GIS and outranking multi criteria analysis for land-use suitability assessment. *International Journal of Geographical Information Science*, *15*(2), 153–174. doi:10.1080/13658810051030487

Kroschel, J., Sporleder, M., Tonnang, H. E. Z. & Simon, R. (2012). Predicting climate-change-caused changes in global temperature on potato tuber moth *Phthorimaea operculella (Zeller)* distribution and abundance using phenology modeling and GIS mapping. *J. Agriculture and Forest Meteorology*. DOI: 10.1016/j.agrformet.2012.06.017

Leclerc, R., & Tilney, M. (2015). *AgTech is the New Queen of Green*. Retrieved October 21, 2015 from http://techcrunch.com/2015/04/01/the-new-queen-of-green/

Lin, H., Wan, Q., Li, X., Chen, J., & Kong, Y. (1997). GIS-based multi criteria evaluation for investment environment. *Environment and Planning. B, Planning & Design*, *24*(3), 403–414. doi:10.1068/b240403

Makropoulos, C., Butler, D., & Maksimovic, C. (2003). A fuzzy logic spatial decision support system for urban water management. *Journal of Water Resources Planning and Management*, *129*(1), 69–77. doi:10.1061/(ASCE)0733-9496(2003)129:1(69)

Malczewski, J. (1999). *GIS and Multi criteria decision analysis*. New York: John Wiley & Sons Inc.

Malczewski, J. (2004). GIS-based land-use suitability analysis: A critical overview. *Progress in Planning*, *62*(1), 3–65. doi:10.1016/j.progress.2003.09.002

Malczewski, J. (2006). GIS-based multi criteria decision analysis: A survey of literature. *International Journal of Geographical Information Science*, *20*(7), 703–726. doi:10.1080/13658810600661508

Malczewski, J., Chapman, T., Flegel, C., Walters, D., Shrubsole, D., & Healy, M. A. (2003). GIS-multicriteria evaluation with ordered weighted averaging (OWA): Case study of developing watershed management strategies. *Environment & Planning A*, *35*(10), 1769–1784. doi:10.1068/a35156

Malczewski, J., & Ogryczak, W. (1996). The multiple criteria location problem: 2. Preference-based techniques and interactive decision support. *Environment & Planning A*, *28*(1), 69–98. doi:10.1068/a280069

Malczewski, J., & Rinner, C. (2005). Exploring multicriteria decision strategies in GIS with linguistic quantifiers: A case study of residential quality evaluation. *Journal of Geographical Systems*, *7*(2), 249–268. doi:10.1007/s10109-005-0159-2

Meyer-Roux, J., & Vossen, P. (1994). The first phase of the MARS project, 1988-1993: overview, methods and results. In *Proceedings of the Conference on the MARS Project: Overview and Perspectives*. Commission of the European Communities.

Milne, B. (2015). *Hackathon Challenges Developers to Make a Difference in Ag*. Retrieved October 23, 2015 from http://techcrunch.com/2015/10/19/hackathon-challenges-developers-to-make-a-difference-in-ag/?ncid=rss

Mohamed, A. B. A. A., Sharifi, M. A., & van Keulen, H. (2000). An integrated agro-economic and agro-ecological methodology for land use planning and policy analysis. *International Journal of Applied Earth Observation and Geoinformation*, *2*(2), 87–103. doi:10.1016/S0303-2434(00)85003-5

Mokarram, M., & Aminzadeh, F. (2009). GIS-based Multicriteria Land Suitability Evaluation using Ordered Weight Averaging with Fuzzy Quantifier: A Case Study in Shavur Plain, Iran. *The International Archives of the Photogrammetry, Remote Sensing and Spatial Information Sciences*, *38*(2), 508–512.

Pandey, N. K., Pandit, A., & Chandran, K. P. (2007). *Potato Statistics – India and world.* Technical Bulletin. No. 81. Central Potato Research Institute.

Rai, A., Sahoo, P. M., & Ahmad, T. (2008). *Geo-informatics in Agricultural Research and Development: An IASRI Perspective*. IASRI. Retrieved October 19, 2015 from http://iasri.res.in/Sovenior/Article_08.pdf

Siddons, P. A., Jones, R. J. A., Hollis, J. M., Hallett, S. H., Huyghe, C., Day, J. M., Scott, T. & Milford, G. F. J. (1994). The use of a land suitability model to predict where autumn-sown determinate genotypes of the white lupin (Lupinus albus) might be grown in England and Wales. *Journal of Agricultural Science, 123,* 199–205.

Sys, C. (1985). *Land evaluation*. Ghent, The Netherlands: State University of Ghent.

Van Ranst, E., Tang, H., Groenemam, R., & Sinthurahat, S. (1996). Application of fuzzy logic to land suitability for rubber production in peninsular Thailand. *Geoderma*, *70*(1), 1–19. doi:10.1016/0016-7061(95)00061-5

Vossen, P. (1995). Early assessment of national yields: the approach developed by the MARSSTAT project on behalf of European Commission. In *Proceedings of the Seminar on Yield Forecasting*.

Wade, G., Mueller, R., Cook, P., & Doraiswamy, P. (1994). AVHRR map products for crop condition assessment: A geographic infonnation systems approach. *Photogrammetric Engineering and Remote Sensing*, *60*(9), 1145–1150.

Yager, R. R. (1988). On ordered weighted averaging aggregation operators in multicriteria decision making. *IEEE Transactions on Systems, Man, and Cybernetics*, *18*(1), 183–190. doi:10.1109/21.87068

Yager, R. R. (1996). Quantifier guided aggregation using OWA operators. *International Journal of Intelligent Systems*, *11*(1), 49–73. doi:10.1002/(SICI)1098-111X(199601)11:1<49::AID-INT3>3.3.CO;2-L

Compilation of References

Abdel-Naby, A., El-Beltagy, M. S., Abou-Hadid, A., Helmy, Y. I., & El-Abd, S. O. (2001). Effect of Salinity Stress Modified by Calcium Application on Potato Plants. *Egyptian Journal of Horticulture*, *28*(4), 519–530.

Abdrabbo, M. A., Farag, A. A., & Abul-Soud, M. (2013). The intercropping effect on potato under net house as adaption procedure of climate change impacts. *Researcher, 5*(6), 48-60. Retrieved from http://www.sciencepub.net/researcher

Abliz, A., Tiyip, T., Ghulam, A., Halik, U., Ding, J. L., Sawut, M., & Abliz, A. et al. (2016). Effects. of shallow groundwater table and salinity on soil salt dynamics in the Keriya Oasis, Northwestern China *Environ.mental. Earth Science*, *75*(3), 1–15. doi:10.1007/s12665-015-4794-8

Abumhadi, N., Todorovska, E., Assenov, B., Tsonev, S., Vulcheva, D., & Vulchev, D. (2012). Agricultural Research in 21 st century : Challenges facing the food security under the impacts of climate change. *Bulgarian Journal of Agricultural Science*, *18*(6), 801–818.

Adejuwon, J. O. (2005). Food crop production in Nigeria. I: Present effects of climate change. *Climate Research*, *32*, 229–245. doi:10.3354/cr032229

Adejuwon, J. O. (2006). Food crop production in Nigeria. II: Potential effects of climate variabilbity. *Climate Research*, *30*, 53–60. doi:10.3354/cr030053

Agarwal, P. K. (2009). *Global Climate change and Indian agriculture: Case studies from ICAR network project*. Indian Council of Agricultural Research.

Agbola, T., & Ijeleye, D. (2007). Climatic change and food crop production in Ibadan, Nigeria. *African Crop Science Conference Proceedings, 8,* 1423-1433.

Aggarwal, P. K. (2008). Climate change: implications for Indian agriculture. In S. S. Lal, P. M. Govindakrishnan, V. K. Dua, J. F. Singh, & K. Pandey (Eds.), *Impact Assessment of Climate Change for Research Priority Planning in Horticultural Crops* (pp. 1–10). Central Potato Research Institute, Shimla. Retrieved August 17, 2016, from http://krishikosh.egranth.ac.in/bitstream/1/2054278/1/CPRI034.pdf

Aggarwal, P. K. (2009). Global climate change and Indian agriculture: Case studies from ICAR network project. Indian Council of Agricultural Research.

Aghaei, K., Ehsanpour, A. A., & Komatsu, S. (2008). Proteome Analysis of Potato under Salt Stress. *Journal of Proteome Research*, *7*(11), 4858–4868. doi:10.1021/pr800460y PMID:18855355

Aghaei, K., Ehsanpour, A. A., & Komatsu, S. (2009). Potato responds to salt stress by increased activity of antioxidant enzymes. *Journal of Integrative Plant Biology*, *51*(12), 1095–1103. doi:10.1111/j.1744-7909.2009.00886.x PMID:20021557

Agoumi, A. (2003). Vulnerability of North African countries to climatic changes: Adaptation and implementation strategies for climatic change. *Developing Perspectives on Climate Change: Issues and Analysis from Developing Countries and Countries with Economies in Transition.* IISD/Climate Change Knowledge Network. Retrieved July 31, 2016, from http://danida.vnu.edu.vn/cpis/files/Papers_on_CC/Vulnerability/Vulnerability%20of%20North%20African%20Countries%20to%20Climatic%20Changes.pdf

Agricultural Statistics at a Glance. (2014). Government of India, Ministry of Agriculture, Department of Agriculture and Cooperation, Directorate of Economics and Statistics.

Agrios, G. N. (1997). Plant pathology. London: Elsevier.

Akita, S. (1993). Respiration: variation and potential for manipulation. In International Crop Science I. Crop Science Society of America.

Alhagdow, M. M., Barthakur, N. N., & Donnelly, D. J. (1999). Salinity stress and sodium-potassium micropropagated potatoes interactions. *Potato Research*, *42*, 73–78. doi:10.1007/BF02358392

Allakhverdiev, S. I., Nishiyama, Y., Miyairi, S., Yamamoto, H., Inagaki, N., Kanesaki, Y., & Murata, N. (2002). Salt stress inhibits the repair of photodamaged photosystem II by suppressing the transcription and translation of psbA genes in synechocystis. *Plant Physiology*, *130*(3), 1443–1453. doi:10.1104/pp.011114 PMID:12428009

Allen, L. H., Jr., & Prasad, P. V. (2004). Crop responses to elevated carbon dioxide. Encyclopedia of Plant and Crop Sci., 346-348.

Allen, E. J., & Scott, R. K. (1992). Principles of agronomy and their application in the potato Industry. In P. M. Harris (Ed.), *The potato crop. The scientific basis for improvement* (pp. 816–881). London: Chapman and Hall. doi:10.1007/978-94-011-2340-2_17

Alpert, P., & Kischka, P. (2008). Quantification of the effect of urbanization on solar dimming. *Geophysical Research Letters*, *35*(8), L08801. doi:10.1029/2007GL033012

Amadi, C. O. (2011). *Breeding potato for Adaptation to Heat stress: Implication for Food Security in Nigeria. Root and Tuber Crops Research for Food Security and Empowerment.* Retrieved January 6, 2016, from http://www.nrcri gov.ng

Anderson, P. K., Cunningham, A. A., Patel, N. G., Morales, F. J., Epstein, P. R., & Daszak, P. (2004). Emerging infectious diseases of plants: Pathogen pollution, climate change and agrotechnology drivers. *Trends in Ecology & Evolution*, *19*(10), 535–544. doi:10.1016/j.tree.2004.07.021 PMID:16701319

Anonymous, . (2006). *Annual Scientific Report, 2005-06*. Shimla: Central Potato Research Institute.

Anonymous. (2008). Annual Scientific Report, 2007-08. Central Potato Research Institute.

Anonymous. (2009). Annual Scientific Report, 2008-09. Central Potato Research Institute.

Anonymous, . (2010). *Annual Scientific Report, 2009-10*. Shimla: Central Potato Research Institute.

Antonie, J., Fischer, G., & Makowski, M. (1997). Multiple criteria land use analysis. *Applied Mathematics and Computation*, *85*(2-3), 195–215. doi:10.1016/S0096-3003(96)00190-7

APN Project Report. (2011). *The Effects of Climate Change on Pests and Diseases of Major Food Crops in the Asia Pacific Region*. Author.

Ardón, M., Helton, A. M., & Bernhardt, E. S. (2016). Drought and saltwater incursion synergistically reduce dissolved organic carbon export from coastal freshwater wetlands. *Biogeochemistry*, *127*(2-3), 411–426. doi:10.1007/s10533-016-0189-5

Arnell, N. W. (1999). Climate change and global water resources. *Global Environmental Change*, *9*(1), 31–49. doi:10.1016/S0959-3780(99)00017-5

Arvind, R. M. (2001). Exchange of gaseous nitrogen compounds between agricultural systems and the atmosphere. *Plant and Soil*, *228*, 17–27. doi:10.1023/A:1004821205442

AU (African Union) Summit. (2008). *Ban Ki-Moon. Secretary-General's address to the Summit of the African Union. Addis Ababa, Ethiopia*. Retrieved 31 January 2008, from http://www.un.org/apps/sg/sgstats.asp?nid=2978

Aydin, G., Yucel, M., Chan, M. T., & Oktem, H. A. (2014). Evaluation of abiotic stress tolerance and physiological characteristics of potato (Solanum tuberosum L. cv. Kennebec) that heterologously expresses the rice Osmyb4 gene. *Plant Biotechnology Reports*, *8*(3), 295–304. doi:10.1007/s11816-014-0322-7

Aziz, I., & Khan, M. A. (2001). Experimental assessment of salinity tolerance of Ceriops tagal seedlings and saplings from the Indus delta, Pakistan. *Aquatic Botany*, *70*(3), 259–268. doi:10.1016/S0304-3770(01)00160-7

Backhausen, J. E., Klein, M., Klocke, M., Jung, S., & Scheibe, R. (2005). Salt tolerance of potato (Solanum tuberosum L. var. Desiree) plants depends on light intensity and air humidity. *Plant Science*, *169*(1), 229–237. doi:10.1016/j.plantsci.2005.03.021

Baker, K. M., Kirk, W. W., Andresen, J., & Stein, J. M. (2004). A problem case study: influence of climatic trends on late blight epidemiology in potatoes. Acta Horticulturae, 638, 37-42. doi:10.17660/ActaHortic.2004.638.3

Banai, R. (1993). Fuzziness in geographical information systems: Contributions from an analytic hierarchy process. *International Journal of Geographical Information Systems*, *7*(4), 315–329. doi:10.1080/02693799308901964

Barredo, J. I. (1996). *Sistemas de Informacion Geografica y Evaluacio Multicriterio en la Ordenacion del Territorio*. Madrid: Ra-Ma.

Barredo, J. I., Benavidesz, A., Hervhl, J., & van Westen, C. J. (2000). Comparing heuristic landslide hazard assessment techniques using GIS in the Tirajana basin, Gran Canaria Island, Spain. *International Journal of Applied Earth Observation and Geoinformation*, *2*(1), 9–23. doi:10.1016/S0303-2434(00)85022-9

Bashi, E., & Rotem, J. (1974). Adaptation of four pathogens to semi-arid habitats as conditioned by penetration rate and germinating spore survival. *Phytopathology*, *64*(7), 1035–1039. doi:10.1094/Phyto-64-1035

Becktell, M. C., Daughtrey, M. L., & Fry, W. E. (2005). Temperature and leaf wetness requirements for pathogen establishment, incubation period, and sporulation of *Phytophthora infestans* on *Petunia* × *hybrida*. *Plant Disease*, *89*(9), 975–979. doi:10.1094/PD-89-0975

Beedasy, J., & Whyatt, D. (1999). Diverting the tourists: A spatial decision support system for tourism planning on a developing island. *J. Appl. Earth Observ. Geoinform.*, *3/4*(3-4), 163–174. doi:10.1016/S0303-2434(99)85009-0

Beevers, L., & Hegeman, R. H. (1969). Nitrate reduction in higher plants. *Annual Review of Plant Physiology*, *20*(1), 495–522. doi:10.1146/annurev.pp.20.060169.002431

Bejat, L., Perfect, E., Quisenberry, V. L., Coyne, M. S., & Haszler, G. R. (2000). Solute Transport as Related to Soil Structure in Unsaturated Intact Soil Blocks. *Soil Science Society of America Journal*, *64*(3), 818. doi:10.2136/sssaj2000.643818x

Belanger, G., Walsh, T. R., Richards, J. E., Milburn, P. H., & Ziadi, N. (2000). Yield responses of two potato cultivars to supplemental irrigation and N fertilizer in New Brunwick. *American Journal of Potato Research*, *77*(1), 11–21. doi:10.1007/BF02853657

Bernstein, L. (1975). Effects of Salinity and Sodicity on Plant Growth. *Annual Review of Phytopathology*, *13*(1), 295–312. doi:10.1146/annurev.py.13.090175.001455

Beukema, H. P., & Van der Zaag, D. E. (1990). *Introduction to potato production*. Wageningen: Pudoc.

Bharali, B., Haloi, B., Chutia, J., & Dey, P. C. (2012). *Atmospheric deposition of Nitrogen in North-east India and productivity of winter cereals.* In The 5th International Conference on Water, Climate and Environment, Ohrid, Macedonia. Retrieved June 5, 2016, from http//www.BALWOIS.Com

Bharali, B., & Bates, J. W. (2002). Soil cations influence Bryophyte susceptibility to bisulphyte. *Annals of Botany*, *90*(3), 337–343. doi:10.1093/aob/mcf192 PMID:12234145

Bharali, B., & Bates, J. W. (2004). Influences of extracellular calcium and iron on membrane sensitivity to bisulphiote in the mosses *Pleurozium schreberi* and *Rhytidiadelphus triquetrus*. *Journal of Bryology*, *26*(1), 53–59. doi:10.1179/jbr.2004.26.1.53

Bharali, B., Haloi, B., Chutia, J., Chack, S., & Hazarika, K. (2015a). Susceptibility of some wheat (*Triticum aestivum* L.) varieties to aerosols of oxidised and reduced Nitrogen. *Advances in Crop Science and Technology, 3*, 4. doi:10.4172/2329-8863.1000182

Bharali, B., Haloi, B., Dey, P. C., Chutia, J., Chack, S., & Hazarika, K. (2015a). Roles of oxidized and reduced nitrogen aerosols on productivity of winter rice (*Oryza sativa* L.). *Intl. J. of Agric. and Environment, 1*, 1–15.

Bhattacharaya, S. K., Shekhawat, G., & Singh, B. P. (1990). *Potato Late Blight. Technical Bulletin No. 27*. Shimla: Central Potato Research Institute.

Bindi, M., Hacour, A., Vandermeiren, K., Craigon, J., Ojanpera, K., Selldén, G., & Fibbi, L. et al. (2002). Chlorophyll concentration of potatoes grown under elevated carbon dioxide and/or ozone concentrations. *European Journal of Agronomy, 17*(4), 319–335. doi:10.1016/S1161-0301(02)00069-2

Biswas, M. K., De, B. K., Nath, P. S., & Mohasin, M. (2004). Influence of different weather factors on the population build up of vectors of potato virus. *Annals of Plant Protection Sciences, 12*(2), 352–355.

Blackman, R. L. (1974). Life-cycle variation of Myzus persicae (Sulz.) (Hom., Aphididae) in different parts of the world, in relation to genotype and environment. *Bulletin of Entomological Research, 63*(04), 595–607. doi:10.1017/S0007485300047830

Bojorquez-Tapia, L. A., Diaz-Mondragon, S., & Ezcurra, E. (2001). GIS-based approach for participatory decision making and land suitability assessment. *International Journal of Geographical Information Science, 15*(2), 129–151. doi:10.1080/13658810010005534

Boko, M., Niang, I., Nnong, A., Vogel, C., Githeko, A., Medany, B., ... Yanda, P. (2007). Africa. In M. L. Parry, O. F. Canziani, J. P. Palutikof, P. J. Van der Linden & C. E. Hanson (Eds.), *Impact, Adaptation and Vulnerability contribution of working group II to the fourth assessment report of the inter-governmental panel on climate change* (pp. 433-467). Cambridge University Press UK. Retrieved July 31, 2016, from https://www.ipcc.ch/publications_and_data/ar4/wg2/en/ch9.html

Boland, G. J., Melzer, M. S., Hopkin, A., Higgins, V., & Nassuth, A. (2004). Climate change and plant diseases in Ontario. *Canadian Journal of Plant Pathology, 26*(3), 335–350. doi:10.1080/07060660409507151

Boonekamp, P. M. (2012). Are plant diseases too much ignored in the climate change debate? *European Journal of Plant Pathology, 133*(1), 291–294. doi:10.1007/s10658-011-9934-8

Borah, M. N., & Milthorpe, F. L. (1959). *The effect of light intensity, length of day and temperature on growth and tuber formation in the potato* (Ph.D. thesis). University of Nottingham, UK.

Bouman, B. A. M., van Dipen, C. A., Vossen, P., & van Der Wal, T. (1997). Simulation and systems analysis tools for crop yield forecasting. In P. S. Teng, M. J. Kropff, H. F. M. ten Berge, J. B. Dent, F. P. Lansigan, & H. H. van Laar (Eds.), *Applications of Systems Approaches at the Farm and Regional Levels* (pp. 325–340). Dordrecht, The Netherlands: Kluwer Academic Publishers. doi:10.1007/978-94-011-5416-1_24

Bowen, W. T. (2003). Water productivity and potato cultivation. In J. W. Kijne, R. Barker & D. Molden (Eds.), *Water Productivity in Agriculture: Limits and Opportunities for Improvement* (pp. 229-238). CAB International. Retrieved April 10, 2016, from http://www.iwmi.cgiar.org/pubs/Book/CA_CABISeries/Water_Productivity/Protected/0 851996698ch14.pdf/

Boxman, A. W., Krabbendam, H., Bellemakers, M. J. S., & Roelofs, G. M. (1991). Effects of ammonium and aluminium on the development and nutrition of *Pinus nigra* in hydroculture. *Environmental Pollution*, *73*(2), 119–136. doi:10.1016/0269-7491(91)90018-R PMID:15092085

Boyer, J. S. (1995). Biochemical and biophysical aspects of water deficits and the predisposition to disease. *Annual Review of Phytopathology*, *33*(1), 251–274. doi:10.1146/annurev.py.33.090195.001343 PMID:18999961

Brevik, E. C. (2012). Soil and Climate Change: Gas Fluxes and Soil Processes. *Soil Horizons*, *53*. Retrieved August 04, 2016, from https://www.researchgate.net/publication/262260059_Soils_and_Climate_Change_Gas_Fluxes_and_Soil_Processes

Burke, M. B., Lobell, D. B., & Guarino, L. (2009). Shifts in African crop climates by 2050, and the implications for crop improvements and genetic resources conservation. *Global Environmental Change*, *19*(3), 317–325. doi:10.1016/j.gloenvcha.2009.04.003

Burt, R. L. (1964). Influence of short periods of low temperature on tuber initiation in the potato. *European Potato Journal*, *7*(4), 197–208. doi:10.1007/BF02368251

Bustan, A., Sagi, M., De Malach, Y., & Pasternak, D. (2004). Effects of saline irrigation water and heat waves on potato production in an arid environment. *Field Crops Research*, *90*(2-3), 275–285. doi:10.1016/j.fcr.2004.03.007

Cao, W., & Tibbitts, T. W. (1997). Starch concentration and impact on specific leaf weight and element concentrations in potato leaves under varied carbon dioxide and temperature. *Journal of Plant Nutrition*, *20*(7-8), 871–881. doi:10.1080/01904169709365302 PMID:11541213

Celebi-Toprak, F., Behnam, B., Serrano, G., Kasuga, M., Yamaguchi-Shinozaki, K., Naka, H., & Watanabe, K. N. et al. (2005). Tolerance to salt stress of the transgenic tetrasomic tetraploid potato, Solanum tuberosum cv. Desiree appears to be induced by the DREB1A gene and rd29A promoter of Arabidopsis thaliana. *Breeding Science*, *55*(3), 311–319. doi:10.1270/jsbbs.55.311

Central Potato Research Institute. (2011-12). *Annual Report 1998-1999*. Central Potato Research Institute, Shimla, India.

Central Potato Research Institute. (2012-13). *Annual Report 1998-1999*. Central Potato Research Institute, Shimla, India.

Central Potato Research Institute. (2014-15). *Annual Report 1998-1999*. Central Potato Research Institute, Shimla, India.

Chadha, S., Rana, S. S., Rameshwar, D., & Chaudhary, D. R. (2006). Effect of split application of N, K and FYM on the productivity of seed potatoes in cold desert region of H.P. *Potato Journal*, *33*, 94–96.

Chaerani, R., & Voorrips, R. E. (2006). Tomato early blight (*Alternaria solani*): The pathogen, genetics, and breeding for resistance. *Journal of General Plant Pathology*, *72*(6), 335–347. doi:10.1007/s10327-006-0299-3

Chakraborty, P. K. (1994). Effect of date of sowing and irrigation and the diurnal variation in physiological process in the leaf of Indian mustard (Brassica juncea). *Journal of Oilseeds Research*, *11*(2), 210–216.

Chakraborty, P. K., Majumdar, A., & Chatterjee, B. N. (1991). Physiological processes in Indian mustard (Brassica juncea) and yellow sarson (B. napus var. glauca) and their agronomic appraisal in mild and short winter prevailing in Gangetic plains of Eastern India. *Indian Journal of Agricultural Science*, *61*(11), 815–821.

Chakraborty, S., Tiedemann, A. V., & Teng, P. S. (2000). Climate change: Potential impact on plant diseases. *Environmental Pollution*, *108*(3), 317–326. doi:10.1016/S0269-7491(99)00210-9 PMID:15092926

Chameides, W.L., Yu, H., Liu, S. C., Bergin, M., Zhou, M., Meanrns, L., … Giorg, F. (1998). Case study of the effects on agriculture: An opportunity to enhance crop yields on China through emission controls. *Inaugural articles*: *National Academy of Science*.

Chaturvedi, R. K., Joshi, J., Jayaraman, M., Bala, G., & Ravindranath, N. H. (2012). Multi-model climate change projections for India under representative concentration pathways. *Current Science*, *103*, 791–802.

Chen, C. C., & McCarl, B. A. (2001). An investigation of the relationship between pesticide usage and climate change. *Climatic Change*, *50*(4), 475–487. doi:10.1023/A:1010655503471

Chen, C. T., & Setter, T. L. (2003). Response of potato tuber cell division and growth to shade and elevated CO_2. *Annals of Botany*, *91*(3), 373–381. doi:10.1093/aob/mcg031 PMID:12547690

Chester, F. D. (1892). Diseases of the round potato and their treatment. *Annual Report of the Delaware College Agricultural Experimental Station*, *5*, 67–70.

Chettri, M., Basu, A., & Konar, A. (2003). Effect of plant growth promoting bacteria on the growth and yield of potato in the plains of West Bengal. *J. Indian Potato Assoc.*, *30*(1-2), 83–84.

Coakley, S. M. (1988). Variation in Climate and prediction of disease in Plants. *Annual Review of Phytopathology*, *26*(1), 163–181. doi:10.1146/annurev.py.26.090188.001115

Coakley, S. M., Scherm, H., & Chakraborty, S. (1999). Climate change and plant disease management. *Annual Review of Phytopathology*, *37*(1), 399–426. doi:10.1146/annurev.phyto.37.1.399 PMID:11701829

Collier, R., Fellows, J., Adams, S., Semenov, M., & Thomas, B. (2008). Vulnerability of horticultural crop production to extreme weather events. Effects of climate change on plants: Implications for agriculture. *Aspects of Applied Biology*, *88*, 3–14.

Collins, M. G., Steiner, F. R., & Rushman, M. J. (2001). Landuse suitability analysis in the United States: Historical development and promising technological achievements. *Environmental Management*, *28*(5), 611–621. doi:10.1007/s002670010247 PMID:11568842

Collins, W. B. (1976). Effect of carbon dioxide enrichment on growth of the potato plant. *HortScience*, *11*, 467–469.

Corwin, D. L., Rhoades, J. D., & Simunek, J. (2007). Leaching requirement for soil salinity control: Steady-state versus transient models. *Agricultural Water Management*, *90*(3), 165–180. doi:10.1016/j.agwat.2007.02.007

Craigon, J., Fangmeier, A., Jones, M., Donnelly, A., Bindi, M., De Temmerman, L., & Ojanpera, K. et al. (2002). Growth and marketable-yield responses of potato to increased CO2 and ozone. *European Journal of Agronomy*, *17*(4), 273–289. doi:10.1016/S1161-0301(02)00066-7

Crawley, M. J. (1995). *Methods in Ecology: GLIM for Ecologists*. London: Oxford Blackwell Scientific Publications.

Crosier, W. (1934). Studies in the biology of *Phytophthora infestans* (Mont.) De Bary. *Cornell Univ. Exp. Stn. Memoir, 155*.

Cutter, E. G. (1992). Structure and development of the potato plant. In P. M. Harris (Ed.), *The potato crop. The scientific basis for improvement* (pp. 65–161). London: Chapman and Hall. doi:10.1007/978-94-011-2340-2_3

D'Odorico, P., Bhattachan, A., Davis, K. F., Ravi, S., & Runyan, C. W. (2013). Global desertification: Drivers and feedbacks. *Advances in Water Resources, Elsevier Ltd*, *51*, 326–344. doi:10.1016/j.advwatres.2012.01.013

Daccache, A., Keay, C., Jones, R. J. A., Weatherhead, E. K. M., Stalham, A., & Knox, J. W. (2011). Climate change and land suitability for potato production in England and Wales: Impacts and adaptation. *The Journal of Agricultural Science*, *150*(2), 1–17. Retrieved from https://dspace.lib.cranfield.ac.uk/bitstream/1826/8188/1/Land%20suitability.pdf

Dadhwal, V. K. (2003). *Crop Growth and Productivity Monitoring and Simulation using Remote Sensing and GIS*. Satellite Remote Sensing and GIS Applications in Agricultural Meteorology.

Daiber, C. C. (1965). Notes on potato aphids and leaf roll spread at different South African localities. *Journal of the Entomological Society of Southern Africa*, *27*, 191–215.

Daiber, C. C., & Schöll, S. E. (1959). Further notes on the overwintering of the green peach aphid, *Myzus persicae* (Sulzer), in South Africa. *Journal of the Entomological Society of Southern Africa, 22*, 494–520.

Dai, F. C., Lee, C. F., & Zhang, X. H. (2001). GIS-based geo-environmental evaluation for urban land-use planning: A case study. *Engineering Geology, 61*(4), 257–271. doi:10.1016/S0013-7952(01)00028-X

Darrall, N. M. (1989). The effects of air pollutants on physiological processes in plants. *Plant, Cell & Environment, 12*(1), 1–30. doi:10.1111/j.1365-3040.1989.tb01913.x

Davidson, D. J., Williamson, T., & Parkins, J. R. (2003). Understanding climate change risk and vulnerability in northern forest-based communities. *Canadian Journal of Forest Research, 33*(11), 2252–2261. doi:10.1139/x03-138

De Jong, C. F., Takken, F. L. W., Cai, X., Dewit, P. J. G. M., & Joosten, M. H. A. J. (2002). Attenuation of Cf-mediated defense responses at elevated temperatures correlates with a decrease in elicitor-binding sites. *Molecular Plant-Microbe Interactions, 15*(10), 1040–1049. doi:10.1094/MPMI.2002.15.10.1040 PMID:12437302

de Souza, D. J., Yuki, V. A., Costa, A. S., & Teixeeira, P. R. M. (1980). Study of the spread of virus diseases in a warm climate as compared to a cold climate, with a view to obtaining seed potatoes with a low rate of virus diseases. *Summa Phytopathologica, 14*, 58–59.

Dennis, A., Johnson, J., Alldredge, R., Hamm, P. B. & Frazier, B. E. (2003). Aerial Photography Used for Spatial Pattern Analysis of Late Blight Infection in Irrigated Potato Circles. *Epidemiology- The American Phytopathological Society, 93*(7), 805-812.

Dieter, P. (1984). Calmodulin and calmodulin-mediated processes in plants. *Plant, Cell & Environment, 7*(6), 371–380. doi:10.1111/j.1365-3040.1984.tb01426.x

Doll, P., & Siebert, S. (2002). Global modeling of irrigation water requirements. *Water Resources Research, 38*(4), 1–10. doi:10.1029/2001WR000355

Donnelly, A., Craigon, J., Black, C. R., Colls, J. J., & Landon, G. (2001). Elevated CO_2 increases biomass and tuber yield in potato even at high ozone concentrations. *The New Phytologist, 149*(2), 265–274. doi:10.1046/j.1469-8137.2001.00015.x

Donnelly, A., Lawson, T., Craigon, J., Black, C. R., Colls, J. J., & Landon, G. (2001). Effects of elevated CO_2 and O_3 on tuber quality in potato (*Solanum tuberosum* L.). *Agriculture, Ecosystems & Environment, 87*(3), 273–285. doi:10.1016/S0167-8809(01)00144-X

Douglas, D. R., & Groskopp, M. D. (1974). Control of early blight in eastern and south central Idaho. *American Potato Journal, 51*(11), 361–368. doi:10.1007/BF02852082

Dowley, L. J., & Sullivan, E. O. (1981). Metalaxyl-resistant strains of *Phytophthora infestons* (Mont.) de Bary in Ireland. *Potato Research, 24*(4), 417–421. doi:10.1007/BF02357324

Dua, V. K. (2013). *Nutrient management in potato production*. Compendium on Summer School on Advances in Quality Potato Production and Post-harvest Management.

Dua, V. K., Singh, B. P., Ahmad, I., & Sharma, S. (2015). Potential impact of climate change on late blight outbreak in western Uttar Pradesh and Punjab using JHULSACAST model. *Potato Journal*, *42*(1), 58–71.

Dua, V. K., Singh, B. P., Govindakrishnan, P. M., Kumar, S., & Lal, S. S. (2013). Impact of climate change on potato productivity in Punjab – a simulation study. *Current Science*, *105*(6), 787–794.

Dua, V. K., Singh, B. P., Govindakrishnan, P. M., Kumar, S., & Lal, S. S. (2013). Impact of climate change on potato productivity in Punjab- A simulation study. *Current Science*, *105*(6), 25–58.

Dutt, B. L. (1979). *Bacterial and Fungal Diseases of potato*. ICAR.

Dzengeleski, S., Da Rocha, A. B., Kirk, W. W., & Hammerschmidt, R. (2003). Effect of soil salinity and Fusarium sambucinum infection on development of potatoes cultivar "Atlantic". *Acta Horticulturae*, *19*(619), 251–261. doi:10.17660/ActaHortic.2003.619.28

Elkhatib, H. A., Elkhatib, E. A., Khalaf Allah, A. M., & El-Sharkawy, A. M. (2004). Yield Response of Salt-Stressed Potato to Potassium Fertilization: A Preliminary Mathematical Model. *Journal of Plant Nutrition*, *27*(1), 111–122. doi:10.1081/PLN-120027550

Ellis, J. B., & Martin, G. B. (1882). Macrosporium solani E&M. *American Naturalist*, *16*, 1003.

Elphinstone, J. G. (1996). Survival and possibilities for extinction of *Pseudomonas solanacearum* (Smith) Smith in cool climates. *Potato Research*, *39*(3), 403–410. doi:10.1007/BF02357946

Emilsson, B., & Lindblom, H. (1963). Physiological mechanisms concerned in sprout growth. In J. D. Ivins & F. L. Milthorpe (Eds.), *Growth of Potato. Proceedings of Tenth Easter School in Agricultural Sciences, University of Nottingham* (pp. 45–62). London Butterworths Press.

Enete, A. A., & Amusa, T. A. (2010). Challenges of Agricultural Adaptation to Climate Change in Nigeria: a Synthesis from the Literature. *Field Actions Science Reports*, *4*, 1-11. Retrieved 31 July 2016, from http://factsreports.revues.org/678

Etehadnia, M., Waterer, D. R., & Tanino, K. K. (2008). The method of ABA application affects salt stress responses in resistant and sensitive potato lines. *Journal of Plant Growth Regulation*, *27*(4), 331–341. doi:10.1007/s00344-008-9060-9

Evans, L. T. (1993). *Crop Evolution, Adaptation and Yield*. Cambridge, UK: Cambridge University Press.

Evers, D., Legay, S., Lamoureux, D., Hausman, J. F., Hoffmann, L., & Renaut, J. (2012). Towards a synthetic view of potato cold and salt stress response by transcriptomic and proteomic analyses. *Plant Molecular Biology*, *78*(4-5), 4–5, 503–514. doi:10.1007/s11103-012-9879-0 PMID:22258187

Ewing, E. E. (1997). Potato. In H. C. Wien (Ed.), *The Physiology of Vegetable crops* (pp. 295–344). Wallingford, UK: CAB International.

Ezekiel, R., Sukumaran, N. P., & Shekhawat, G. S. (1999). *Potato: A Wholesome Food.* Tech. Bull. No. 49. CPRI.

Ezekiel, R., & Bhargava, S. C. (2000). Potato top growth as influenced by temperature under short days. *Indian Journal of Plant Physiology*, *5*(1), 85–88.

Fagelfors, H., Wivstad, M., Eckersten, H., Holstein, F., Johansson, S., & Verwijst, T. (2009). *Strategic analysis of Swedish agriculture. Production systems and agricultural landscapes in a time of change*. Swedish University of Agricultural Sciences. Retrieved September 5, 2010, from http://pub-epsilon.slu.se:8080/1626

Fakihri, A. B., & Sombroek, W. G. (2010). Global climate change and agricultural production: An assessment of current knowledge and critical gaps. *FAO Corporate Document Repository*, 1-17.

Falloon, P., & Betts, R. (2010). Climate impacts on European agriculture and water management in the context of adaptation and mitigation: The importance of an integrated approach. *The Science of the Total Environment*, *408*(23), 5667–5687. doi:10.1016/j.scitotenv.2009.05.002 PMID:19501386

Fangmeier, A., De Temmerman, L., Black, C., Persson, K., & Vorne, V. (2002). Effects of elevated CO_2 and/or ozone on nutrient concentrations and nutrient uptake of potatoes. *European Journal of Agronomy*, *17*(4), 353–368. doi:10.1016/S1161-0301(02)00071-0

FAO (Food and Agricultural Organisation of the Unites Nations). (2001). *Farming systems and Poverty: Improving farmer's livelihoods in a Challenging world.* FAO Rome.

FAO (Food and Agricultural Organization of the Unites Nations). (2003). Responding to Agricultural and Food Insecurity Challenges Mobilizing Africa to Implement NEPAD Programmes. Conference of ministers of agriculture of the African union, Maputo, Mozambique.

FAO. (1983). *Guidelines: Land evaluation for rainfed agriculture. Soils Bulletin 52.* Rome: FAO.

FAO. (2004). *Agricultural data. Production and Indices Data Crop Primary.* Retrieved February 11, 2016, from http://www.fao.org/

FAO. (2005). *FAOSTAT Agricultural Data. Agricultural production, crops, primary.* Retrieved February 10, 2016, from http://faostat.fao.org/faostat/collections?subset=agriculture

FAO. (2009). Advances in the Assessment and Monitoring of Salinization and Status of Biosaline Agriculture. *World Soil Resources Reports, 104.*

Fatma, S., Moursy, I., Sadek, S.M., Abolmaaty, & Youssef, M. A . (2013). Identification and quantify of the main climatic factors affecting potato growth and yield. *Journal of Horticultural Science & Ornamental Plants, 5*(3), 177-182.

Finnan, J. M., Donnelly, A., Burke, J. I., & Jones, M. B. (2002). The effects of elevated concentrations of carbon dioxide and ozone on potato (*Solanum tuberosum* L.) yield. *Agriculture, Ecosystems & Environment, 88*(1), 11–22. doi:10.1016/S0167-8809(01)00158-X

Fischer, G. S., Shah, M., & Uelthuizen, H. Z. (2002). Climate change and agriculture vulnerability: international institute for applied systems analysis. Report prepared under UN institutional contract agreement 1113 for world summit on sustainable development, Luxemburg.

Fischer, G., & Heilig, G. K. (1997). Population momentum and the demand on land and water resources. *Philosophical Transactions of the Royal Society of London. Series B, Biological Sciences, 352*(1356), 869–889. doi:10.1098/rstb.1997.0067

Fischnich, O. (1959). Quoted by Emilsson and Lindblom, Physiological mechanisms concerned in sprout growth. In J. D. Ivins & F. L. Milthorpe (Eds.), *Growth of Potato* (pp. 45–62). London Butterworths Press.

Fleisher, D. H., Timlin, D. J. & Reddy, V. R. (2006). Temperature influence on potato leaf and branch distribution and on canopy photosynthetic rate. *Agronomy Journal, 98,* 1442-1452.

Frank, G. (2007). *Potato Wart. Online.* APSnet Features. doi:PSnetFeature-2007-060710.1094/A

Galloway, B. T. (1891). The new potato disease. *Garden and Field, Adelaide, Australia, 16*, 158.

Gao, H. J., Yang, H. Y., Bai, J. P., Liang, X. Y., Lou, Y., Zhang, J. L., & Chen, Y. L. et al. (2015). 'Ultrastructural and physiological responses of potato (Solanum tuberosum L) plantl.ets to gradient saline stress. *Frontiers in Plant Science*, *5*, 1–14. doi:10.3389/fpls.2014.00787

Garrett, K. A., Dendy, S. P., Frank, E. E., Rouse, M. N., & Travers, S. E. (2006). Climate change effects on plant disease: Genomes to ecosystems. *Annual Review of Phytopathology*, *44*(1), 489–509. doi:10.1146/annurev.phyto.44.070505.143420 PMID:16722808

Geriesh, M. H., Balke, K. D., El-Rayes, A. E., & Mansour, B. M. (2015). Implications of climate change on the groundwater flow regime and geochemistry of the Nile Delta, Egypt. *Journal of Coastal Conservation, 19*(4), 589–608. doi:10.1007/s11852-015-0409-5

Ghini, R., Hamada, E., & Bettiol, W. (2008). Climate change and plant diseases. *Scientia Agricola (Piracicaba, Braz.), 65,* 98-107.

Ghosh, S. C., Asanuma, K., Kusutani, A., & Toyota, M. (2000). Effect of temperature at different growth stages on nonstructural carbohydrate, nitrate reductase activity and yield of potato (*Solanum tuberosum*).[Japan]. *Environment Control in Biology, 38*, 197–206. doi:10.2525/ecb1963.38.197

Godwin, D. C., & Singh, U. (1998). Nitrogen balance and crop response to nitrogen in upland and lowland cropping systems. In G.Y. Tsuji, G. Hoogenboom, & P. K. Thornton (Eds.), Understanding options of agricultural production (pp. 55-77). Kluwer Academic Publishers. doi:10.1007/978-94-017-3624-4_4

Goodwin, P. B. (1967). The control of branch growth in potato tubers.II. The pattern of sprout growth. *Journal of Experimental Botany, 18*(1), 87–89. doi:10.1093/jxb/18.1.87

Goodwin, S. B., Cohen, B. A., & Fry, W. E. (1994). Panglobal distribution of a single clonal lineage of the Irish potato famine fungus. *Proceedings of the National Academy of Sciences of the United States of America*, *91*(24), 11591–11595. doi:10.1073/pnas.91.24.11591 PMID:7972108

Gornall, J., Betts, R., Burke, E., Clark, R., Camp, J., Willett, K., & Wiltshire, A. (2010). Implications of climate change for agricultural productivity in the early twenty-first century. *Philosophical Transactions of the Royal Society of London. Series B, Biological Sciences*, *365*(1554), 2973–2989. doi:10.1098/rstb.2010.0158 PMID:20713397

Govindakrishnan, P. M. (2013). Climate requirement for ware & seed potato cultivation. Compendium on Summer School on Advances in Quality Potato Production and Post-harvest Management, Shimla, India.

Granell, C., Casteleyn, S., & Atzberger, C. (2015). Geospatial Data Capturing, Processing, Analysis, and Visualization in Agro-Geoinformatics. *Frontiers in Environmental Science*. Retrieved 10 December 2015 from http://journal.frontiersin.org/article/10.3389/fenvs.2015.00076/full

Grenfell, S. E., Callaway, R. M., Grenfell, M. C., Bertelli, C. M., Mendzil, A. F., & Tew, I. (2016). Will a rising sea sink some estuarine wetland ecosystems? *Science of the Total Environment. Elsevier B.V.*, *554-555*, 276–292.

Grewal, J. S., Trehan, S. P., & Sharma, R. C. (1991). *Phosphorus and potassium nutrition of potato*. Technical Bulletin No. 31. Central Potato Research Institute.

Grewal, J. S., Sharma, R. C., & Saini, S. S. (1992). *Agro techniques for intensive potato cultivation in India*. New Delhi: ICAR Publication.

Grewal, J. S., Sharma, R. C., & Sud, K. C. (1981). Effect of continuous application of FYM and P K fertilizers on potato yield and some soil properties. *Journal of the Indian Society of Soil Science*, *28*, 29–31.

Grewal, J. S., & Sud, K. C. (1990). Potato response to phosphorus in northern India. *Fertilizer News*, *35*, 25–30.

Grewal, J. S., & Trehan, S. P. (1979). Direct, residue and cumulative effects of phosphorus and potassium applied through organic and inorganic sources on potato (Solanum tuberosum L.) in acidie brown bill soils. *J. Indian Potato Assoc.*, *6*, 41–51.

Grewal, J. S., & Trehan, S. P. (1984). Long-term effects of phosphorus and potassium fertilizers, farmyard manure and their combinations on potato yield, nutrient uptake and soil fertility. *Journal of the Indian Society of Soil Science*, *32*, 384–386.

Grewal, J. S., & Trehan, S. P. (1988). Results of use of P and K fertilizers and farmyard manure on potato yield, nutrient uptake and P and K status of soil. *Journal of Potassium Research*, *4*, 24–30.

Grewal, J. S., Verma, R. S., & Bist, B. S. (1979). Method, time and level of nitrogen application on growth, yield and quality of potato grown on acidic brown hill soils of Shimla. *Indian Journal of Agricultural Sciences*, *49*, 683–688.

Guhathakurta, P., & Rajeevan, M. (2008). Trends in rainfall pattern over India. *International Journal of Climatology*, *28*(11), 1453–1469. doi:10.1002/joc.1640

Gupta, N. K. (2007). *On spatial prediction modeling* (Unpublished Ph.D. thesis). P. G. School IARI, New Delhi, India.

Haas, B. J., Kamoun, S., Zody, M. C., Jiang, R. H., Handsaker, R. E., Cano, L. M., & Nusbaum, C. et al. (2009). Genome. sequence and analysis of the Irish potato famine pathogen *Phytophthora infestans Letters. Nature*, *461*(7262), 393–398. doi:10.1038/nature08358 PMID:19741609

Hamdi, M. M., Bettaieb, T., Harbaoui, Y., Mougou, A. A., & Jardin, P. (2009). Insight into the role of catalases in salt stress in potato (Solanum tuberosum L.). *Wild*, *13*(3), 373–379.

Hanjra, M. A., & Qureshi, M. E. (2010). Global water crisis and future food security in an era of climate change. *Food Policy, Elsevier Ltd*, *35*(5), 365–377. doi:10.1016/j.foodpol.2010.05.006

Han, S., Evans, R. G., Hodges, T., & Rawlins, S. L. (1995). Linking geographic information system with a potato simulation model for site-specific crop management. *Journal of Environmental Quality*, *24*(4), 772–777. doi:10.2134/jeq1995.00472425002400040031x

Hanson, B. R., Grattan, S. R. & Fulton, A. (2006). Agricultural Salinity and Drainage, Division of Agriculture and Natural Resources, *Publication, 3375*.

Harrington, R. (1994). Aphid Layer. *Antenna*, *18*, 50–51.

Harrington, R., Clark, S. J., Welham, S. J., Verrier, P. J., Denholm, C. H., Hullé, M., & Cocu, N. et al. (2007). Environmental change and the phenology of European aphids. *Global Change Biology*, *13*(8), 1550–1564. doi:10.1111/j.1365-2486.2007.01394.x

Harrison, M. D., Livingston, C. H., & Oshima, N. (1965a). Control of early blight in Colorado. I. Fungicidal spray schedules in relation to the epidemiology of the disease. *American Potato Journal*, *42*(11), 319–327. doi:10.1007/BF02869697

Harrison, M. D., Livingston, C. H., & Oshima, N. (1965b). Control of early blight in Colorado. II. Spore traps as a guide for initiating applications of fungicides. *American Potato Journal*, *42*(11), 333–340. doi:10.1007/BF02869699

Harrison, M. D., & Venette, J. R. (1970). Control of potato early blight and its effect on potato yield. *American Potato Journal*, *47*(3), 81–86. doi:10.1007/BF02864808

Hartz, T. K., & Moore, F. D. (1978). Prediction of potato yield using temperature and insolation data. *American Potato Journal*, *55*(8), 431–436. doi:10.1007/BF02852146

Hassanein, A. M. (1999). Alterations in protein and esterase patterns of peanut in response to salinity stress. *Biologia Plantarum*, *42*(2), 241–248. doi:10.1023/A:1002112702771

Haverkort, A. J. & Verhagen, A. (2008). Climate Change and Its Repercussions for the Potato Supply Chain. *Potato Research, 51*, 223–237. doi:10.1007/s11540-008-9107-0

Haverkort, A. J. (1990). Ecology of potato cropping systems in relation to latitude and altitude. *Agricultural Systems, 32*(3), 251–272. doi:10.1016/0308-521X(90)90004-A

Haverkort, A. J., & Verhagen, A. (2008). Climate Change and Its Repercussions for the Potato Supply Chain. *Potato Research, 51*(3-4), 223–237. doi:10.1007/s11540-008-9107-0

Headford, D. W. R. (1961). *Sprout growth of the potato* (Ph.D. thesis). University of Nottingham, UK.

Heagle, A. S., Heck, W. W., Rawlings, J. O., & Philbeck, R. B. (1983). Effects of Chronic Doses of O_3 and Sulfur Dioxide on Injury and Yield of Soybeans in Open-Top Field Chambers. *Crop Science, 23*(6), 1184. doi:10.2135/cropsci1983.0011183X002300060037x

Heagle, A. S., Miller, J. E., & Pursley, W. A. (2003). Growth and yield responses of potato to mixtures of carbon dioxide and ozone. *Journal of Environmental Quality, 32*(5), 1603–1610. doi:10.2134/jeq2003.1603 PMID:14535300

Heath, L. R., & Packer, L. (1968). Photo-oxidation in isolated chloroplast. I kinetics and stochiomentry of fatty acid peroxidation. *Archives of Biochemistry and Biophysics, 125*, 189–198. doi:10.1016/0003-9861(68)90654-1 PMID:5655425

Heldt, H.W., Flügge, U.I. and Stitt, M. (1986). Kohlenhydratstoffwechsel der pflanzlichen photosynthese. Biologie in unserer Zeitschrift, 16: 97-105.

Heuer, B., & Nadler, A. (1998). Physiological response of potato plants to soil salinity and water deficit. *Plant Science, 137*(1), 43–51. doi:10.1016/S0168-9452(98)00133-2

Heywood, I., Oliver, J., & Tomlinson, S. (1995). Building an exploratory multi-criteria modelling environment for spatial decision support. In P. Fisher (Ed.), *Innovations in GIS* (vol. 2, pp. 127–136). London: Taylor & Francis.

Hightower, R., Badon, L., Penzes, E., Lund, P., & Dunsumuir, F. (1991). Expression of antifreeze proteins in transgenic plants. *Plant Molecular Biology, 17*(5), 1013–1021. doi:10.1007/BF00037141 PMID:1932678

Hijmans, R. J. (2003). The Effect of Climate Change on Global Potato Production. *American Journal of Potato Research, 80,* 271–280. doi:10.1007/bf02855363

Hijmans, R. J. (2003). The Effect of Climate Change on Global Potato Production. *American Journal of Potato Research, 80*(4), 271–280. doi:10.1007/BF02855363

Hiremath, D. B., Patil, N. R., & Dasgupta, A. (2013). Geospatial technique for potato cold storage allocation. *Journal of Geomatics, 7*(1), 13–17.

Hisamatsu, S., Nihira, J., Takeuchi, Y., Satoh, S., & Kondo, N. (1988). NO_2 suppression of light-induced nitrate reductase in squash cotyledons. *Plant & Cell Physiology, 29*, 395–401.

Holden, N. M., Brereton, A. J., Fealy, R., & Sweeney, J. (2003). Possible change in Irish climate and its impact on barley and potato yields. *Agricultural and Forest Meteorology*, *116*(3-4), 181–196. doi:10.1016/S0168-1923(03)00002-9

Hopkins, L. (1977). Methods for generating land suitability maps: A comparative evaluation. *Journal of the American Institute of Planners*, *34*(1), 19–29.

Horton, D. E., & Anderson, J. L. (1992). Potato production in the context of the world and farm economy. In P. M. Harris (Ed.), *The potato crop* (pp. 794–15). Chapman and Hall London. doi:10.1007/978-94-011-2340-2_16

Houghton, J. T., Ding, Y., Griggs, D. J., Noguer, M., van der Linden, P. J., Xiaosu, D., … Johnson, C. A. (Eds.). (2001). *Climate Change 2001. The Scientific Basis.* Contribution of Working Group I to the Third Assessment Report of the Intergovernmental Panel on Climate Change (IPCC). Cambridge University Press. Retrieved January 09, 2016 from http://cpri.ernet.in/

Houghton, J. T., Ding, J., Griggs, D. J., Noguer, M., van der Linden, P. J., & Xiaosu, D. et al. (2001). *The Scientific Basis. Contribution of Working Group I to the Third Assessment Report of the Intergovernmental Panel on Climate Change.* Cambridge University.

Hughes, L., & Bazzaz, F. A. (2001). Effects of elevated CO_2 on five plant-aphid interactions. *Entomologia Experimentalis et Applicata*, *99*, 87–96.

Ifeanyi-obi, C. C., Etuk, U. R. & Jike-wai, O. (2012). Climate Change, Effects and Adaptation Strategies; Implication for Agricultural Extension System in Nigeria. *Greener Journal of Agricultural Sciences, 2*(2), 53-60.

Ifenkwe, O. P. (1981). A review of the status of potato production in Plateau State Nigeria. Academic Press.

Indian Horticulture Database. (2014). *National Horticulture Board, Ministry of Agriculture.* Govt. of India.

Indian Institute of Spices Research. (2012). *PhytoFura consolidated report 2009-12*. Kozhikode, Kerala: Indian Institute of Spices Research.

Intergovernmental Panel on Climate Change. (2007). *Climate change: Impact, adaptation and vulnerability, contribution of WG II to the third assessment report of the Intergovernmental Panel on climate Change*. Cambridge Univ. Press. Retrieved from www.ipcc-wg2.org

IPCC. (1995) Climate Change 1994. Cambridge University Press.

IPCC. (2001). *Third assessment report of the Intergovernmental Panel on climate change*. WMO, UNEP.

IPCC. (2001). *Third Assessment Report*. Cambridge, UK: Intergovernmental Panel on Climate Change Cambridge University Press.

IPCC. (2007), Contribution of working group 11. Intergovernmental Panel on climate change. Cambridge University Press.

IPCC. (2007). Climate Change 2007, Mitigation of Climate Change. In B. Metz, O. R. Davidson, P. R. Bosch, R. Dave, & L. A. Meyer (Eds.), *Contribution of Working Group III to the Fourth Assessment Report of the Intergovernmental Panel on Climate Change, 2007*. Cambridge University Press. Retrieved July 31, 2016, from https://www.ipcc.ch/pdf/assessment-report/ar4/wg3/ar4_wg3_full_report.pdf

IPCC. (2007). Climate Change 2007, The Physical Science Basis. In S. Solomon, D. Quin, M. Manning, Z. Chen, M. Marquis, K. B. Averut, & H. L. Miller et al. (Eds.), *Contribution of Working Group – I to the Fourth Assessment Report of the IPCC*. Cambridge University Press.

IPCC. (2007). *Fourth assessment report of the Intergovernmental Panel on climate change*. WMO, UNEP.

IPCC. (2007). *The Physical Science Basis Contribution of Working Group I to the Fourth Assessment Report of the Intergovernmental Panel on Climate Change*. Cambridge, UK: Cambridge University Press.

IPCC. (2007). The Physical Science Basis. In S. Solomon., D. Qin, M. Manning, Z. Chen, M. Marquis, K. B. Averyt, M. Tignor, & H. L. Miller (Eds.), Contribution of Working Group I to the Fourth Assessment Report. Cambridge University Press.

IPCC. (2012). *Managing the risks of extreme events and disasters to advance climate change adaptation IPCC 2012*. Retrieved October 10, 2016, from https://www.ipcc.ch/pdf/special-reports/srex/SREX_Full_Report.pdf

IPCC. (2014). Climate Change 2014: Synthesis Report. In R. K. Pachauri & L. A. Meyer (Eds.), *Contribution of Working Groups I*. Geneva, Switzerland: II and III to the Fifth Assessment Report of the Intergovernmental Panel on Climate Change.

IPCC. (2014). Impacts, Adaptation and Vulnerability. *Summary for Policy Makers*. Working Group II Contribution to the Fifth Assessment Report of the Intergovernmental Panel on Climate Change.

Jaarsma, R., de Vries, R. S. M., & de Boer, A. H. (2013). Effect of Salt Stress on Growth, Na+ Accumulation and Proline Metabolism in Potato (Solanum tuberosum) Cultivars. *PLoS ONE*, *8*(3), e60183. doi:10.1371/journal.pone.0060183 PMID:23533673

Jacobsen, T., & Adams, R. M. (1958). Salt and silt in ancient Mesopotamian Agriculture. *Science*, *128*(3334), 1251–1258. doi:10.1126/science.128.3334.1251 PMID:17793690

Jankowski, P. (1995). Integrating geographical information systems and multiple criteria decision-making methods. *International Journal of Geographical Information Science*, *9*, 251–273.

Jankowski, P., & Richard, L. (1994). Integration of GIS-based suitability analysis and multicriteria evaluation in a spatial decision support system for route selection. *Environment & Planning B*, *21*(3), 326–339.

Jatav, M. K., Sud, K. C., & Dua, V. K. (2007). Nutrient status of soils from high hills of potato growing areas of Shimla district in Himachal Pradesh. *Potato Journal*, *34*(3-4), 216–220.

Jatav, M. K., Sud, K. C., & Trehan, S. P. (2010). Effect of organic and inorganic sources of P and K on their different fraction in potato-radish sequence in Brown Hill Soils. *Journal of the Indian Society of Soil Science*, *58*(4), 388–393.

Jesus, J. M., Danko, A. S., Fiúza, A., & Borges, M. T. (2015). Phytoremediation of salt-affected soils: A review of processes, applicability, and the impact of climate change. *Environmental Science and Pollution Research International*, *22*(9), 6511–6525. doi:10.1007/s11356-015-4205-4 PMID:25854203

Jiang, H., & Eastman, J. R. (2000). Application of fuzzy measures in multi-criteria evaluation in GIS. *International Journal of Geographical Information Systems*, *14*(2), 173–184. doi:10.1080/136588100240903

Joerin, F. (1995). Me ́thode multicrite ̀re d'aide a ̀ la de ́cision et SIG pour la recherche d'un site. Rev. *Int. Ge ́omatique*, *5*(1), 37–51.

Joerin, F., Thériault, M., & Musy, A. (2001). Using GIS and outranking multi criteria analysis for land-use suitability assessment. *International Journal of Geographical Information Science*, *15*(2), 153–174. doi:10.1080/13658810051030487

Jones, L. R. (1912). *Potato diseases in Wisconsin and their control.* University ofWisconsin Experimental Station Circular of Information 36.

Jones, L. R. (1893). The new potato disease or early blight. *Vermont Agricultural Experimental Station Bulletin*, *6*, 66–70.

Jones, P. G., & Thornton, P. K. (2002). *Croppers to livestock keepers: Livelihood transition to 2010 in Africa due to climate change. Global Environmental Change*. Geneva, Switzerland: World Health Organization.

Jones, R. A. C. (2009). Plant virus emergence and evolution: Origins, new encounter scenarios, factors driving emergence, effects of changing world conditions, and prospects for control. *Virus Research*, *141*(2), 113–130. doi:10.1016/j.virusres.2008.07.028 PMID:19159652

Joshi, K. C., Grewal, S. S., Mishra, J. B., & Verma, S. C. (1982). Discoloration of potato tubers in relation to K fertilization. In B. B. Nagaich, G. S. Shekhawat., P. C. Gaur, & S. C. Verma (Eds.), Potato in Developing Countries (pp. 265-69). Indian Potato Association.

Katny, M. A. C., Hoffmann, T. G., Schrier, A. A., Fangmeier, A., & Jager, H. J. (2005). Increase of photosynthesis and starch in potato under elevated CO2 is dependent on leaf age. *Journal of Plant Physiology*, *162*(4), 429–438. doi:10.1016/j.jplph.2004.07.005 PMID:15900885

Kaufman, Z., Nachmias, A., Livescu, L., Meiri, A., & Tibor, M. (1990). Verticillium wilt of potatoes under irrigation with saline water. *Hassadeh*, *70*(6), 898–901.

Kaukoranta, T. (1996). Impact of global warming on potato late blight: Risk, yield loss, and control. *Agriculture Food Science Finland*, *5*, 311–327.

Khan, I. A., Deadman, M. L., Al Nabhani, H. S., & Al Habsi, K. A. (2003). Interactions between temperature and yield components in exotic potato cultivars grown in Oman. *Acta Horticulturae*, *619*(619), 353–359. doi:10.17660/ActaHortic.2003.619.41

Khavari-Nejad, R. A., & Chaparzadeh, N. (1998). The effects of NaCl and $CaCl_2$ on photosynthesis and growth of alfalfa plants. *Photosynthetica*, *35*(3), 461–466. doi:10.1023/A:1006928721986

Khrais, T., Leclerc, Y., & Donnelly, J. D. (1998). Relative Salinity Tolerance of Potato Cultivars Assessed By I n Vitro Screening Multiplication of In Vitro Material. *American Journal of Potato Research*, *75*, 207–210. doi:10.1007/BF02854214

Kim, S., Yang, C., Jeong, J., Choi, W., Lee, K., & Kim, S. (2013). Physiological Response of Potato Variety to Soil Salinity. *Korean Journal of Crop Science*, *58*(2), 85–90. doi:10.7740/kjcs.2013.58.2.085

Kirk, W., da Rocha, A., Hollosy, S., Hammerschmidt, R., & Wharton, P. (2006). Effect of Soil Salinity on Internal Browning of potato Tuber Tissue in Two Soil Types. *American Journal of Potato Research*, *83*(3), 223-232. Retrieved 30, 2016, from http://link.springer.com/article/10.1007/BF02872158

Kitamura, Y., Yano, T., Honna, T., Yamamoto, S., & Inosako, K. (2006). Causes of farmland salinization and remedial measures in the Aral Sea basin-Research on water management to prevent secondary salinization in rice-based cropping system in arid land. *Agricultural Water Management*, *85*(1-2), 1–14. doi:10.1016/j.agwat.2006.03.007

Klopping, H. L., & Delp, C. J. (1980). 2-cyano-N [(ethylamino) carbonyl]-2-(methoxylmino) acetamide, a new fungicide. *Journal of Agricultural and Food Chemistry*, *28*(2), 467–468. doi:10.1021/jf60228a010 PMID:7391384

Kocmánková, E., Miroslav, T., Jan, J., Martin, D. D. S., Martin, M., & Zdeněk, Ž. (2009). Impact of climate change on the occurrence and activity of harmful organisms. *Plant Protection Science*, *45*(Special Issue), S48–S52.

Kooman, P. L., & Haverkort, A. J. (1995). Modelling development and growth of the potato crop influenced by temperature and daylength: Lintul-Potato. In A. J. Haverkort & D. K. L. MacKerron (Eds.), *Potato ecology and modelling of crops under conditions limiting growth* (pp. 41–60). Dordrecht: Kluwer Academic Publishers. doi:10.1007/978-94-011-0051-9_3

Kooman, P. L., & Rabbinge, R. (1996). An analysis of the Relation between Dry Matter Allocation on the Tuber and Earliness of a Potato Crop. *Annals of Botany*, *77*(3), 235–242. doi:10.1006/anbo.1996.0027

Kroschel, J., Sporleder, M., Tonnang, H. E. Z. & Simon, R. (2012). Predicting climate-change-caused changes in global temperature on potato tuber moth *Phthorimaea operculella (Zeller)* distribution and abundance using phenology modeling and GIS mapping. *J. Agriculture and Forest Meteorology*. DOI: 10.1016/j.agrformet.2012.06.017

Kudela, V. (2009). Potential impact of climate change on geographic distribution of plant pathogenic bacteria in Central Europe. *Plant Protection Science*, *45*, S27–S32.

Ku, G., Edwards, E., & Tanner, C. B. (1977). Effects of light, carbon dioxide and temperature on photosynthesis, oxygen inhibition of photosynthesis and transpiration in *Solanum tuberosum*. *Plant Physiology*, *59*(5), 868–872. doi:10.1104/pp.59.5.868 PMID:16659958

Kumar, K. (2009). *Impact of climate change on India's monsoon climate and development of high resolution climate change scenarios for India*. Presented at MoEF, New Delhi, India. Retrieved Jan. 2015, from http: moef.nic.

Kumar, M., Rawat, S., Jatav, M. K., Dua, V. K., & Lal, S. S. (2009c). GIS mapping of spatial variability of available nutrients for site specific nutrient management in potato. In P. M. Govindakrishnan, J. P. Singh, S. S. Lal, V. K. Dua, S. Rawat, & S. K. Pandey (Eds.), Information Technology Application in Horticultural Crops (pp. 99–105). Academic Press.

Kumar, D., & Lal, S. S. (2004). Evaluation of potato based cropping systems in Bihar plains. *Potato Journal*, *31*(1-2), 49–53.

Kumar, M. (2011). North East India: Soil and water management imperatives for food security in a changing climate? *Current Science*, *101*, 1119.

Kumar, M., Jatav, M. K., Trehan, S. P., & Lal, S. S. (2009a). Integrated nutrient management in potato based cropping systems for eastern Indo-Gangetic plains of India. *Potato Journal*, *36*, 136–142.

Kumar, M., Trehan, S. P., Jatav, M. K., & Lal, S. S. (2009b). Efficacy of potato (Solenum tuberosum) cultivars under varying levels of nitrogen and growth duration in eastern Indo-Gangetic plains. *Indian Journal of Agronomy*, *54*, 63–68.

Kumar, S. (2006). Climate change and crop breeding objectives in the twenty first century. *Current Science*, *90*, 1053–1054.

Lal, M., & Yadav, S. (2014). *Efficacy of fungicides against Phytophthora infestans in the era of global climate change*. Global conference on technological challenges and human resources for climate smart horticulture-issues and strategies, NAU Navsari.

Lal, S. S. (2013). *Potato based cropping systems and weed management in potato production*. Compendium on Summer School on Advances in Quality Potato Production and Post-harvest Management, Shimla, India.

Lal, R. (2001). Future climate change: Implications for India Summer monsoon and its variability. *Current Science*, *81*, 1205–1207.

Lal, R. (2004). Soil carbon sequestration in India. *Climatic Change*, *65*(3), 277–296. doi:10.1023/B:CLIM.0000038202.46720.37

Larsson, D., & Kershaw, K. A. (1975). Measurement of CO_2 exchangeable in lichens, a new method. *Canadian Journal of Botany*, *53*(15), 1535–1541. doi:10.1139/b75-184

Latham, J. (1990). Control of Global warming? *Nature*, *347*(6291), 339–340. doi:10.1038/347339b0 PMID:2215647

Latijnhouwers, M., Ligterink, W., Vleeshouwers, V. G., VanWest, P., & Govers, F. (2004). A Gα subunit controls zoospore mobility and virulence in the potato late blight pathogen Phytophthora infestans. *Molecular Microbiology*, *51*(4), 925–936. doi:10.1046/j.1365-2958.2003.03893.x PMID:14763970

Lawson, T., Craigon, J., Tulloch, A. M., Black, C. R., Colls, J. J., & Landon, G. (2001). Photosynthetic responses to elevated CO2 and O3 in field-grown potato (*Solanum tuberosum*). *Journal of Plant Physiology*, *158*(3), 309–323. doi:10.1078/0176-1617-00105

Leclerc, R., & Tilney, M. (2015). *AgTech is the New Queen of Green*. Retrieved October 21, 2015 from http://techcrunch.com/2015/04/01/the-new-queen-of-green/

Legge, R. L., Thompson, J. E., Baker, J. E., & Lieberman, M. (1982). The effect of calcium on the fluidity of phase properties of microsomal membranes isolated from postclimacteric golden delicious apples. *Plant & Cell Physiology*, *23*, 161–169.

Lenka, D. M., Okonkwo, J. C., Amadi, C. O., Asumgha, G. N., & Nwosu, K. I. (2008). *Potato production, storage and marketing in Nigeria. In Farmer's perspectives and practice* (p. 33). National Root Crops Research Institute Umudike.

Lennefors, B. L., Lindsten, K., & Koenig, R. (2000). First record of A and B type *Beet necrotic yellow vein virus* in sugar beets in Sweden. *European Journal of Plant Pathology*, *106*(2), 199–201. doi:10.1023/A:1008780812215

Levy, D., Fogelman, E. & Itzhak, Y. (1988). The effect of water salinity on potatoes (Solanum tuberosum L.): Physiological indices and yielding capacity. *Potato Research, 31*(4), 601–610.

Levy, D. (1992). The response of potatoes (Solanum tuberosum L.) to salinity: Plant growth and tuber yields in the arid desert of Israel. *Annals of Applied Biology*, *120*(1), 547–555. doi:10.1111/j.1744-7348.1992.tb04914.x

Levy, D., Coleman, W. K., & Veilleux, R. E. (2013). Adaptation of Potato to Water Shortage: Irrigation Management and Enhancement of Tolerance to Drought and Salinity. *American Journal of Potato Research*, *90*(2), 186–206. doi:10.1007/s12230-012-9291-y

Levy, D., Fogelman, E., & Ytzhak, Y. (1993). Influence of water and soil salinity on emergence and early development of potato (Solanum tubersum L.) cultivars and effect of physiological age of seed tubers. *Potato Research*, *36*(4), 335–340. doi:10.1007/BF02361800

Levy, D., & Tai, G. C. C. (2013). Differential response of potatoes (Solanum tuberosum L.) to salinity in an arid environment and field performance of the seed tubers grown with fresh water in the following season. *Agricultural Water Management. Elsevier B.V.*, *116*, 122–127.

Li, H., Yi, J., Zhang, J., Zhao, Y., Si, B., Hill, R., … Liu, X. (2015). Modeling of Soil Water and Salt Dynamics and Its Effects on Root Water Uptake in Heihe Arid Wetland Gansu, China. *Water, 7*(5), 2382–2401. Retrieved July 30, 2016, from http://www.mdpi.com/2073-4441/7/5/2382

Lilley, R. M., Fitzgerald, M. P., Rienits, K. G., & Walker, D. A. (1975). Criteria of intactness and the photosynthetic activity of spinach chloroplast preparations. *The New Phytologist*, *75*(1), 1–10. doi:10.1111/j.1469-8137.1975.tb01365.x

Lin, H., Wan, Q., Li, X., Chen, J., & Kong, Y. (1997). GIS-based multi criteria evaluation for investment environment. *Environment and Planning. B, Planning & Design*, *24*(3), 403–414. doi:10.1068/b240403

Lipton, S. A., Yun-Beom, C. H., Pan, Z. H., Le, S. Z., Vincent Chen, H. S., & Sucher, N. J. (1993). A redox-based mechanism for the neuroprotective and neurodestructive effects of nitric oxide and related nitroso-compounds. *Nature*, *364*(6438), 626–632. doi:10.1038/364626a0 PMID:8394509

Lobell, D. B., Burke, M. B., Tebaldi, C., Mastrandrea, M. D., Falcon, W. P., & Naylor, R. L. (2008). Prioritizing climate change adaptation needs for food security in 2030. *Science*, *319*(5863), 607–610. doi:10.1126/science.1152339 PMID:18239122

Lobell, D. B., & Field, C. B. (2007). Global scale climate-crop yield relationships and the impacts of recent past. *Environmental Research Letters*, *2*(1), 1–7. doi:10.1088/1748-9326/2/1/014002

Londhe, S. (2016). Cultivation of potato on abiotic stress-affected soils of India. *Current Science, 111*(1), 21-22. Retrieved July 30, 2016, from http://www.currentscience.ac.in/Volumes/111/01/0021.pdf

Loomis, R. S., & Connor, D. J. (1992). *Crop ecology: productivity and management in agricultural systems*. Cambridge, UK: Cambridge University Press. doi:10.1017/CBO9781139170161

Lopata, W. D., & Ullrich, H. (1975). Untersuchungen zu stoffichen und strukturellen Veränderungen an Pflanzen under NO_2-Einfluss. *Staub Reinhaltung der Luft*, *35*, 196–200.

Ludlow, L., & Smit, B. (1987). Assessing the Implications of Environmental Change for Agricultural Production: The Case of Acid Rain in Ontario, Canada. *Journal of Environmental Management*, *25*, 27–44.

Madden, L., Pennypacker, S. P., & MacNab, A. A. (1978). FAST, a forecasting system Alternaria solani on tomato. *Phytopathology*, *68*(9), 1354–1358. doi:10.1094/Phyto-68-1354

Magliulo, V., Bindi, M., & Rana, G. (2003). Water use of irrigated potato (*Solanum tuberosum* L.) grown under free air carbon dioxide enrichment in central Italy. *Agriculture, Ecosystems & Environment, 97*(1-3), 65–80. doi:10.1016/S0167-8809(03)00135-X

Makropoulos, C., Butler, D., & Maksimovic, C. (2003). A fuzzy logic spatial decision support system for urban water management. *Journal of Water Resources Planning and Management, 129*(1), 69–77. doi:10.1061/(ASCE)0733-9496(2003)129:1(69)

Malczewski, J. (1999). *GIS and Multi criteria decision analysis*. New York: John Wiley & Sons Inc.

Malczewski, J. (2004). GIS-based land-use suitability analysis: A critical overview. *Progress in Planning, 62*(1), 3–65. doi:10.1016/j.progress.2003.09.002

Malczewski, J. (2006). GIS-based multi criteria decision analysis: A survey of literature. *International Journal of Geographical Information Science, 20*(7), 703–726. doi:10.1080/13658810600661508

Malczewski, J., Chapman, T., Flegel, C., Walters, D., Shrubsole, D., & Healy, M. A. (2003). GIS-multicriteria evaluation with ordered weighted averaging (OWA): Case study of developing watershed management strategies. *Environment & Planning A, 35*(10), 1769–1784. doi:10.1068/a35156

Malczewski, J., & Ogryczak, W. (1996). The multiple criteria location problem: 2. Preference-based techniques and interactive decision support. *Environment & Planning A, 28*(1), 69–98. doi:10.1068/a280069

Malczewski, J., & Rinner, C. (2005). Exploring multicriteria decision strategies in GIS with linguistic quantifiers: A case study of residential quality evaluation. *Journal of Geographical Systems, 7*(2), 249–268. doi:10.1007/s10109-005-0159-2

Martinez, C. A., Maestri, M., & Lani, E. G. (1996). In vitro salt tolerance and proline accumulation in Andean potato (Solanum spp.) differing in frost resistance. *Plant Science, 116*(2), 177–184. doi:10.1016/0168-9452(96)04374-9

Marwaha, R. S. (1989). Changes in chemical constituents of potato tuber, with the application of Zinc. *Proceedings of the National Academy of Sciences of the United States of America, 59*, 229–233.

Marwaha, R. S., & Sandhu, S. K. (2002). Yield, growth components and processing quality of potatoes as influenced by crop maturity under short and long days. *Advances in Horticultural Sciences, 16*, 47–52.

Mayton, H., Forbes, G. A., Mizubuti, E. S. G., & Fry, W. E. (2001). The roles of three fungicides in the epidemiology of potato late blight. *Plant Disease, 85*(9), 1006–1012. doi:10.1094/PDIS.2001.85.9.1006

Mc Clean, C. J., van den Berg, L. J. L., Ashmore, M. R., & Preston, C. D. (2011). Atmospheric nitrogen deposition explains patterns of plant species loss. *Global Change Biology, 17*(9), 2882–2892. doi:10.1111/j.1365-2486.2011.02462.x

McDonald, R. I., & Girvetz, E. H. (2013). Two Challenges for U.S. Irrigation Due to Climate Change: Increasing Irrigated Area in Wet States and Increasing Irrigation Rates in Dry States. *PLoS ONE*, *8*(6), 1–10. doi:10.1371/journal.pone.0065589 PMID:23755255

McElrone, A. J., Sherald, J. L., & Forseth, I. N. (2001). Effects of water stress on symptomatology and growth of *Parthenocissus quinquefolia* infected by *Xylella fastidiosa*. *Plant Disease*, *85*(11), 1160–1204. doi:10.1094/PDIS.2001.85.11.1160

Meena, H. B., Sharma, R. P., & Rawat, U. S. (2006). Status of macro and micronutrients in some soils of Tonk district of Rajasthan. *Journal of the Indian Society of Soil Science*, *54*, 508–512.

Mehree, I., & Mohr, H. (1989). Ammonium toxicity:description of the syndrome in Sinapsis alba and the search for its causation. *Physiologia Plantarum*, *77*(4), 545–554. doi:10.1111/j.1399-3054.1989.tb05390.x

Mello, A. F. S., Olarte, R. A., Gray, S. M., & Perry, K. L. (2011). Transmission efficiency of Potato virus Y strains $PVY^{O}$and PVY^{N-Wi} by five aphid species. *Plant Disease*, *95*(10), 1279–1283. doi:10.1094/PDIS-11-10-0855

Mendelsohn, R., Nordhaus, W. D., & Shaw, D. (1994). The Impact of Climate Change on Agriculture. A Ricardian Analysis. *The American Economic Review*, *84*(4), 753–771. Retrieved from http://www.jstor.org/stable/2118029?origin=JSTOR-pdf&seq=1#page_scan_tab_contents

Mercure, P. (1998). *Early Blight and Late Blight of Potato*. University of Connecticut, Integrated Pest Management. Retrieved December 15, 2015, from www.hort.uconn.edu/IPM/VEG/HTMS/BLTPOT.HTM

Meyer, R. D., & Marcum, D. B. (1998). Potato yield, petiole nitrogen and soil nitrogen response to water and nitrogen. *Agronomy Journal*, *90*(3), 420–429. doi:10.2134/agronj1998.00021962009000030017x

Meyer-Roux, J., & Vossen, P. (1994). The first phase of the MARS project, 1988-1993: overview, methods and results. In *Proceedings of the Conference on the MARS Project: Overview and Perspectives*. Commission of the European Communities.

Midmore, D. J., & Prange, R. K. (1992). Growth responses of two *Solanum* species to contrasting temperatures and irradiance levels: Relations to photosynthesis, dark respiration and chlorophyll fluorescence. *Annals of Botany*, *69*, 13–20.

Miglietta, F., Magliulo, V., Bindi, M., Cerio, L., Vaccari, F. P., Loduca, V., & Peressotti, A. (1998). Free air CO2 enrichment of potato (*Solanum tuberosum* L.), development, growth and yield. *Global Change Biology*, *4*(2), 163–172. doi:10.1046/j.1365-2486.1998.00120.x

Milne, B. (2015). *Hackathon Challenges Developers to Make a Difference in Ag*. Retrieved October 23, 2015 from http://techcrunch.com/2015/10/19/hackathon-challenges-developers-to-make-a-difference-in-ag/?ncid=rss

Milthorpe, F. L. (1963). Some aspects on plant growth an introductory survey. In J. D. Ivins & F. L. Milthorpe (Eds.), *Growth of Potato. Proceedings of Tenth Easter School in Agricultural Sciences, University of Nottingham* (pp. 3–16). London Butterworths Press.

Minhas, J. S., Kumar, D., Joseph, T. A., Raj, B. T., Paul Khurana, S. M., Pandey, S. K., & Naik, P. S. et al. (2006). Variety suitable for early planting in north-western plains, peninsular India and processing into french fries and chips. *Potato Journal*, *33*(1-2), 35–43.

Mitchell, C. E., & Power, A. G. (2003). Release of invasive plants from fungal and viral pathogens. *Nature*, *24*, 421. PMID:12571594

Mitchell, C. E., Reich, P. B., Tilman, D., & Groth, J. V. (2003). Effects of elevated CO_2, nitrogen deposition, and decreased species diversity on foliar fungal plant disease. *Global Change Biology*, *9*(3), 438–451. doi:10.1046/j.1365-2486.2003.00602.x

Mizubuti, E. S. G., & Fry, W. E. (1998). Temperature effects on developmental stages of isolates from three clonal lineages of *Phytophthora infestans. Phytopathology*, *88*(8), 837–843. doi:10.1094/PHYTO.1998.88.8.837 PMID:18944891

Mohamed, A. B. A. A., Sharifi, M. A., & van Keulen, H. (2000). An integrated agro-economic and agro-ecological methodology for land use planning and policy analysis. *International Journal of Applied Earth Observation and Geoinformation*, *2*(2), 87–103. doi:10.1016/S0303-2434(00)85003-5

Mohan, M., & Kumar, S. (1998). Review of Acid rain potential in India: Future threats and remedial measures. *Current Science*, *75*(6), 579–593.

Mokarram, M., & Aminzadeh, F. (2009). GIS-based Multicriteria Land Suitability Evaluation using Ordered Weight Averaging with Fuzzy Quantifier: A Case Study in Shavur Plain, Iran. *The International Archives of the Photogrammetry, Remote Sensing and Spatial Information Sciences*, *38*(2), 508–512.

Monneveux, P., Quiroz, R., Posadas, A., & Kleinwechter, U. (2012). *Facing climate change effects on potato cultivation-an integrative approach.* In *International conference on managing soils for food security and climate change, adaptation and mitigation*, Vienna, Austria.

Moorby, J., & Milthorpe, F. L. (1975). Potato. In L. T. Evans (Ed.), *Crop Physiology some case histories* (pp. 225–257). London: Cambridge University Press.

Moss, R. H., Edmonds, J. A., Hibbard, K. A., Manning, M. R., Rose, S. K., Vuuren, D. P., & Wilbanks, T. J. et al. (2010). The next generation of scenarios for climate change research and assessment. *Nature*, *463*(7282), 747–756. doi:10.1038/nature08823 PMID:20148028

Moya, J. L., Gómez-Cadenas, A., Primo-Millo, E., & Talon, M. (2003). Chloride absorption in salt-sensitive Carrizo citrange and salt-tolerant Cleopatra mandarin citrus rootstocks is linked to water use. *Journal of Experimental Botany*, *54*(383), 825–833. doi:10.1093/jxb/erg064 PMID:12554725

Mukherjee, D. (2004b). Influence of stresses on crop efficiency. *Kisan World, 31*(3), 40.

Mukherjee, D. (2010a). Studies on climate change effect on hill agro-ecosystem. In *Abstract National Conference of Plant Physiology on "Physiological and Molecular Approaches for Crop Improvement under Changing Environment"* (pp. 1-2). Department of Plant Physiology, Institute of Agricultural Sciences, Banaras Hindu University, Varanasi.

Mukherjee, D. (2010b). *Lead Lecture on "Resource conservation through indigenous farming system in hills of west Bengal"*. Workshop on ITK in Modern Agriculture (Souvenir), Kalyani, India.

Mukherjee, D. (2013a). Nutrient use efficiency for maximization of crop productivity. In Advances in Plant Physiology. Scientific Publishers.

Mukherjee, D. (2014c). Irrigation water quality in agriculture. In H. Rodriguez, N. C. Sarkar, R. Ramanjaneyulu, & R. Maity (Eds.), Advances in Agro-Technology: A text book (vol. 2, pp. 146-166). Puspa Publishing House.

Mukherjee, D. (2014d). Nutrient and its management: Prospect and challenges under the changing environment scenario. In A. Hemantaranjan (Ed.), Advances in Plant Physiology (vol. 15, pp. 413-442). Scientific Publishers.

Mukherjee, D. (2014e). Natural resource management under changing forest ecosystem. *MFP Newsletter, 24*(2), 15-22.

Mukherjee, D. (2002). Food security coping with nutritional challenge ahead. *Agriculture Today, 8*, 30–32.

Mukherjee, D. (2004a). Herbicides: Safe and appropriate option for weed control. *Agriculture Today, 11*, 26–27.

Mukherjee, D. (2007). Weed management strategy in rice -A Review. *Agricultural Review, 27*(4), 247–257.

Mukherjee, D. (2011). Global warming mitigation and agricultural practices.*Proceedings of 57th Orientation Course* (pp 8-9). Banaras Hindu University.

Mukherjee, D. (2013b). Sustainable uses of water resources and quality with respect to agriculture. In S. R. Maloo & L. L. Somani (Eds.), *Advances in Experimental Agriculture* (Vol. 3, pp. 5–48). Udaipur, India: Agrotech Publishing Academy.

Mukherjee, D. (2014a). Climate change and its impact on Indian agriculture. In S. Nehra (Ed.), *Plant disease management and microbes* (pp. 193–206). Jaipur, India: Aavishkar Publishers.

Mukherjee, D. (2014b). Water harvesting and resource conservation technique of Darjeeling Himalaya. *Himalayan Research Journal, 2*(1), 72–79.

Mukherjee, D. (2015a). Current environment issue related to hill agriculture. *Himalayan Times, 4*(10), 11–13.

Mukherjee, D. (2015b). Microbial diversity for soil sustainability and crop productivity. In B. K. Sharam & A. Singh (Eds.), *Microbial empowerment in Agriculture: A key to sustainability and crop productivity* (pp. 237–268). New Delhi: Biotech publishers.

Mukherjee, D., & Singh, R. P. (2002). Herbicide bating low-dose options. *Agriculture Today, 12*, 44–45.

Murray, A. J., & Wellborn, A. R. (1985). Differences in nitrogen metabolism between cultivars of tomato and pepper during exposure to glasshouse atmospheres containing oxides of nitrogen. *Environmental Pollution, 39*(4), 303–316. doi:10.1016/0143-1471(85)90112-6

Muthukumarasamy, M., Gupta, D. S., & Panneerselvam, R. (2000). Enhancement of peroxidase, polyphenol oxidase and superoxide dismutase activities by triadimefon in NaCl stressed Raphanus sativus L. *Biologia Plantarum, 43*(2), 317–320. doi:10.1023/A:1002741302485

Nachmias, A., Kaufman, Z., Livescu, L., Tsror, L., Meiri, A., & Caligari, P. D. S. (1993). Effect of salinity and its interactions with disease incidence on potatoes grown in hot climates. *Phytoparasitica, 21*(3), 245–255. doi:10.1007/BF02980946

Nandekar, D. N., Sawarkar, S. D., & Naidu, A. K. (2006). Effect of bio fertilizers and NPK on the growth and yield of potato in Satpura plateau. *Potato Journal, 33*(3-4), 168–169.

Nathan, C. (1995). Natural resistance and nitric oxide. *Cell, 82*(6), 873–876. doi:10.1016/0092-8674(95)90019-5 PMID:7553846

Nayyar, H., Malik, C. P., Singh, P., Parmar, U., Grewal, M., & Kaur, S. (1990). Diurnal variation in photosynthetic parameters in peanut. *Photosynthetica, 24*, 276–274.

NBSS & LUP. (2008). *Harmonization of Degraded lands/Wasteland Dataset of India- GIS- based Approach.* Nagpur: NBSS&LUP.

Nearing, M. A. (2001). Potential changes in rainfall erosivity in the U.S. with climate change during the 21st century. *Journal of Soil and Water Conservation, 56*, 229–232.

Newman, J. A., Gibson, D. J., Parsons, A. J., & Thornley, J. H. M. (2003). Hoe predictable are aphid population responses to elevated CO_2? *Journal of Animal Ecology, 72*(4), 556–566. doi:10.1046/j.1365-2656.2003.00725.x

Ngaira, J. K. W. (2009). Challenges of water resource management and food production in a changing climate in Kenya. *Journal of Geography and Regional Planning, 2*(4), 97-103.

Nonhebel, S. (1993). *The importance of weather data in crop growth sim- ulation models and assessment of climatic change effects* (Ph.D. thesis). Wageningen Agricultural University, Wageningen, The Netherlands.

Nwakor, F. N., Anyaegbunam, H. N., & Ekedo, T. O. (2013). Effects of Climate change on root and tuber crop production among farmers in southeastern Nigeria. *International Journal of Applied Research and Technology, 2*(6), 42–47. Retrieved from http://www.esxpublishers.com/images/ijrt-0613-0220.pdf

Obioha, I. B. (2002). The Late Onset of the Rains in the Coastal Area of Nigeria. *Weather*, *5*(6), 222–227.

Odjugo, P. A. O. (2010). General Overview of Climate Change Impacts in Nigeria. *Journal of Human Ecology (Delhi, India)*, *29*(1), 47–55.

Okonkwo, J. C. Amadi, C. O. & Nwosu, K.I. (2009). *Potato production, storage, processing and utilization in Nigeria.* National Root Crop Research Institute, Umudike, Nigeria.

Okonkwo, J. C., Ene, L. S. O., & Okoli, O. O. (1995). Potato production in Nigeria. National Root Crops Research Institute, Umudike, Nigeria.

Olivo, N., Martinez, C. A., & Oliva, M. A. (2002). The photosynthetic response to elevated CO2 in high altitude potato species (*Solanum curtilobum*). *Photosynthetica*, *40*(2), 309–313. doi:10.1023/A:1021370429699

Opena, G. B., & Porter, G. A. (1999). Soil management and supplemental irrigation effects on potato II. Root growth. *Agronomy Journal*, *91*(3), 426–431. doi:10.2134/agronj1999.00021962009100030011x

Osswald, W. F., Fleischmann, F., & Heiser, I. (2006). Investigations on the effect of ozone, elevated CO_2 and nitrogen fertilization on host- parasite interactions. *Summa Phytopathologica*, *32S*, S111–S113.

Oustani, M., Halilat, M. M. T., & Chenchouni, H. (2014). Effect of poultry manure on the yield and nutriments uptake of potato under saline conditions of arid regions. *Emirates Journal of Food and Agriculture*, *27*(1), 106–120. doi:10.9755/ejfa.v27i1.17971

Pallas, J. E. Jr, & Samish, Y. B. (1974). Photosynthetic response of peanut. *Crop Science*, *14*(3), 478–482. doi:10.2135/cropsci1974.0011183X001400030042x

Pandey, N. K., Pandit, A., & Chandran, K. P. (2007). *Potato Statistics – India and world.* Technical Bulletin. No. 81. Central Potato Research Institute.

Pant, K. P. (2009). Effects of Agriculture on Climate Change: A Cross Country Study of Factors Affecting Carbon Emissions. *J. Agric. Envi.*, *10*, 72–88.

Parida, A. K., & Das, A. B. (2005). Salt tolerance and salinity effects on plants: A review. *Ecotoxicology and Environmental Safety*, *60*(3), 324–349. doi:10.1016/j.ecoenv.2004.06.010 PMID:15590011

Parya, M. (2009). *Studies on the variation in physical environment and its effect on the growth and yield of wheat cultivars (Triticum aestivum L.) under different dates of sowing* (Ph. D. Thesis). B.C.K.V., Mohanpur, India.

Parya, M., Dutta, S. K., Jena, S., Nath, R., & Chakraborty, P. K. (2010). Effect of thermal stress on wheat productivity in West Bengal. *Journal of Agrometerology*, *12*(2), 217–220.

Paulo, C. F. S., Heder, B., Camilo, B., & Paulo, R. C. (2010). Economic optimum nitrogen fertilizer rates and nitrogen fertilizer rate effects on tuber characteristics of potato cultivars. *Potato Research*, *53*(3), 167–179. doi:10.1007/s11540-010-9160-3

Peet, M. M., & Wolfe, D. W. (2000). Crop ecosystem responses to climate change: Vegetable crops. In K. R. Reddy & H. F. Hodges (Eds.), *Climate Change & Global Crop Production* (pp. 213–243). CAB International. doi:10.1079/9780851994390.0213

Pérombelon, M. C. M. (1992). Potato blackleg: Epidemiology, host–pathogen interaction and control. *European Journal of Plant Pathology*, *98*, 135–146.

Plessl, M., Elstner, E. F., Rennenberg, H., Habermeyer, J., & Heiser, I. (2007). Influence of elevated CO_2 and ozone concentrations on late blight resistance and growth of potato plants. *Environmental and Experimental Botany*, *60*(3), 447–457. doi:10.1016/j.envexpbot.2007.01.003

Prasad, B. J., & Rao, D. N. (1980). Alteration in metabolic pools of nitrogen dioxide exposed wheat plants. *Indian Journal of Experimental Biology*, *18*, 879–882.

Prasad, P. V., & Potluri, S. D. P. (1996). Influence of proline and hydroxyproline on salt-stressed axillary bud cultures of two varieties of potato (Solanum tuberosum). *In Vitro Cellular & Developmental Biology. Plant*, *32*(1), 47–50. doi:10.1007/BF02823013

Pryor, W. A., & Lightsey, J. W. (1981). Mechanisms of NO_2 reaction: Initiation of lipid peroxidates and the production of nitrous acid. *Science*, *214*, 435–437. doi:10.1126/science.214.4519.435 PMID:17730242

Qadir, M., & Oster, J. D. (2004). Crop and irrigation management strategies for saline-sodic soils and waters aimed at environmentally sustainable agriculture. *The Science of the Total Environment*, *323*(1-3), 1–19. doi:10.1016/j.scitotenv.2003.10.012 PMID:15081713

Radcliffe, E. B. (1982). Insect pests of potato. *Annual Review of Entomology*, *27*(1), 173–204. doi:10.1146/annurev.en.27.010182.001133

Radcliffe, E. B., & Ragsdale, D. W. (2002). Aphid-transmitted potato viruses: The importance of understanding vector biology. *American Journal of Potato Research*, *79*(5), 353–386. doi:10.1007/BF02870173

Rahnama, H., & Ebrahimzadeh, H. (2004). The effect of NaCl on proline accumulation in potato seedlings and calli. *Acta Physiologiae Plantarum*, *26*(3), 263–270. doi:10.1007/s11738-004-0016-9

Rahnama, H., & Ebrahimzadeh, H. (2005). The effect of NaCl on antioxidant enzyme activities in potato seedlings. *Biologia Plantarum*, *49*(1), 93–97. doi:10.1007/s10535-005-3097-4

Rahnama, H., Vakilian, H., Fahimi, H., & Ghareyazie, B. (2011). Enhanced salt stress tolerance in transgenic potato plants (Solanum tuberosum L.) expressing a bacterial mtlD gene. *Acta Physiologiae Plantarum*, *33*(4), 1521–1532. doi:10.1007/s11738-010-0690-8

Rai, A., Sahoo, P. M., & Ahmad, T. (2008). *Geo-informatics in Agricultural Research and Development: An IASRI Perspective*. IASRI. Retrieved October 19, 2015 from http://iasri.res.in/Sovenior/Article_08.pdf

Rajasthan Agriculture Statistics at a Glance. (2012-13). Commissionerate of Agriculture, Rajasthan, Jaipur (Statistical Cell).

Ramakrishna, Y. S., Rao, G. G. S. N., Rao, V. U. M., Rao, A. V. M. S., & Rao, K. V. (2007). Water management- water use in rainfed regions of India. In M. V. K. Sivakumar & R. P. Motha (Eds.), *Managing Weather and Climate Risks in Agriculture* (pp. 245–262). Springer. doi:10.1007/978-3-540-72746-0_15

Ramanathan, V., & Carmichael, G. (2008). Global and regional climate changes due to black carbon. *Nature Geoscience*, *1*(4), 221–227. doi:10.1038/ngeo156

Rambabu, E., Chandini, R., & Pradeepthi, L. (2016). Climate change challenges in vegetables and their adaptations through bio-diversity. *Readers Shelf*, *12*, 9–10.

Ram, S. (1987). Performance of zinc sulphate and lime combination for controlling late blight in the north-east hills. *Journal of Hill Research*, *1*, 96–101.

Rattan, R. K. (2014). Soil processes and climate change. *Journal of the Indian Society of Soil Science*, *62*(Supplement), s5–s24.

Rengasamy, P. (2006). World salinization with emphasis on Australia. *Journal of Experimental Botany*, *57*(5), 1017–1023. doi:10.1093/jxb/erj108 PMID:16510516

Reynolds, M. P., & Ewing, E. E. (1989a). Effects of High Air and Soil Temperature Stress on Growth and Tuberization in *Solanum tuberosum*. *Annals of Botany*, *64*(3), 241–247. Retrieved from http://aob.oxfordjournals.org/content/64/3/241.short

Reynolds, M. P., Ewing, E. E., & Owens, T. G. (1990). Photosynthesis at high temperature in tuber bearing *Solanum species*. *Plant Physiology*, *93*(2), 791–797. doi:10.1104/pp.93.2.791 PMID:16667538

Richardson, K. V. A., Wetten, A. C., & Caligari, P. D. S. (2001). Cell and nuclear degradation in root meristems following exposure of potatoes (Solanum tuberosum L.) to salinity. *Potato Research*, *44*(4), 389–399. doi:10.1007/BF02358598

Riediger, J., Breckling, B., Nuske, R. S., & Schröder, W. (2014). Will climate change increase irrigation requirements in agriculture of Central Europe? A simulation study for Northern Germany. *Environmental Sciences Europe*, *26*(18), 1–13.

Rikko, L.S. (2010). Emerging issues from world climate trends and technologies for climate change adaptation. *Journal of Sustainable Development*, *7*(2), 27-33.

Roberts, S., & Mc Dole, R. E. (1985). Potassium Nutrition of Potatoes. In R. S. Munson (Ed.), *Potassium in Agriculture* (pp. 800–818). Madison, WI: ASA-CSSA-SSSA.

Robert, Y., Woodford, J. A. T., & Ducray-Bourdin, D. G. (2000). Some epidemiological approaches to the control of aphid-borne virus diseases in seed potato crops in northern Europe. *Virus Research*, *71*(1-2), 33–47. doi:10.1016/S0168-1702(00)00186-6 PMID:11137160

Rodrigues, M. A., Coutinho, J., Mertins, F., & Arrobas, M. (2005). Quantitative sidedress nitrogen recommendations for potatoes based upon crop nutritional indices. *European Journal of Agronomy*, *23*(1), 79–88. doi:10.1016/j.eja.2004.10.001

Romero-Aranda, R., Soria, T., & Cuartero, J. (2001). Tomato plant-water uptake and plant-water relationships under saline growth conditions. *Plant Science*, *160*(2), 265–272. doi:10.1016/S0168-9452(00)00388-5 PMID:11164598

Rotem, J. (1994). The Genus Alternaria: Biology, Epidemiology, and Pathogenicity. APS Press.

Roudier, P., Sultan, B., Quirion, P., & Berg, A. (2011). The impact of future climate change on West African crop yields: What does the recent literature say? *Global Environmental Change*, *1*(3), 1073–1083. doi:10.1016/j.gloenvcha.2011.04.007

Rowland-Bamford, A. J., Lea, P. J., & Wellburn, A. R. (1989). NO_2 flux into leaves of nitrite reductase deficient barley mutants and corresponding changes in nitrate reductase activity. *Environmental and Experimental Botany*, *29*(4), 439–444. doi:10.1016/0098-8472(89)90022-1

Roy, S. K., Sharma, R. C., & Trehan, S. P. (2001). Integrated nutrient management by using farmyard manure and fertilizers in potato-sunflower–paddy rice rotation in the Punjab. *The Journal of Agricultural Science*, *137*(03), 271–278. doi:10.1017/S0021859601001472

Rydén, K., Eriksson, B., & Insunza, V. (1986). Rostringar hos potatis orsakade av potatismopptoppvirus (PMTV). *Växtskyddsnotiser*, *50*, 97–102.

S. J. V. (2007). *Swedish Board of Agriculture. En meter i timmen—klimatförändringarnas påverkan på jordbruket i Sverige*. (English summary). Report no. 16. ISSN 1102–3007.

Sajid, Z. A., & Aftab, F. (2009). Amelioration of salinity tolerance in Solanum tuberosum L. by exogenous application of ascorbic acid. *In Vitro Cellular & Developmental Biology. Plant*, *45*(5), 540–549. doi:10.1007/s11627-009-9252-4

Sajid, Z., & Aftab, F. (2012). Role of Salicylic Acid in Amelioration of Salt Tolerance in Potato (Solanum Tuberosum L.) Under in Vitro Conditions. *Pakistan Journal of Botany*, *44*, 37–42.

Sale, P. J. M. (1979). Growth of potatoes (*Solanum tuberosum* L) to the small tuber stage as related to soil temperature. *Australian Journal of Agricultural Research*, *30*, 667–675.

Sandgren, M. (1995). Potato mop-top virus (PMTV): Distribution in Sweden, development of symptoms during storage and cultivar trials in field and glasshouse. *Potato Research*, *38*(4), 387–397. doi:10.1007/BF02357743

Santala, J., Samuilova, O., Hannukkala, A., Latvala, S., Kortemaa, H., Beuch, U., & Valkonen, J. P. T. et al. (2010). Detection, distribution and control of *Potato mop-top virus*, a soil-borne virus, in northern Europe. *Annals of Applied Biology*, *157*(2), 163–178. doi:10.1111/j.1744-7348.2010.00423.x

Scehepers, H. T. A. M. (1996). Effect of rain on efficacy of fungicides deposits potato against *Phytophthora infestans. Potato Research*, *39*(4), 541–550. doi:10.1007/BF02358473

Schapendonk, A. H. C. M., Oijen van, N., Dijkstra, P., Pot, C. S., Jordi, W. J. R. M., & Stoopen, G. M. (2000). Effects of elevated CO_2 concentration on photosynthetic acclimation and productivity of two potato cultivars grown in open-top chambers. *Australian Journal of Plant Physiology*, *27*, 1119–1130.

SEI. (2008). *Climate change and adaptation in African agriculture*. Rockefeller Foundation. Retrieved July 31, 2016 from https://www.sei-international.org/mediamanager/documents/Publications/Climate/Using_Science_To_Improve_Adaptation_Africa.pdf

Sharma, A. K. (2013). *Practices for Quality Seed Potato Production*. Compendium on Summer School on Advances in Quality Potato Production and Post-harvest Management, Shimla, India.

Sharma, R. C., & Upadhayay, N. C. (1993). Nitrogen nutrition of potato. In K. L. Chadha & J. S. Grewal (Eds.), Advances in Horticultural (vol. 7, pp. 231-59). Malhotra Publishing House.

Sharma, S. (2013). *Fungal diseases of potato and their management*. Compendium on Summer School on Advances in Quality Potato Production and Post-harvest Management, Shimla, India.

Sharma, P. D., & Singh, M. V. (2012). State of health of Indian soils. In N. N. Goswami, B. Singh, D. K. Pal, R. K. Rattan, & D. L. N. Rao (Eds.), *Soil Science in the Service of Nation* (pp. 191–213). New Delhi: Indian Society of Soil Science.

Sharma, R. C., Grewal, J. S., & Sud, K. C. (1976). Relative suitability of different phosphatic sources for potato on brown hill soils of Shimla. *Journal of the Indian Society of Soil Science*, *24*, 95–97.

Sharma, R. C., Trehan, S. P., Roy, S. K., & Kumar, D. (1999). Nutrient management in potato. *Indian Farming*, *49*(9), 52–54.

Sharma, R. P., Singh, M., & Sharma, J. P. (2003). Correlation studies on DTPA – extractable micronutrients vis-à-vis soil properties in some soils of Nagaur district in semi-arid region of Rajasthan. *Journal of the Indian Society of Soil Science*, *51*, 522–527.

Sharma, U. C., & Singh, K. (1989). Nutrient balance sheet and economics of three potato-based cropping systems in acidic soils. *Journal of the Indian Society of Soil Science*, *37*, 754–758.

Shekhawat, G. S., Gadewar, A. V., & Chakrabarti, S. K. (2000). *Potato bacterial wilt in India*. Technical Bulletin No. 38 (Revised). C.P.R.I.

Sherf, A. F., & MacNab, A. A. (1986). Vegetable Disease and Their Control. John Wiley and Sons.

Shtienberg, D., & Fry, W. E. (1990). Influence of host resistance and crop rotation on initial appearance of potato early blight. *Plant Disease*, *74*(11), 849–852. doi:10.1094/PD-74-0849

Siddons, P. A., Jones, R. J. A., Hollis, J. M., Hallett, S. H., Huyghe, C., Day, J. M., Scott, T. & Milford, G. F. J. (1994). The use of a land suitability model to predict where autumn-sown determinate genotypes of the white lupin (Lupinus albus) might be grown in England and Wales. *Journal of Agricultural Science, 123,* 199–205.

Sidi, M. S. (2012). *The Impact of the 2012 Floods on Agriculture and Food Security in Nigeria Using GIS.* United Nations International Conference on Space-based Technologies for Disaster Management -Risk Assessment in the Context of Global Climate Change, Beijing, China.

Silva, J. A. B., Otoni, W. C., Martinez, C. A., Dias, L. M., & Silva, M. A. P. (2001). Microtuberization of Andean potato species (Solanum spp.) as affected by salinity. *Scientia Horticulturae*, *89*(2), 91–101. doi:10.1016/S0304-4238(00)00226-0

Singh, B. (2013). *Potato scenario- past, present and future.* Compendium on Summer School on Advances in Quality Potato Production and Post-harvest Management, Shimla, India.

Singh, B. P., Dua, V. K., Govindakrishnan, P. M., & Sharma, S. (2013). Impact on climate change on potato. In H. P. Singh et al. (Eds.), Climate-Resilient Horticulture: Adaptation and Mitigation Studies (pp. 125-135). . doi:10.1007/978-81-322-0974-4_12

Singh, J. P., & Trehan, S. P. (1993). *A case study of soil and irrigation water related constraints in potato crop production: Cause and correction.* National Seminar on 'Developments in Soil Science, 58th Annual convention, Dehradun.

Singh, J. P., Govindakrishnan, P. M., Lal, S. S., & Aggarwal, P. K. (2008). Infocrop-Potato a Model for Simulating Growth and Yield of Potato in the Sub-Tropics. Central Potato Research Institute, Shimla.

Singh, J. P., Jaiswal, V. P., & Trehan, S. P. (1994). Potato: Present and Future. GS Indian Potato Association.

Singh, J. P., Lal, S. S., & Pandey, S. K. (2009). Effect of climate change on potato production in India. Central Potato Research Institute, Shimla, Newsletter, 40, 17-18.

Singh, J. P., Lal, S. S., & Sharma, R. C. (2001). Productivity and sustainability of potato based cropping systems with reference to fertilizer management in north-western plains of India. In *Proceedings of Global Conference on Potato.* Indian Potato Assoc.

Singh, J. P., Lal, S. S., Govindakrishnan, P. M., Dua, V. K., & Pandey, S. K. (2008). Impact of climate change on potato. In S. S. Lal, P. M. Govindakrishnan, V. K. Dua, J. F., Singh, & K. Pandey (Eds.), *Impact Assessment of Climate Change for Research Priority Planning in Horticultural Crops* (pp. 125–136). Central Potato Research Institute, Shimla. Retrieved August 17, 2016, from http://krishikosh.egranth.ac.in/bitstream/1/2054278/1/CPRI034.pdf

Singh, N. (2013). *Micro irrigation system and water management in potato production.* Compendium on Summer School on Advances in Quality Potato Production and Post-harvest Management, Shimla, India.

Singh, R. P., Singh, R. K., & Mukherjee, D. (2004). Effect of weed interference on efficiency of crop. In Abstract Seminar on biotic and abiotic stress on agro-ecosystems W.B. Institute of Agricultural Sciences (P.S.B).

Singh, B. P., & Bhat, M. N. (2010). Impact assessment of climate change on potato diseases and insect pests. In H. P. Singh, J. P. Singh, & S. S. Lal (Eds.), *Challenges of climate change Indian horticulture* (pp. 178–184). New Delhi: Westville Publishing house.

Singh, J. P., Govindakrishnan, P. M., Lal, S. S., & Aggarwal, P. K. (2005). Increasing the efficiency of agronomy experiments in potato using INFOCROP-POTATO model. *Potato Research, 48*(3-4), 131–152. doi:10.1007/BF02742372

Singh, J. P., & Grewal, J. S. (1995). Requirement of potassium to potato crop and effect of some agrotechniques on yield and fertilizer use efficiency. *J. Potassium Res., 11*, 160–165.

Singh, J. P., Lal, S. S., & Pandey, S. K. (2009). Effects of climate change on potato production in India. Central Potato Research Institute. *Shimla Newsletter, 40*, 17–18.

Singh, J. P., Salaria, A., Singh, K., & Gangwar, B. (2008a). Efficiency of diversified rice-wheat cropping systems including potato, vegetable peas and groundnut crops in trans-gangetic plains. *Potato Journal, 35*(1-2), 53–60.

Singh, J. P., Trehan, S. P., & Sharma, R. C. (1997). Crop residue management for sustaining the soil fertility and productivity of potato based cropping systems in Punjab. *Journal of Indian Potato Association, 24*, 85–89.

Singh, J. P., Trehan, S. P., Upadhayay, N. C., & Lal, S. S. (2008b). Potato based cropping system. In S. K. Pandey & S. K. Chakrabarti (Eds.), *Twenty steps towards hidden treasure* (pp. 105–125). Shimla, India: Central Potato Research Institute.

Singh, K. (2000). Effect of inoculation with Azotobacter and Phosphobacteria on potato (Solanum tuberosum) in north- eastern hills. *Indian Journal of Agricultural Sciences, 70*(6), 385–386.

Singh, K. (2001). Effect of phosphobacteria, nitrogen and phosphorus on the tuber yield of potato (Solanum tuberosum) under East Khasi hill conditions of Meghalaya. *Indian Journal of Agronomy, 46*(2), 375–379.

Singh, M. V. (2006). Micro- and secondary nutrients and pollutant elements research in India. Coordinator Report-AICRP Micro and secondary nutrients and pollutant elements in soil and plants. *IISS, Bhopal., 30*, 1–110.

Singh, N. P., Ojha, A. P., & Kumar, V. (1986). Response of potato (Solanum tuberosum L.) varieties on various fertility levels. *Progressive Horticulture, 18*, 114–117.

Singh, N. P., & Raghav, M. (2000). Response of potato to nitrogen and potassium fertilization under U.P. tarai conditions. *Journal of Indian Potato Association*, *27*, 47–48.

Singh, N., Sood, M. C., & Lal, S. S. (2005). Evaluation of potato based cropping sequences under drip, sprinkler and furrow methods of irrigation. *Potato Journal*, *32*(3-4), 175–176. Retrieved from WWW.agriexchange.apeda.gov.in/production/

Singh, S. K. (2002). Effect of inoculation with Azotobacter and Phosphobacteria on potato (Solanum tuberosum) in north- eastern hills. *Indian Journal of Agronomy*, *47*(2), 273–277.

Singh, S. N., Singh, B. P., Singh, O. P., Singh, R., & Singh, R. K. (2007). Effect of nitrogen application in conjunction with bio-inoculants on the growth, yield and quality of potato under indo-gangetic plain region. *Potato Journal*, *34*(1-2), 103–104.

Singh, S. N., Trehan, S. P., & Manoj, K. (2008c). Response of potato cultivars to nitrogen in the plains of Bihar. *Potato Journal*, *35*(3-4), 158–160.

Sivakumar, M. V. K. (2007). Interactions between climate and desertification. *Agricultural and Forest Meteorology*, *142*(2-4), 143–155. doi:10.1016/j.agrformet.2006.03.025

Smit, B., Ludlow, L., & Brklacich, M. (1988). Implications of a Global Climatic Warming for Agriculture: A Review and Appraisal. *Journal of Environmental Quality*, *17*(4), 517–527. doi:10.2134/jeq1988.00472425001700040001x

Somani, A. K. (1986). Non-hazardous chemical control of black scurf of potato. *Indian Journal of Agricultural Sciences*, *56*, 366–369.

Somani, A. K. (1988). Control of black scurf (*Rhizoctonia solani*) and common scab (*Streptomyces scabies*) of potato (*Solanum tuberosurn* L.) with boric acid. Control of black scurf (*Rhizoctonia solani*) and common scab (*Streptomyces scabies*) of potato (*Solanum tuberosurn* L.) with boric acid. *Indian Journal of Agricultural Sciences*, *58*, 693–698.

Son, S. W., Kim, H. Y., Choi, G. J., Lim, H. K., Jang, K. S., Lee, S. O., & Kim, J. C. et al. (2008). Bikaver.in and fusaric acid from Fusarium oxysporum show antioomycete activity against P infesta. ns. *Journal of Applied Microbiology*, *104*(3), 692–698. doi:10.1111/j.1365-2672.2007.03581.x PMID:17927749

Sood, M. C., & Sharma, R. C. (2001). Value of growth promoting bacteria, vermicompost and Azotobacter on potato production in Shimla hills. *Journal of Indian Potato Association*, *28*(1), 52–53.

Sparrow, L. A., & Chapman, K. S. R. (2003). Effects of nitrogen fertilizer on potato (Solanum tuberosum L.) in Jasmanica 1. Yield and quality. *Australian Journal of Experimental Agriculture*, *43*, 631–641. doi:10.1071/EA02099

Srivastava, H. S., & Ormrod, D. P. (1984). Effect of nitrogen dioxide and nitrate nutrition on growth and assimilation in bean leaves. *Plant Physiology*, *76*(2), 418–423. doi:10.1104/pp.76.2.418 PMID:16663857

Stainforth, D. A., Thomas, E. D., Richard, W., Ana, L., & Mark, N. (2007). Issues in the interpretation of climate model ensembles to inform decisions. *Philosophical Transactions of the Royal Society A: Mathematical, Physical and Engineering Sciences*, *365*(1857), 2163–2177. doi:10.1098/rsta.2007.2073 PMID:17569657

Staubli, B., Wenger, R., & Wymann Von Dach, S. (2008). *Potatoes and climate change*. Retrieved March 6, 2016 from http://www.inforesources.ch/pdf/focus08_1_e.pdf

Sterrett, S. B., Lee, G. S., Henninger, M. R., & Lentner, M. (1991). Predictive model for onset and development of internal heat necrosis of 'Atlantic' potato. *Journal of the American Society for Horticultural Science*, *116*, 701–705.

Stol, W, de Koning, G. H. J., Haverkort, A. J., Kooman, P.L., van Keulen, H. & Penning de Vries, F. W. T. (1991). *Agro-ecological characterization for potato production. A simulation study at the request of the International Potato Center (CIP), Lima, Peru.* CABO-DLO, Report 155.

Struik, P. C., Geertsema, J., & Custers, C. H. M. G. (1989). Effects of shoot, root and stolon temperatureon the development of potato plant. III. Development of tubers. *Potato Research*, *32*, 151–158. doi:10.1007/BF02358227

Sud, K. C., & Grewal, J. S. (1994). Evaluation of Nitrogen sources and levels on potato nutrition and their effect on soil fertility under potato-fallow-potato crop sequence in Shimla hills. In Potato: Present and Future. Indian Potato Association.

Sud, K. C., Upadhayay, N. C., Trehan, S. P., Jatav, M. K., Kumar, M., & Lal, S. S. (2008). *Soil and Plant Tests for Judicious Use of Fertilizer in Potato*. Technical Bulletin No. 91. Central Potato Research Institute.

Sud, K. C. (1996). Effect of phosphorus and sulphur on potato nutrition in acid soils of Shimla. *Journal of the Indian Society of Soil Science*, *44*, 440–444.

Sud, K. C. (2006). Anionic balance of nitrogen, phosphorus and sulphur in potato leaves under rainfed conditions in Shimla hills. *Journal of the Indian Potato Association*, *33*, 68–73.

Sud, K. C., & Grewal, J. S. (1991). Effect of time of K application on potato nutrition in Shimla hills. *Journal of Potassium Research*, *7*, 266–276.

Sud, K. C., & Grewal, J. S. (1992). Evaluation of methods of potassium application in potato production on acidic soils of Shimla. *Journal of Potassium Research*, *8*, 137–143.

Sud, K. C., Grewal, J. S., & Sharma, R. C. (1991). Efficient use of urea by potato (*Solanum tuberosum* L.) in Shimla hills. *Indian Journal of Agricultural Sciences*, *61*, 389–394.

Sud, K. C., & Jatav, M. K. (2007). Response of potato to phosphorus and phosphorus solubilizing bacteria in brown hill soils of Shimla. *Potato Journal*, *34*, 109–110.

Sud, K. C., Sharma, R. C., & Verma, B. C. (1996). Evaluation of levels and sources of sulphur on potato nutrition in Shimla hills. *Journal of the Indian Potato Association*, *23*, 134–139.

Sud, K. C., Trehan, S. P., & Lal, S. S. (2007). Performance of organic and inorganic in potato –garlic sequence under zero tillage in the mid hills of Himachal Pradesh. *Potato Journal*, *34*, 203–208.

Sullivan, C. Y., & Rose, M. W. (1972). Selection for draught and resistance in grain sorghum In H. Mussel & R. Staples (Eds.), Stress Physiology in Crop Plants (pp. 263-289). Wiley.

Sys, C. (1985). *Land evaluation*. Ghent, The Netherlands: State University of Ghent.

Takahashi, K., Matsuoka, Y., & Harasawa, H. (1998). Impacts of climate change on water resources, crop production and natural ecosystem in the Asia and Pacific region. *Journal of Global Environmental Engineering*, *11*, 91–103.

Takemura, T., Hanagata, N., Sugihara, K., Baba, S., Karube, I., & Dubinsky, Z. (2000). Physiological and biochemical responses to salt stress in the mangrove, Bruguiera gymnorrhiza. *Aquatic Botany*, *68*(1), 15–28. doi:10.1016/S0304-3770(00)00106-6

Teixeira, J., & Fidalgo, F. (2009). Salt stress affects glutamine synthetase activity and mRNA accumulation on potato plants in an organ-dependent manner. *Plant Physiology and Biochemistry. Elsevier Masson SAS*, *47*(9), 807–813.

Teixeira, J., & Pereira, S. (2007). High salinity and drought act on an organ-dependent manner on potato glutamine synthetase expression and accumulation. *Environmental and Experimental Botany*, *60*(1), 121–126. doi:10.1016/j.envexpbot.2006.09.003

Teixeira, J., Pereira, S., Queirós, F., & Fidalgo, F. (2006). Specific roles of potato glutamine synthetase isoenzymes in callus tissue grown under salinity: Molecular and biochemical responses. *Plant Cell, Tissue and Organ Culture*, *87*(1), 1–7. doi:10.1007/s11240-006-9103-5

Teng, P. S., & Bissonnette, H. L. (1985). Estimating potato yield responses for chemical control of early blight in Minnesota. *American Potato Journal*, *62*(11), 595–606. doi:10.1007/BF02854434

Trehan, S. P. (2010). Nitrogen and Potassium use efficiency vis-à-vis N & K interaction in different potato cultivars/hybrid. Presented in Round Table Discussion on "Enhancing Nutrient Use Efficiency through Potassium fertilization" being held by Fertilizer Association of India and International Potash Institute Switzerland on 9th Feb., 2010 at the Fertilizer Association of India House, New Delhi, India.

Trehan, S. P., & Sharma, R. C. (1999). *Micronutrients Requirement of potato*. Technical Bulletin 23 (revised). CPRI.

Trehan, S. P., Upadhayay, N. C., Sud, K. C., Kumar, M., Jatav, M. K., & Lal, S. S. (2008). *Nutrient Management in Potato*. Technical Bulletin No. 90. Central Potato Research Institute.

Trehan, S. P. (2003). Evaluation of nitrogen efficiency of different potato cultivars. *Journal of the Indian Potato Association*, *30*, 63–64.

Trehan, S. P. (2006). Genetic control of different potato cultivars in the manipulation of nitrogen uptake from green manured soil. *Advances in Horticultural Science, 20*, 199–207.

Trehan, S. P. (2009). Improving nutrient use efficiency by exploiting genetic diversity of potato. *Potato Journal, 36*, 121–135.

Trehan, S. P., & Grewal, J. S. (1991). *Micronutrients management in potato. In Annual Scientific Report* (p. 65). Jalandhar, India: Central Potato Research Station.

Trehan, S. P., & Grewal, J. S. (1995). Response of zinc, copper and iron and their critical levels for potato. *Journal of the Indian Society of Soil Science, 43*, 89–91.

Trehan, S. P., Sharma, K. K., & Grewal, J. S. (1995). Effect of soil and foliar application of micronutrients and fungicides on late blight (*Phytophthora infestans*) and tuber yield of potato (*Solanum tuberosum* L.). *Indian Journal of Agricultural Sciences, 65*, 836–841.

Trenberth, K. E. (2011). Changes in precipitation with climate change. *Climate Research, 47*(1-2), 123–138. doi:10.3354/cr00953

United States Salinity Laboratory. (1954). Diagnosis and Improvement of Saline and Alkali Soils. *Agricultural Handbook, 60.*

Upadhayay, N. C., & Sharma, R. C. (2002). Fertilizer prescription equations for N, P and K in potato. In S. M. Paul Khurana, G. S. Shakawat, S. K. Pandey & B. P. Singh (Eds.), Potato, Global Research & Development (vol. 2, pp. 838-842). Indian Potato Association.

Upadhyaya, C. P., Venkatesh, J., Gururani, M. A., Asnin, L., Sharma, K., Ajappala, H., & Park, S. W. (2011). Transgenic potato overproducing l-ascorbic acid resisted an increase in methylglyoxal under salinity stress via maintaining higher reduced glutathione level and glyoxalase enzyme activity. *Biotechnology Letters, 33*(11), 2297–2307. doi:10.1007/s10529-011-0684-7 PMID:21750996

Upadhyay, N. C., Sharma, R. C., & Singh, N. (1999). *Effect of organic and biofertilizer economy in green manure-potato-okra system.Global Conference on Potato*, New Delhi, India.

Usharani, K. S., Surendernath, B., Paul-Khurana, S. M., Garg, I. D., & Malathi, V. G. (2004). Potato leaf curl – a new disease of potato in northern India caused by a strain of Tomato leaf curl New Delhi virus. *Plant Pathology, 53*(2), 235. doi:10.1111/j.0032-0862.2004.00959.x

Vaccari, F. P., Miglietta, F., Giuntoli, A., Magliulo, V., Cerio, L., & Bindi, M. (2001). Free air CO2 enrichment of potato (*Solanum tuberosum* L.). Photosynthetic capacity of leaves. *Italian Journal of Agronomy, 5*, 3–10.

Van de Geijn, S. C., & Dijkstra, P. (1995). Physiological effects of changes in atmospheric carbon dioxide concentration and temperature on growth and water relations of crop plants. In A. J. Haverkort & D. K. L. MacKerron (Eds.), *Potato ecology and modelling of crops under conditions limiting growth* (pp. 89–100). Dordrecht: Kluwer Academic Publishers. doi:10.1007/978-94-011-0051-9_6

Van der Waals, J. E. (2002). *A review of early blight on potatoes*. University of Pretoria.

Van der Waals, J. E., Korsten, L., Aveling, T. A. S., & Denner, F. D. N. (2003). Influence of environmental factors on field concentrations of *Alternaria solani* conidia above a South African potato crop. *Phytoparasitica*, *31*(4), 353–364. doi:10.1007/BF02979806

Van Keulen, H., & Stol, W. (1995). Agro-ecological zonation for potato production. In A. J. Haverkort & D. K. L. MacKerron (Eds.), *Potato Ecology and Modelling of Crops under Conditions Limiting Growth* (pp. 357–372). Dordrecht: Kluwer Academic Publishers. doi:10.1007/978-94-011-0051-9_23

Van Ranst, E., Tang, H., Groenemam, R., & Sinthurahat, S. (1996). Application of fuzzy logic to land suitability for rubber production in peninsular Thailand. *Geoderma*, *70*(1), 1–19. doi:10.1016/0016-7061(95)00061-5

Vandermeiren, K., Black, C., Lawson, T., Casanova, M. A., & Ojanpera, K. (2002). Photosynthetic and stomatal responses of potatoes grown under elevated CO2 and/or O3 - results from the European CHIP-programme. *European Journal of Agronomy*, *17*(4), 337–352. doi:10.1016/S1161-0301(02)00070-9

Varallyay, G. (2007). Potential impacts of climate change on agro-ecosystems. *Review Agriculturae Conspectus Scientificus*, *72*, 1–8.

Venette, J. R., & Harrison, M. D. (1973). Factors affecting infection of potato tubers by Alternaria solani in Colorado. *American Potato Journal*, *59*, 313–325.

Verbeek, M., Piron, P. G. M., Dullemans, A. M., Cuperus, C., & Van der Vlugt, R. A. A. (2010). Determination of aphid transmission efficiencies for N, NTN and Wilga strains of Potato virus Y. *Annals of Applied Biology*, *156*(1), 39–49. doi:10.1111/j.1744-7348.2009.00359.x

Vergeer, P., Van Den Berg, L. L. J., Bulling, M. T., Ashmore, M. R., & Kunin, W. E. (2008). Geographical variation in the response to nitrogen deposition in Arabidopsis lyrata petraea. *The New Phytologist*, *179*(1), 129–141. doi:10.1111/j.1469-8137.2008.02445.x PMID:18422899

Verma, R. S., & Grewal, J. S. (1978). Evaluation of levels and methods of phosphatic fertilizer application to potatoes in acidic brown hill soil. *Journal of the Indian Potato Association*, *5*, 76–82.

Vicente-Serrano, S. M., Lopez-Moreno, J. I., Beguería, S., Lorenzo-Lacruz, J., Sanchez-Lorenzo, A., García-Ruiz, J. M., . . . Espejo, F. (2014). Evidence of increasing drought severity caused by temperature rise in southern Europe. *Environmental Research Letters*, *9*(4), 1-9. Retrieved July 30, 2016 from http://digital.csic.es/bitstream/10261/95049/1/BegueriaS_EnvResLett_2014.pdf

Visser, J. C., & Bellstedt, D. U. (2009). An assessment of molecular variability and recombination patterns in South African isolates of Potato virus Y. *Archives of Virology*, *154*(12), 1891–1900. doi:10.1007/s00705-009-0525-3 PMID:19862472

Vorne, V., Ojanpera, K., De Temmerman, L., Bindi, M., Högy, P., Jones, M. B., & Persson, K. et al. (2002). Effects of elevated carbon dioxide and ozone on potato tuber quality in the European multiple-site experiment CHIP-project. *European Journal of Agronomy*, *17*(4), 369–381. doi:10.1016/S1161-0301(02)00072-2

Vos, J. (1995). Nitrogen and the growth of potato crops. In A. J. Haverkort & D. K. L. MacKerron (Eds.), *Potato ecology and modelling of crops under conditions limiting growth* (pp. 115–128). Dordrecht: Kluwer Academic Publishers. doi:10.1007/978-94-011-0051-9_8

Vossen, P. (1995). Early assessment of national yields: the approach developed by the MARSSTAT project on behalf of European Commission. In *Proceedings of the Seminar on Yield Forecasting*.

Wade, G., Mueller, R., Cook, P., & Doraiswamy, P. (1994). AVHRR map products for crop condition assessment: A geographic infonnation systems approach. *Photogrammetric Engineering and Remote Sensing*, *60*(9), 1145–1150.

Waggoner, P. E., & Parlange, J. Y. (1975). Slowing of spore germination with changes between moderately warm and cool temperatures. *Phytopathology*, *65*(5), 551–553. doi:10.1094/Phyto-65-551

Wang, K., Dickinson, R. E., & Liang, S. (2012). Global atmospheric evaporative demand over land from 1973 to 2008. *Journal of Climate*, *25*(23), 8353–8361. doi:10.1175/JCLI-D-11-00492.1

Wang, Y., Bao, Z., Zhu, Y., & Hua, J. (2009). Analysis of temperature modulation of plant defense against biotrophic microbes. *Molecular Plant-Microbe Interactions*, *22*(5), 498–506. doi:10.1094/MPMI-22-5-0498 PMID:19348568

Wang, Y., & Nii, N. (2000). Changes in chlorophyll, ribulose bisphosphate carboxylase-oxygenase, glycine betaine content, photosynthesis and transpiration in Amaranthus tricolor leaves during salt stress. *The Journal of Horticultural Science & Biotechnology*, *75*(6), 623–627. doi:10.1080/14620316.2000.11511297

Watson, R. T., Patz, J., Gubler, D. J., Parson, E. A., & Vincent, J. H. (2005). Environmental health implications of global climate change. *Journal of Environmental Monitoring*, *7*(9), 834–843. doi:10.1039/b504683a PMID:16121261

Wellborn, A. R. (1990). Why atmospheric oxides of nitrogen usually phytotoxic and not alternative fertilizers? *The New Phytologist*, *115*, 395–429. doi:10.1111/j.1469-8137.1990.tb00467.x

Wellborn, A. R., Mayernik, O., & Wellborn, F. (1972). Effect of SO_2 and NO_2 polluted air upon the ultrastructure of chloroplasts. *Environmental Pollution*, *3*(1), 37–49. doi:10.1016/0013-9327(72)90016-X

Wharton, P., & Kirk, W. (2012). *Early Blight. Potato Disease*. Michigan State University. Retrieved January, 05, 2016, from http://www.potatodiseases.org/earlyblight.html

Wheeler, R. M., Tibbitts, T. W., & Fitzpatrick, A. H. (1991). Carbon dioxide effects on potato growth under different photoperiods and irradiance. *Crop Science*, *31*(5), 1209–1213. doi:10.2135/cropsci1991.0011183X003100050026x PMID:11537629

Wikipedia. (2010). Climate change and agriculture. Author.

Wikipedia. (2016). *Climate change and potato.* Author.

Williams, E. K., & Rosenheim, B. E. (2015). What happens to soil organic carbon as coastal marsh ecosystems change in response to increasing salinity? An exploration using ramped pyrolysis. *Geochemistry Geophysics Geosystems, 16*(7), 2322–2335. doi:10.1002/2015GC005839

Wolfe, D. W., Fereres, E., & Voss, R. E. (1983). Growth and yield response of two potato cultivars to various levels of applied water. *Irrigation Science, 3*(4), 211–222. doi:10.1007/BF00272837

Workman, M., Harrison, M. D. & Franc, G. D. (1983). *Effects of nitrogen fertilization and fungicide treatments on the Monona potato cultivar.* Colorado State University Experimental Station Bulletin.

WWAP. (2015). *The United Nations World Water Development Report 2015: Water for a Sustainable World.* Paris: UNESCO.

Yadav, S., Irfan, M., Ahmad, A., & Hayat, S. (2011). Causes of salinity and plant manifestations to salt stress: A review. *Journal of Environmental Biology, 32*(5), 667–685. PMID:22319886

Yager, R. R. (1988). On ordered weighted averaging aggregation operators in multi-criteria decision making. *IEEE Transactions on Systems, Man, and Cybernetics, 18*(1), 183–190. doi:10.1109/21.87068

Yager, R. R. (1996). Quantifier guided aggregation using OWA operators. *International Journal of Intelligent Systems, 11*(1), 49–73. doi:10.1002/(SICI)1098-111X(199601)11:1<49::AID-INT3>3.3.CO;2-L

Yang, X. B., & Scherm, H. (1997). El Niño and infectious disease. *Science, 275*(5301), 7399. doi:10.1126/science.275.5301.737c PMID:9036530

Yaycili, O., & Alikamanoglu, S. (2012). Induction of salt-tolerant potato (Solanum tuberosum L.) mutants with gamma irradiation and characterization of genetic variations via RAPD-PCR analysis. *Turkish Journal of Biology, 36*(4), 405–412.

Ying, J., Peng, S., He, Q., Yang, H., Yang, C., Visperas, R. M., & Cassman, K. G. (1998). Comparison of high-yield rice in tropical and subtropical environments I. Determinants of grain and dry matter yields. *Field Crops Research, 57*(1), 71–84. doi:10.1016/S0378-4290(98)00077-X

Yoshida, S., & Douglas, A., Forno, J., Cock Kwanchai, A., & Gomez, A. (1976). *Laboratory Manual for Physiological Studies of Rice.* Intl. Rice Res. Inst.

Yuan, F. M. & Bland, W. (2004). Light and temperature modulated expolinear growth model of potato (Solanum tuberosum L.). *Agricultural and Forest Meterology,* (121), 141-151.

Ywa, N. S., Walling, L., & McCool, P. M. (1995). Influence of elevated CO_2 on disease development and induction of PR proteins in tomato roots by *Phytophthora parasitica. Plant Physiololgy, 85*(Supplement), 113. Retrieved November 4, 2015, from http://www.climateandforming.org

Zeevaart, A. Z. (1976). Some effects of fumigating plants for short periods with NO_2. *Environmental Pollution*, *11*(2), 97–108. doi:10.1016/0013-9327(76)90022-7

Zhang, Z., Mao, B., Li, H., Zhou, W., Takeuchi, Y., & Yoneyama, K. (2005). Effect of salinity on physiological characteristics, yield and quality of microtubers in vitro in potato. *Acta Physiologiae Plantarum*, *27*(4), 481–489. doi:10.1007/s11738-005-0053-z

Zoellick, R. B. (2009). *A Climate Smart Future. In The Nation Newspapers*. Lagos, Nigeria: Vintage Press Limited.

About the Contributors

Sunil Lalasaheb Londhe is associated as Soil Health and Geoinformatics Scientist with International Centre for Research in Agroforestry Centre (ICRAF) popularly known as World Agroforestry Centre, New Delhi, India. He earned PhD degree from Birla Institute of Technology, Ranchi and received M.Sc. in Agriculture in the subject of Agricultural Chemistry and Soil Science with specialization in Land Resources Management. He is having 17 years of professional experience in agriculture research and development. He was associated as professional scientist with Centre for Development of Advanced Computing (C-DAC), Pune, India, Jharkhand Space Applications Center (JSAC), Ranchi and National Bureau of Soil Survey and Land Use Planning, Nagpur. His current research interest includes Land Health, Agroforestry, Pedology, Climate Change, Land Degradation, Land Resources Inventory, Land Use Planning, Wasteland Change Analysis, Geoinformatics, and Spectral Reflectance of Soils.

* * *

P. Bandyopadhyay worked in crop production aspects involving principal crops of Lower Gangetic Plains. Contributed to national networks of Allied fibres and Forage groups. Wide exposure in research and development works in drought prone areas of the state. Managed externally funded research projects with both government and corporate support. Themes covered in programs include natural resource management and sustainable production, crop weather interaction, ground water pollution. Contributed to education management at the university level. Have wide work experience in agriculture extension with farm science centers and stakeholders. Experience in organizing state and national level training and agribusiness courses. Teach UG & PG students, supervise M.Sc., Ph.D students, and review papers of national and international journals.

Bhagawan Bharali obtained B.Sc.(Agri,), M.Sc.(Agri) from Assam Agricultural University, Jorhat; and , Ph.D.(Biology) from London University, U.K. He has been holding the position of Professor & Head (Crop Physiology) in Assam Agricultural University. His research field is Environmental Stress Physiology.

Sonbeer Chack mastered in Crop Physiology from Assam Agricultiural university, Jorhat. Currently Mr. Sonbeer has been pursuing Ph.D. degree in the in the area of nitrogen use efficiency of crops in the Dept. of Crop Physiology, Assam Agricultural University.

Joyashree Chutia is MSc.(Agri) in Plant Breeding and Genetics. She worked as JRF in the Research Project entitled 'Atmospheric deposition of nitrogen and productivity of winter cereal crops in Northeast India', at the Department of Crop Physiology, Assam Agricultural University, Jorhat (Assam).

V. K. Dua is working on a project climate change impact and adaptation strategies for potato. He is a principal scientist in CPRI, Shimla. He is a principal scientist in ICAR.

Upasana Dutta is a Senior Technical Officer in the Emerging Solutions and e-Governance Group with Centre for Development of Advanced Computing (C-DAC) (Under DeITY, Ministry of Communications and Information Technology, Govt. of India) Pune since 2005, Ms. Dutta has over 10 yrs experience in the field of Geo-informatics & Environment Science. During her tenure in the Geomatics Division of C-DAC Pune, she was involved in a number of major projects of national repute (including the prestigious PMGSY project), and is also responsible for the quality control activities of the group at C-DAC Pune. Prior to joining C-DAC, she had worked as JRF in IIT Guwahati & as a research assistant in a Biodiversity Conservation NGO. MSc. in Environment Science, her major work areas include Natural Resource Management & Analysis, Infrastructure Management, EIA Preparation and Analysis, Environment Modeling, Disaster Management and Mitigation using GIS and Remote Sensing. She was part of the team conferred with Geo-spatial Excellence award and also was a finalist in the eNorthEast award in the Innovative ICT Category in the year 2012. Ms. Dutta has also authored a number of research publications of national & international repute. She is also in the executive panel of a number of prestigious scientific societies.

P. M. Govindakrishnan is Project Coordinator AICPIP, CPRI, Shimla.

Moni Gupta is Associate Professor in Div. of Biochemistry, SKUAST-J. Her main area of research is biochemistry of plant pathogens.

Sachin Gupta is serving as Assistant Professor in Division of Plant Pathology. His main area of interest is Mushroom Biology and Plant Disease Management.

Bhupendra Haloi worked in various capacities in the area of research and teaching in the Department of Crop Physiology. Dr. Haloi accomplished his post graduate degree from Jawaharlal Nehru Krishi Vishwa Vidyalaya, Jabalpur, Madhya Pradesh and Ph.D. from IARI, New Delhi. Now he has retired from the post of Professor & Head, Department of Crop Physiology, Assam Agricultural University.

Lokesh Jain passed his post graduation in Agronomy from SKRAU Bikaner Rajasthan with honours and receipt gold medal. he has a vast experience of research (10 years as SRF in AICRP on Dryland Agriculture), extension (3 years as Subject Matter Specialist at KVK Barmer) and teaching as Assistant Professor since last 2 years to B.Sc. Ag. students of College of Agriculture Sumerpur. He has published a lot of publication like 1 book as co-author, contributed 9 chapters in books, 5 manuals and booklets, 50 research paper along with a lot of technical and popular articles. He is also working as Master Trainer of Agronomy for zone IIb of Rajasthan and a member of seed certification committee.

M. K. Jatav, Senior Scientist (Soil Science), Serving ICAR since 2003. He has research specialization in area of potato production technologies, integrated nutrient management in potato based cropping systems.

Sarika Jena completed her doctoral degree in Agronomy, with a specialisation in light interactions under intercropping, from Bidhan Chandra Krishi Viswa Vidyalaya, West Bengal. Presently, working as a research scientist on low cost fibre production technology. at Jute Research Station, Kendrapara. She has an experience of four years in research and extension activities and published 26 research papers in National and International levels.

Surinder Kumar Kaushik, Principal Scientist, ICAR-NBPGR, New Delhi born on 29th June, 1967, had completed his Ph. D (Ag.) Plant Breeding (1993) from CCS Haryana Agricultural University, Hisar. He joined CPRI, Shimla in 1994 and held various scientific positions including Joint Director, CPRI Campus, Modipuram. At present, he is working as Principal Scientist (Plant Breeding), Germplasm Evaluation Division, ICAR-NBPGR, New Delhi. He contributed immensely not only for research but also for infrastructure development at CPRI, Shimla and its regional stations located at Kufri and Modipuram. His research contributions include development of three late blight resistant varieties, elite 'triplex' PVY resistant line, hybrids having combined resistances to late blight, PVY and cyst nematodes, development of mapping population and initiation of marker assisted selection for late blight resistance in potatoes, broadened the genetic base of cultivated potato by incorporating gene pool from wild/semi cultivated Solanum species through development of meiotic tetraploids etc. Besides he has developed and introduced more than a dozen high yielding late blight and virus resistant hybrids into AICRP (P) for multi-location testing. He has published 117 research articles including 40 research papers in national and inter-national journals. He has been awarded BOYSCAST Fellowship and several best poster paper awards. He is the fellow of the Indian Potato Association (IPA) and Confederation of Horticulture Associations of India (CHAI). He is the Vice-President (Region IV), Indian Potato Association. He has organized several training programmes for farmers/ agricultural/horticultural officers and kisan melas.

Manoj Kumar is a principal scientist & Head in ICAR-Central Potato Research Station.

Mehi Lal has worked seven years on potato diseases for diversity and management.

Dhiman Mukherjee is working as an Associate Professor (Agronomy), under the aegis of Bidhan Chandra Krishi Viswavidayalaya. He is officer incharge of All India Coordinated Research Project on Wheat and Barley improvement project, at Kalyani. He is PI of two project and Co-PI of one research assignment. He has an experience in research and extension work in agronomy in various ways for last 16 years. His contribution mainly confined to resource conservation and crop husbandry practices of various field crop mainly potato, rapeseed, wheat etc. He was also associated with high altitude medicinal plant project from 2005 to 2014, and registered 42 valuable germplasm of high altitude medicinal plant. He has been honored with Indira Gandhi Shiromany award and Bharat Jyoti award in 2011 for outstanding achievement and contribution in the field of agricultural sciences. He received Bharat Shiksha Ratan award in the year 2012, for his colossal involvement in the field of research and teaching activity. He received CWSS Young Scientist Award in 2013 for valuable contribution in the field of agronomy. He was honored with "Outstanding Scientist Award-2015" from Venus International Foundation. He has so far published more than 84 papers in national and international journals. He has attended more than 33 international and national conferences and presented his research work as key note speaker. He has published 8 review papers, 16 book chapters and 24 popular articles. Dr. Mukherjee is a Fellow of Society of Hind Agri-Horticultural Society (India). He has also authored four popular books mainly on medicinal plant, seed science, organic agriculture and agrometeorology. He is a member of 25 professional scientific societies, including society of oilseed and brassica research, agronomy society, weed science etc. He is editorial board member of three national reputed journals and one journal of international level.

R. K. Naitam born on 1st July 1978 at Kurul in Maharashtra state in India. Presently he is working as Scientist in the field of Pedology at ICAR-National Bureau of soil survey and land use planning, Nagpur since 2010.

Ratneswar Poddar works as an Assistant Professor in Agronomy of AICRP on Irrigation Water Management under Directorate of Research, Bidhan Chandra Krishi Viswavidyalaya, Gayeshpur, Nadia (W.B.), India. He has a brilliant academic carrier got JRF, SRF of ICAR. Mr. Poddar has carrying out important field works and significant dedicated researcher, primarily on Weed Management, Soil Fertility and Ground water contaminated irrigation management. Now he is engaged in research and teaching activities.

R. P. Sharma Scientist (Soil Science), serving ICAR since 1999. He has research specialization in area of vegetable production, nutrient management, land resource inventorization, soil survey and land use planning in India.

B. P. Singh has worked more than 35 years on potato diseases. he had been joint- director of ICAR-Central Potato Research Institute Meerut and Director of ICAR-Central Potato Research Institute Shimla.

R. Singh is the Principal Scientist at NBSS&LUP, Regional Centre, Udaipur.

S. Singh is Director at ICAR-NBSS&LUP, Nagpur.

Baby Summuna is Ph.D scholar in Division of Plant Pathology. Presently she is doing her research on plant pathogens.

Zafar Ullah obtained B.Sc. (Agri), M.Sc. (Agri) & Ph.D. from Assam Agricultural University, Jorhat, Assam. Currently Dr. Ullah has been working as Principal Scientist on AICRP (Potato) at AAU, Jorhat, Assam.

Saurabh Yadav has worked six years on late blight of potato on diversity and management aspects.

Index

P

R

S

T

V

W

Y

www.ingramcontent.com/pod-product-compliance
Lightning Source LLC
Chambersburg PA
CBHW081213180926
51613CB00003B/228